The Primate Postcranial Skeleton
Studies in Adaptation and Evolution

Edited by

ELIZABETH STRASSER and MARIAN DAGOSTO

City University of New York and Northwestern University

ACADEMIC PRESS

Harcourt Brace Jovanovich, Publishers

London   San Diego   New York   Boston
Sydney   Tokyo   Toronto

ACADEMIC PRESS LIMITED
24–28 Oval Road
London NW1 7DX
(Registered Office)

United States edition published by
ACADEMIC PRESS INC.
San Diego, CA 92101

(Reprinted from *Journal of Human Evolution,* Vol. 17, No. 1/2, 1988)

ISBN 0-12-673005-9

Printed in Great Britain by Black Bear Press Limited, Cambridge

List of Contributors

K. C. BEARD, *Department of Cell Biology and Anatomy, Johns Hopkins University, School of Medicine, 725 North Wolfe Street, Baltimore, MD 21205, U.S.A.*

H. H. COVERT, *Department of Anthropology, University of Colorado, Boulder, CO 80309-0233, U.S.A.*

M. DAGOSTO, *Departments of Anthropology and Cell Biology & Anatomy, Northwestern University, Evanston, IL 60208, U.S.A.*

S. M. FORD, *Department of Anthropology, Southern Illinois University, Carbondale, IL 62901, U.S.A.*

D. L. GEBO, *Department of Anthropology, Northern Illinois University, DeKalb, IL 60115, U.S.A.*

L. R. GODFREY, *Department of Anthropology, University of Massachusetts, Amherst, MA 01003, U.S.A.*

M. GODINOT, *Laboratoire de Paléontologie, Université des Sciences et Techniques du Languedoc, Place Eugène-Bataillon, F-34060 Montpellier, France*

W. L. JUNGERS, *Department of Anatomical Sciences, School of Medicine, State University of New York at Stony Brook, Stony Brook, NY 11794-8081, U.S.A.*

M. D. ROSE, *Department of Anatomy, New Jersey Medical School, 100 Bergen Street, Newark, NJ 07103, U.S.A.*

E. STRASSER, *Department of Anthropology, Graduate School, City University of New York, 33 West 42nd Street, New York, NY 10036, U.S.A.*

F. S. SZALAY, *Department of Anthropology, Hunter College, City University of New York, 695 Park Avenue, New York, NY 10021, U.S.A.*

Preface

Most of the papers included in this volume were originally presented at a symposium entitled "Evolution and Diversification of the Postcranium in Primates" which was held at the Fifty-Sixth Annual Meeting of the American Association of Physical Anthropologists in New York on April 4, 1987. The purpose of the symposium was to address the relationship between the acquisition of novel locomotor strategies and the origin and radiation of major groups of primates. Primates exhibit a wide range of locomotor and postural behaviors and a correlated diversity in the morphology of the postcranial skeleton which is perhaps unparalleled in any other order of mammals. Thus, studies of primate locomotion and postcranial anatomy have always been important in the field of primatology, but have received increased attention in the past two decades (e.g., Jenkins, 1974; Morbeck *et al.*, 1979). The symposium also marked the twentieth anniversary of the publication of one of the most influential and controversial papers in the field of primate locomotion, Napier & Walker's (1967) "Vertical clinging and leaping—a newly recognized category of locomotor behavior of primates".

One primary area of emphasis in past studies of the primate locomotor system has been the understanding of the functional-adaptive importance of characters and the use of these for inferring the behavior of fossil primates. Another more recently developed focus has been the incorporation of postcranial features in phylogenetic analyses of primates. Following the pioneering efforts of Szalay (e.g., Szalay & Decker, 1974), we encouraged the symposium participants to synthesize the results of these two approaches.

As data accumulate from these studies it has become increasingly apparent that unique sets of postcranial features characterize primates at almost all taxonomic levels. This means that adaptations of the postcranial skeleton, and thus those most likely related to positional behavior, have had an impact on the origin of most groups of primates. Our object in organizing this symposium was to examine such questions in greater detail. We asked the participants to identify those features that distinguish phylogenetically related groups of primates, discuss the functional significance of these features, explore the relationships between these features and behavioral and ecological factors, and thus identify possible explanations for the origin and subsequent radiation of some primate taxa. We are extremely pleased at how many of these lofty goals were met by the contributors.

The papers can be organized loosely into three types. The first are those which delineate the autapomorphous characters (and their behavioral implications) in a particular group of primates. Dagosto discusses the origin of Euprimates, Ford the origin of Platyrrhini,

i

Strasser the origin and diversification of cercopithecids, Gebo & Dagosto the origin of indriids, and Godfrey the diversification of extant and subfossil Malagasy lemurs. The phylogenetic orientation of these studies is primarily cladistic. The approach to relating morphology to behavior is basically analogic or associational, but with an eye toward bolstering such hypotheses with a "mechanical" understanding of features.

In the second group of papers, intensive scrutiny is given to a particular anatomical region in a wide range of taxa, discussing important functional consequences of variations in anatomy, and also commenting upon the possible phylogenetic implications of these features. In this group Szalay & Dagosto address the question of the anatomical bases of grasping in Primates, Rose carefully analyzes the mechanics and function of the anthropoid elbow, and Jungers uses a variety of regression techniques to examine joint size-body size relationships in hominoids from which he makes some important comments on hominid evolution.

A third group of papers describes new adapid postcranial fossils and discusses the implications of this material for the phylogenetic relationships within Euprimates (Covert) and within Strepsirhini (Beard & Godinot).

Two contributions to the original symposium do not appear here. One was published in a previous issue of this journal (Gebo, 1987). The other paper was an interesting discussion by Mark Birchette concerning the limits of behavioral reconstruction in fossil primates. Two of the contributions published here (Covert and Gebo & Dagosto) were not part of the original symposium, but because of their relevance to this topic, the authors were invited to contribute to this published version.

We hope that this first attempt at a synthesis of primarily functionally oriented with more phylogenetically directed studies of the primate postcranium will provide a basis for future, more completely understood, hypotheses of evolution and diversification in primates.

E. Strasser
M. Dagosto
New York and Chicago
January, 1988

References

Gebo, D. L. (1987). Anthropoid origins—the foot evidence. *J. hum. Evol.* **15**, 421–430.

Jenkins, F. A., Jr. (1974). *Primate Locomotion.* New York: Academic Press.

Morbeck, M. E., Preuschoft, H. & Gomberg, N. (1979). *Environment, Behavior, and Morphology: Dynamic Interactions in Primates.* New York: Gustav Fischer.

Napier, J. H. & Walker, A. C. (1967). Vertical clinging and leaping—a newly recognized category of locomotor behavior of primates. *Folia primatol.* **6**, 204–219.

Szalay, F. S. & Decker, R. L. (1974). Origins, evolution and function of the tarsus in late Cretaceous eutherians and Paleocene Primates. In (Jenkins, F. A., Jr., Ed.) *Primate Locomotion,* pp. 223–259. New York: Academic Press.

Contents

Strasser, E. & Dagosto, M. Preface i

Szalay, F. S. & Dagosto, M. Evolution of hallucial grasping in the primates . . . 1

Dagosto, M. Implications of postcranial evidence for the origin of euprimates 35

Covert, H. H. Ankle and foot morphology of *Cantius mckennai:* adaptations and phylogenetic implications 57

Beard, K. C. & Godinot, M. Carpal anatomy of *Smilodectes gracilis* (Adapiformes, Notharctinae) and its significance for lemuriform phylogeny . . 71

Godfrey, L. R. Adaptive diversification of Malagasy strepsirrhines . . 93

Gebo, D. L. & Dagosto, M. Foot anatomy, climbing, and the origin of the Indriidae . 135

Ford, S. M. Postcranial adaptations of the earliest platyrrhine . . 155

Rose, M. D. Another look at the anthropoid elbow 193

Strasser, E. Pedal evidence for the origin and diversification of cercopithecid clades 225

Jungers, W. L. Relative joint size and hominoid locomotor adaptations with implications for the evolution of hominid bipedalism . . 247

Frederick S. Szalay

*Department of Anthropology, and
Graduate Programs in Anthropology
and Ecology and Evolutionary
Biology, Hunter College, C.U.N.Y.,
695 Park Avenue, New York, NY
10021, U.S.A.*

Marian Dagosto

*Department of Cell Biology and
Anatomy, Johns Hopkins University,
Medical School, Baltimore, MD
21205, U.S.A.**

Received 15 December 1986
Revision received 15 October
1987 and accepted 18 December
1987

Publication date June 1988

Keywords: Pedal morphology,
grasping, first metatarsal,
entocuneiform, marsupials,
erinaceids, archontans; primate,
euprimate, and anthropoid
origins.

Evolution of hallucial grasping in the primates

Homology of the adaptive solutions of grasping, like other attributes of the postcranial skeleton, have long been assumed for marsupials, early eutherians, and euprimates. Evidence is presented which contradicts this view. The origin of grasping is documented and discussed in the order Primates, the semiorder Euprimates, the suborders Strepsirhini and Haplorhini, and the semisuborder Anthropoidea. Grasping may have been primitive in the cohort Archonta, but the euprimate grasp appears to be related not only to climbing but to a saltatory, graspleaping, locomotor mode of the common ancestor. The origin of anthropoid modifications involves a reduced emphasis on the pedal grasp. The "prehallux" hypothesis for the explanation of the sesamoid in the entocuneiform–hallucial articulation of anthropoids cannot be corroborated by either topographical, developmental, or functional evidence.

Journal of Human Evolution (1988) **17,** 1–33

Introduction

Any student who has ever examined either the morphology, the mechanics, or the behaviors associated with primate extremities cannot but be struck by the pedal-grasping adaptations compared to other mammals. In Cartmill's (1985) recent account of climbing solutions in vertebrates, the phenomenon of pedal grasping is placed in its proper perspective in relation to arboreal locomotion and its complex attendant problems. In spite of abundant references in the literature, however, there has been no deeply focused effort which concentrated on the evolutionary history of pedal grasping within the order, encompassing morphological (form-function), adaptive, and genealogical perspectives. By analyzing the morphology and mechanics of the first metatarsal, the entocuneiform, and selected associated soft anatomy in living and fossil primates, and behaviors related to this region, we hope to provide a better understanding of the transformation of this critical area in primates.

As the construction of taxon phylogeny is properly based on well-understood charactercline transformations, so is the understanding of the evolutionary transformation of a character complex grounded on detailed morphological, functional, adaptive, and paleontological knowledge. In analyzing this transformation we accept the simple assumption that the mechanics and subsequently the possible range of environmentally related activities (bioroles; see especially Bock & von Wahlert, 1965) is dependent on the mechanical relationships of the musculoskeletal system and that this is mediated and reflected in the articular surfaces of the bones involved.

* Current address: Department of Anthropology and Cell Biology & Anatomy, Northwestern University, Evanston, IL 60201, U.S.A.

0047–2484/88/01/20001 + 33 $03.00/0

Based on the assumption that the grasping foot of the most primitive living marsupials, the didelphids, is representative of the ancestors of eutherians, Matthew's (1904) influential paper paved the road for the acceptance of a putative protoeutherian which was arboreal and consequently assumed to be a power grasper (see Szalay, 1984, for a detailed discussion). For example, the most recent and one of the most detailed comparative pedal anatomical studies by Lewis (1980*a–c*) assumed incontrovertible evidence for the arboreal habits of the first eutherians. With this long history of assumptions about the protoeutherians, the origin of primates was almost always assumed to have been from an already arboreal and grasping stock of animals. A notable exception was Haines (1958) who argued for protoplacental terrestriality based on the comparative anatomy of the manus.

This paper documents and analyses the form and mechanics of the entocuneiform–first metatarsal joint (EMt1J) in selected marsupials and nonprimate eutherians, and in Paleogene and living primates. Based on this, we will assess: (1) the disputed background for the origin of primate pedal grasping (and therefore aspects of primate origins in general); (2) the primate–euprimate transition; and (3) the origin of the anthropoid pedal grasping complex. For this study we emphasize the articular relationship and consequent mechanics of the entocuneiform–first metatarsal joint (EMt1J), the entocuneiform–mesocuneiform joint (EMcJ), and the entocuneiform–second metatarsal joint (EMt2J). We attempt throughout the paper to connect causally these data with the known locomotor behavior of the living species.

The comparative accounts of nonprimates provided here are not descriptively all-inclusive; the statements will emphasize those characteristics which diagnose their morphology compared to primate, particularly euprimate, characters. More detailed treatment of the metatherians in particular is found in Szalay (in prep.). The abbreviations used in the text and figures are listed in Table 1.

Our descriptions of joint form and hypotheses of joint motion are based on the concept that form and mechanical function inseparably mirror each other, and we particularly follow the morphological-functional classification of MacConaill (1946, 1953, 1966, 1973). In this system, joints are classed into four types, and the geometry of each determines allowable kinds of directions of motion.

Our study, however, should not be regarded as a conventional, functionally oriented description. We supplement the functional accounts with observations of bony morphology which indicate, through comparison, phylogenetically meaningful relationships. We hope to have judiciously mixed comparative morphological, functional, and phylogenetic perspectives to make this a study in evolutionary morphology.

Form-function of saddle joints

Most of the joints we describe are sellar, or are derived from saddle joints. It is important, therefore, to present briefly the salient features of such joints. These points about saddle joints are:

1. They are biaxial, with opposing concavo-convex surfaces which allow minimal or no translation, making saddle joints more stable that other joint types which facilitate equivalent primary movements.

2. Movements, primarily around the two axes, occur in two planes, and the swing component of a movement is accompanied by conjunct rotation of the bone.

Table 1   **Abbreviations used in the figures and in the text**

Abbreviation	Meaning
AMNH	Department of Vertebrate Paleontology, American Museum of Natural History
AMNH-M	Department of Mammalogy, American Museum of Natural History
As	astragalus
Ca	calcaneus
Do	dorsal
Ec	ectocuneiform
EMc	entocuneiform–mesocuneiform facet
EMcd	distal entocuneiform–mesocuneiform articular facet
EMcp	proximal entocuneiform–mesocuneiform articular facet
EMcJ	entocuneiform–mesocuneiform joint
EMJ	entocuneiform–mesocuneiform joint
EMt1	entocuneiform–first metatarsal articular facet
EMt1d	distal entocuneiform–first metatarsal articular facet
EMt1J	entocuneiform–first metatarsal joint
EMt1p	proximal entocuneiform–first metatarsal articular facet
EMt2J	entocuneiform–second metatarsal joint
En	entocuneiform
EPh	entocuneiform–prehallux facet
KNM	Kenya National Museum
lp	lateral process
Me	medial
Mc	mesocuneiform
Mt1	first metatarsal
Na	navicular
NEn	naviculo–entocuneiform articular facet
ph	prehallux
Pl	plantar
Plp	plantar process of entocuneiform
ppl	process for peroneus longus
UCMP	University of California Museum of Paleontology
USNM–VZ	United States National Museum, Vertebrate Zoology/Mammals

3. Stability and the range of excursion about each axis is determined by either one or all three of the following factors: (a) ligamentous restraints; (b) differences in the area of articulating male and female surfaces; (c) surface curvature.

4. Surface curvature is significant in determining whether, with slight separation, an independent spin, slide (involving minimal translation), or roll may occur in various directions. In general, the possibility for adjunct axial rotations (independent movements which provide an additional degree of freedom) is largely influenced by surface curvature. Thus, slight surface curvature in a saddle joint, while perhaps less resistant to loads, allows a greater range of adjustability by either muscular or passive forces.

Form-function of metatherian pedal grasping

In connection with a recent survey of marsupial pedal morphology (Szalay, in prep.) the osseous anatomy of the entocuneiform and hallux of living and all known and available fossil genera of metatherians were studied. The morphology of the articular ends of the entocuneiform and first metatarsals which form the entocuneiform–hallucial joint (EMt1J) of a representative didelphid (Figure 1A) and representatives of two distinct groups of

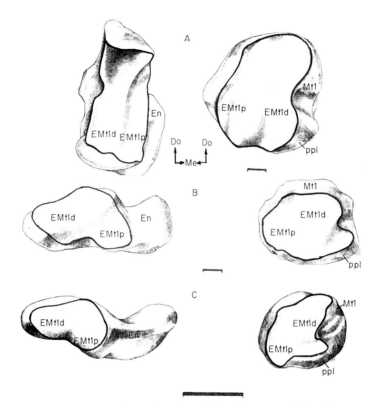

Figure 1. Comparisons, in equivalent orientation, of the right entocuneiform–first metatarsal joint (EMtlJ) in the major groups of arboreal metatherians, the ameridelphian didelphids (*Didelphis*, above) and the australidelphian phalangeriforms (Australian *Pseudocheirus*, middle) and australidelphian microbiotheres (Patagonian–Chilean *Dromiciops*, below). The distal view of the entocuneiform (left) and the proximal view of the first metatarsal (right) illustrate the characteristic morphology of the two major groups (abbreviations as in Table 1).

Scales represent 1 mm; arrows depict anatomical directions *dorsal* and *medial* for a quadruped, for the left and right drawings, respectively. These conventions are followed in all figures, unless otherwise noted.

australidelphians (Figures 1B and C) illustrate the two basic patterns of this joint relevant in this study. These patterns faithfully mirror the morphotypic condition of the order Didelphida and the cohort Australidelphia (see Szalay, 1982 for discussion of marsupial relationships).

Didelphida

Unlike in most eutherians, articular contact between the entocuneiform and the hallucial metatarsal in all arboreal metatherians, like didelphids, is offset slightly medially from the distomedial section of the entocuneiform. Although superficially this may appear similar to the condition in euprimates, scrutiny shows that the euprimate condition is a modification of a distinctive eutherian ancestral pattern.

The essential uniformity of the EMtlJ in all of its identifiable details among living didelphids is mirrored in *Didelphis*. The EMtl on the entocuneiform is an attenuated quasi-cylindrical (or perhaps better described as ridge-like) facet on the lateral side of the joint. It articulates with a corresponding female conarticular facet on the first metatarsal. Continuous with this articular contact is a flattish facet on the metatarsal which articulates with another female surface continuous with and medial to the cylinder on the entocuneiform. Although the lateral surfaces are sellar in configuration, the joint itself cannot be considered a sellar one by any conventional standard of shape.

The medial conarticular area is best regarded as a stop or buttress which bears the bulk of the load during extremes of hallucial abduction. Adduction from this position is minimal since these animals have a habitually abducted and relatively rigidly constrained hallux even when walking on level ground. During adduction hallucial movement is strictly guided by the lateral, distally protruding articular areas (EMt1d on Figure 1A). The articular contact of the entocuneiform with the mesocuneiform (EMc facet) is single, unlike in archontans. The facet is dorsal and not parallel to the naviculoentocuneiform (NEn) facet as in eutherians, but adjacent and proximal to the second metatarsal facet.

The total morphological configuration and the manner of the mechanics it dictates is distinct from all Eutheria, and does not bear meaningful resemblance to any archontan or primate pattern.

Australidelphia

The polyprotodont microbiotheriid *Dromiciops* from South America and the diprotodont phalangeriforms share, in addition to other numerous tarsal synapomorphies (Szalay, 1982), a unique and functionally sellar entocuneiform–hallucial joint. Although the joint is functionally similar to that found between the human trapezium and metacarpal, the details of its topography (see Figures 1B and C) leave no doubt that it was derived from a pattern seen in didelphids. The derived metatherian condition of the EMt1J (along with the single articular contact of the entocuneiform with the mesocuneiform) render the australidelphian common ancestor even more different from any known eutherian or archontan than is the didelphid pattern.

Although the morphotypic primate EMt1J is also sellar in shape, all its details, as will become apparent in the discussion below, are constructed differently. Principally, the primate (and other eutherian) EMt1J is on the distal end of the entocuneiform, and not slightly medial as in marsupials. The medial extension of the euprimate joint is from the distally located condition of the joint. In metatherians the whole joint is offset to the medial side of the entocuneiform, and part of the distal end of the bone, unlike in eutherians, is nonarticular.

In comparing this joint between metatherians and eutherians, we see what has also been documented for the rest of the tarsus (Szalay, 1984). Arboreally related adaptations of the foot in the two great groups in general, and primates in particular, have been independently attained. There is no justification to assume that pedal grasping in primates and in arboreal marsupials is homologous.

Form-function of some nonprimate eutherian entocuneiform–hallucial joints

We have selected here a number of eutherians which may shed light on the configuration of this joint in the protoeutherian. We have paid particular attention to groups suspected to have possible close ties to the primates, particularly members of the great arctocyonid derived radiations (Matthew, 1937), the earliest rodents (Szalay, 1985), carnivorans and creodonts (Matthew, 1937), and the dermopterans. In Paleogene members of these groups form-function of the hallux, even in highly arboreal forms, is fundamentally different from what we see among the primates. The tupaiids occupy an interesting position, as discussed below.

The EMtlJ in the early Paleogene eutherians, and therefore possibly in the protoeutherian, although slightly sellar, is relatively flat, mediolaterally narrow, and restricted to the distal end of the entocuneiform, as in living *Echinosorex* or *Tenrec* (see Figure 2). It appears to have been involved in receiving either compressive or tensile loads with only a modicum of hallucial swing on the entocuneiform. In such undoubtedly arboreal forms as the early paramyid rodents (see Szalay, 1985) the joint reflects a modest range of movements, probably in tension, which allowed some abduction of the hallux.

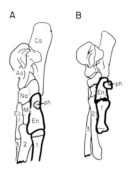

Figure 2. Position of the prehallux and the shape of the entocuneiform and its articulation with the first metatarsal in some living Insectivora. A, the erinaceid *Echinosorex* and B, the tenrecoid *Tenrec* (abbreviations as in Table 1).

Insectivora

The tenrecoids display a unique solution of hallucial mechanics, and their configuration does not resemble that of any other eutherian we have sampled. The ball and socket configuration coupled with a ridge-in-trough pattern is almost certainly *sui generis* in the Tenrecoidea, as traces of this pattern are suggested even in the aquatic *Potamogale*

A probably primitive condition for erinaceoids, displayed by *Echinosorex*, is strikingly different (Figure 3) from the tenrecoid condition. We will compare this below with that of tupaiids.

Scandentia

The tupaiids, as expected, show an intriguing morphological pattern. Although the joint is restricted, as in other eutherians, to the distal end of the entocuneiform, its conically shaped sellar configuration, in both *Tupaia* and *Ptilocercus* (see Figure 4), could have been derived from a pattern which may have been antecedent to that described for didelphids. The EMc facet is double in tupaiids, and these (EMcp and EMcd) are contiguous on the proximal end of the entocuneiform adjacent to the NEn facet.

The postcranially most primitive living erinaceoid, the terrestrial *Echinosorex* (Figure 3), displays the EMtlJ pattern seen in tupaiids. This, of course, suggests that this pattern is not a tupaiid one but that the shared morphology is inherited from the last common ancestor of these two groups. That such a form was similar to the protoeutherian is likely,

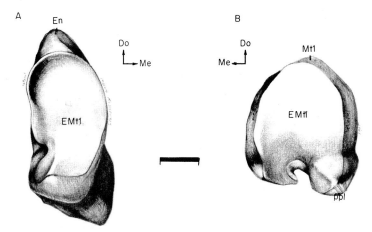

Figure 3. Articular facets of the right EMtlJ in the erinaceid *Echinosorex gymnurus* (abbreviations as in Table 1).

but unfortunately it is not adequately corroborated by detailed studies on early eutherians and living forms.

What is this eutherian pattern? Its most distinguishing attribute is the relatively wide and distally extending male surface of the entocuneiform EMt1 facet. This latter narrows plautarly into a ridge, then, as it turns dorsally again it slightly fans medially and laterally, somewhat asymmetrically. This morphology is complemented by the corresponding female conarticulation on the proximal hallucial metatarsal. It is the conical configuration of the male facet which gives its unique attribute rather than its sellariform configuration. In tupaiids the area opposite the peroneal process of the first metatarsal is relatively wider than in *Echinosorex*.

The possible homologous similarity of this pattern to a pre-didelphid, primitive metatherian condition lies in the shared presence of a slightly sellar ridge and the dorsally flared medial and lateral sides. The tupaiid and *Echinosorex* patterns (in *Erinaceus*, due to the great reduction of the hallux, this joint is very poorly developed) may well reflect both an ancient therian solution for moderate hallucial mobility as well as the primitive eutherian

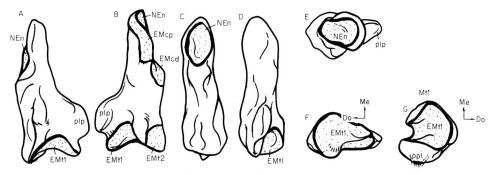

Figure 4. The right entocuneiform (A–F) and the proximal end of first metatarsal (G) in *Ptilocercus lowii*. Medial (A), lateral (B), dorsal (C), ventral (D), proximal (E), and distal (F) views, respectively, of the entocuneiform, and proximal view of Mt1 (G) (abbreviations as in Table 1).

pattern. The metatherian condition may be derived from this possibly primitive eutherian one (compare Figures 1, 3, and 4). The Cretaceous evidence will be critical in the further understanding of this problem.

In the multituberculates (see Krause & Jenkins, 1983) the joint is constructed functionally somewhat similarly to those of the eutherians described, and this has some bearing on the assessment of primitive therian morphology. It is a near certainty that eutherians and metatherians (= Theria, *sensu stricto*) are sister-groups. The entocuneiform of the multituberculate *Eucosmodon* (AMNH 16325) is quite dissimilar to those of therians in that it is short and proximally wide. A sellar pattern on its distal end is not specifically similar to any metatherian or eutherian. Since such a sellar joint is responsible for a relatively simple and straightforward form-function of hallucial abduction–adduction and flexion–extension, it is not unlikely that it evolved convergently. Additional, early fossil evidence will be critical to examine this problem. While possible, a synapomorphy in this feature between multituberculates and therians (dating to the Jurassic), does not appear likely in the light of the total tarsal evidence.

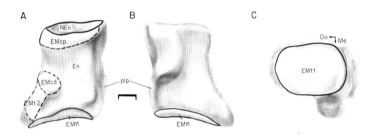

Figure 5. Left entocuneiform of the dermopteran *Cynocephalus philippiensis*. Lateral (A), medial (B) and distal (C) views, respectively (abbreviations as in Table 1).

Dermoptera

As expected in a group of mammals which use their feet primarily for tensile loading, the living dermopterans show a rather featureless entocuneiform-hallucial joint (Figure 5). The mediolaterally somewhat broad facet on the distal entocuneiform is nearly flat, displaying a curvature of only 5–10° of included angle. The hallux, like the rest of the cheiridia, is usually slightly abducted, and under tension by the pedally hanging animal. There is no special similarity of the entocuneiform or first metatarsal of colugos to the primates, unlike those noted in the proximal tarsals (Szalay & Drawhorn, 1980). However, we find the double articulation of the entocuneiform and mesocuneiform, and a small but well-defined plantar process, in dermopterans, in archaic primates, and in the early euprimates; this feature may be of some phylogenetic significance.

Form-function of primate pedal grasping

Archaic primates

The known record of the entocuneiform and first metatarsal of archaic primates is extremely poor. The entocuneiforms of *Plesiadapis tricuspidens* from the French late Paleocene are almost certainly correctly allocated, and a nearly identical specimen recovered from the Bison Basin Saddle locality (AMNH 92011) is equally certainly that of

Plesiadapis (including *Nannodectes*) (compare Figures 6 and 7). The morphology of these three specimens is the sum total of the fossil record from which we may reconstruct the function and biorole of the hallux in archaic primates.

A unique aspect of these specimens is the large size of a process on the plantar surface of the bone. This may be an expanded area of insertion for the tibialis anterior and posterior, powerful invertors of the foot, for ligaments, and/or to serve as a medial wall for the tunnel for powerful pedal flexors (flexor tibialis and flexor fibularis). A suggestion of its homolog can be seen on the entocuneiform in colugos and tupaiids, as well as in nonanthropoid euprimates. As in tupaiids and euprimates, the entocuneiform is narrow proximally and deep distally in the dorsoplantar dimension. The medial surface is nondescript, with no

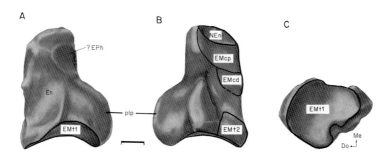

Figure 6. Right entocuneiform of *Plesiadapis* sp., AMNH 92011, from the middle Paleocene Bison Basin Saddle locality. Medial (A), lateral (B), and proximal (C) views, respectively (abbreviations as in Table 1).

apparent articular surface for the prehallux (see p. 35 on the homology of the anthropoid "prehallux"). Nevertheless, the presence of such a therian prehallux (see Figure 2) in archaic primates cannot be excluded, and in fact the "appropriate" indentation (where the prehallux would have fitt) on the proximomedial side of the entocuneiform (see Figure 6) supports this suggestion. This bone is prominent in didelphid and phalangeroid marsupials, edentates, many groups of insectivorans, and can also be suspected, judged from the entocuneiform conformations, in many Paleogene groups. As far as we know, this prehallux is absent from euprimates as neither adapids, omomyids, living strepsirhines, or *Tarsius* show any indication of it.

The lateral side of the archaic primate entocuneiform displays a prominent proximo-distally aligned tuberosity, presumably for a robust ligament binding the bone to the second metatarsal. The articulation with the mesocuneiform, unlike the pattern in either didelphids or phalangeroids, is a double articulation with the more proximal of the two being offset from the longitudinal plane of the entocuneiform. The double entocuneiform–mesocuneiform pattern of plesiadapids is found also in all euprimates. In the latter the distal articular facet (EMcd) is closer to the facet articulating with the second metatarsal (EMt2) than to the more proximal entocuneiform–mesocuneiform facet (EMcp).

The navicular facet on the proximal end of the entocuneiform (NEn) is concave, facing laterally and dorsally. This appears to be a common therian pattern, found also in didelphids, but altered in the phalangeroids to a convex facet on the entocuneiform. Unlike in dermopterans (see Figure 5), the plesiadapid, tupaiid, and euprimate facet is relatively small. The plesiadapid entocuneiform EMt1 facet, although primarily on the distal end of the bone but with a telling slight extension onto the medial surface, is a joint surface strikingly sellar in shape, even though the concave–convex dimensions cannot yield more than 20° of arc. This articular surface (we do not know the metatarsal facet) is similar to that of *Ptilocercus*, and to some degree to other tupaiids, except for being relatively wider.

The hypertrophied plantar process and the shape of the EMt1 in plesiadapids and tupaiids certainly suggests that of these characters at least the hypertrophied plantar process is an archontan synapomorphy.

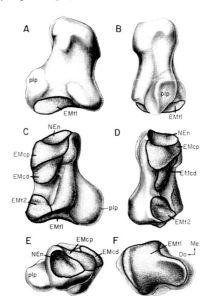

Figure 7. Left entocuneiform of *Plesiadapis tricuspidens*, from the late Paleocene of France. Medial (A), ventral (B), lateral (C), dorsal (D), proximal (E), and distal (F) views, respectively (abbreviations as in Table 1).

The plesiadapids are so far the only representatives of the semiorder Paromomyiformes which allow a glimpse at this joint. Evidence from the latter indicates that a hallux which, although unlike the permanently abducted one of many marsupials, and strepsirhine and tarsiiform primates, was capable of a reasonable range of abduction. The range of movement was probably similar to that described for *Ptilocercus* (Clark, 1926), which, in fact, both grasps with and is capable of reversing the hindfoot when descending a branch. Using Napier & Napier's (1967) terminology for the hand, the foot of plesiadapids could probably thus be described as being of the prehensile but non-opposable type.

Euprimates
Entocuneiforms and first metatarsals are known in a range of taxa available for study. Specimens of adapids and omomyids, primarily from the early and middle Eocene of North America, were sampled and described. Information for the primarily European Adapinae, other than what can be gleaned from the first metatarsal, is lacking. The pattern of

similarity which emerges from the combined appraisal of the living primates and these Paleogene taxa, however, leaves no serious objection to the hypothesis that the basic pattern described for the adapids below (and the nailed condition attributed to them by Gregory, 1920, and Koenigswald, 1979) is the morphotypic pattern for the Euprimates.

The transformation of conditions related to pedal grasping from an archaic primate to the ancestral euprimate species is a rather drastic reorganization in a single lineage, a unique phylogenetic event. These attributes in subsequent euprimates, therefore, are not an expression of some ill defined series of trends, independently attained. This cannot be sufficiently emphasized in light of the literature-entrenched notions of various "trends" in primates. The joint complex and attendant form-functions of pedal grasping represent an advanced primate condition, as well as the primitive euprimate character complex. This taxon specific constraint is inherited and it is either retained or tellingly modified in subsequent descendants. An understanding of these transformations is the basis for choosing a direction for the change in a morphocline, and its subsequent use in establishing taxon phylogeny.

Adapiformes

The condition described for the entocuneiform–first metatarsal and entocuneiform–mesocuneiform articulations in *Pelycodus*, *Notharctus*, and *Smilodectes* is probably the representative euprimate morphotypic state. We will, under the various subheadings, present arguments for the as yet unrefuted nature of this hypothesis. The basic reason for this stand is that all Eocene and post-Eocene differences from these also incorporate, or build on, the pattern which we see first among these primates.

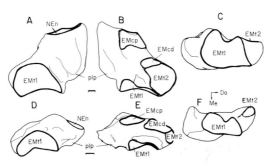

Figure 8. Comparison of the right entocuneiforms of *Notharctus* sp., AMNH 13050, middle Eocene of North America (A, B, C), and *Perodicticus potto*, AMNH-M 52717 (D, E, F). Medial (A, D), lateral (B, E), and distal (C, F) views, respectively (abbreviations as in Table 1).

Notharctinae. On the entocuneiform the plantar process, like in plesiadapids, is prominent (Figures 8A and B). The distal articulation with the first metatarsal is much larger than that with the navicular, and the former is extended onto the medial third of the entocuneiform surface. The EMt1 facet on the entocuneiform is convex in a mediolateral direction and covers about 180° of arc. In the dorsoplantar direction the surface is gently concave; the whole joint thus forms an unmodified sellar surface. The dorsoplantar concavity is characteristic of both the dorsal and medial components of the joint. The extensive medial migration of the articular area remains the basis for grasping related mechanics in other euprimates.

At such an unmodified sellar surface three movements are allowed. The two slides (in dorsoplantar and mediolateral directions) are greatly limited by ligamentous constraints and by the overlap of the metatarsal and entocuneiform facets. The predominant movement at this joint is an arcuate swing, which must be accompanied by a conjunct rotation. Thus, the metatarsal rotates slightly laterally and dorsally as it is adducted. As MacConaill (1946) has shown, a sellar joint is the more efficient system in those cases where, as in the opposition of the pollex and hallux, in addition to a swing, a rotation of the bone is required. Due to aspects of the geometry of sellar surfaces, a swing upon a sellar surface produces more conjunct rotation than does a swing of the same magnitude upon an ovoid surface. Therefore, the necessary rotation can be accomplished more efficiently at a sellar joint than at an ovoid one. The advantage of a sellar joint, like the euprimate EMt1J, lies in this efficiency with which conjunct rotation can be produced. Since motion is largely restricted to a single arcuate swing, the joint is relatively stable, a point of particular importance in a highly mobile articulation. The same morphology however, also has the possible disadvantage of not allowing adjunct (independent) rotation of the hallux. Similarly, in the case of the australidelphian marsupial condition discussed above, it is probably these factors which guided the evolution of a more open and less stable pattern from the more restrictive ameridelphian one.

On the lateral side of the entocuneiform in notharctines the EMt1 facet is extensive on the plantar half of the bone. The conarticular surface on the metatarsal is primarily concave with the noted sellar component for the joint. The distolateral facet on the entocuneiform articulates with a broadly based narrowing extension onto the strong and plantarly twisted peroneal process of the first metatarsal. The notharctine EMt1J is very similar to that of *Lemur* (see Figure 9 for adapines, and Figures 10–11 for the articulation of the distal tarsals in some living strepsirhines). There seems to be one close-packed position when the hallux is completely abducted. In the extreme of adduction a slight spin and dorsoventral adjustment of the hallucial metatarsal seems possible. The plantad twist of the peroneal process is an indication of the power of the grasp. Also, we believe, it is a suggestive clue that the medial migration of this joint, in order to facilitate extremes of abduction, was brought about with the aid of the tightly bracing facet on the dorsal side of the peroneal process (see Figures 8 and 9).

The double articulation of the entocuneiform with the mesocuneiform, as noted before, is universal in euprimates. The proximal and distal facets are separated by a nonarticular gap so that the EMcp facet is contiguous with the NEn facet, and the EMcd facet is contiguous with the EMt2 facet. The distal entocuneiform–mesocuneiform (EMcd) facet is only slightly offset from the entocuneiform–second metatarsal (EMt2) articulation in notharctines. This offset is slightly more exaggerated in living lemuriforms and in some anthropoids, in which, as a rule, this offset is small (e.g., *Alouatta*, see below). Even in *Daubentonia* (where slow moving generates different loads in the joints from those of leapers) the EMcd and EMt2 facets are offset from one another. This condition reaches an extreme expression in lemurids and indriids (offset about 90°). This character complex reflects changes in the position and/or relative size of the mesocuneiform (which is smaller in lemuroids than in notharctines or anthropoids). The relationship between these aspects of the tarsus and the abducted hallux, or other aspects of hallucial function or biorole, is not yet clear.

Although the EMcd and EMt2 facets are not drastically offset from one another in notharctines or in omomyids, the medial position of the EMt1 facet on the entocuneiform leaves no question that habitual abduction of the hallux, although not as extreme as in living lemuroids, was clearly an advanced primate attribute of the first known euprimates (i.e., a primitive character in this taxon).

It is likely that the large peroneal process of the first metatarsal of adapids (Figure 9) reflects an equally hypertrophied peroneus longus involved in the grasping and leaping adaptation of the first euprimates. This muscle, an everter of the pes and adductor of the hallux, is an important part of the grasping mechanism in most modern primates (especially lemuriforms). It probably served the same bioroles in the first euprimates. The elongation of the process may be related to the resistance of tensile loads placed upon it by

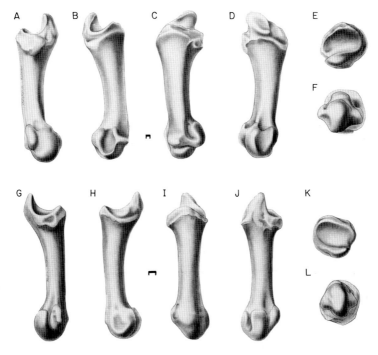

Figure 9. First right metatarsal of the late Eocene adapine adapid *Leptadapis magnus* (A–F) and left metatarsal of *Adapis parisiensis* (G–L) from the phosphorites of Quercy, Montauban collection. Posterior (A, G), anterior (B, H), dorsomedial (C, I), plantolateral (D, J), proximal (E, K), and distal (F, L) views, respectively. The designated anatomical positions are those of a quadrupedal lemur (and not of a human) grasping a horizontal branch with its pes (abbreviations as in Table 1; scales represent 1 mm).

the peroneus longus and to the lever advantage for the peroneus longus while rotating the hallux during adduction and leaping. The load generated during pushoff was probably resisted at the distal end of the tarsals, and thus involved the EMt1J. A quick look at the relative size differences in the peroneal process in *Microcebus* compared to the vertical clinging and graspleaping *Avahi* appears to support this causal connection (see Figure 10). Allometry is clearly not a factor here as the large adapines have relatively small processes.

The heavy layer of bone plantad to the articulation suggests to us that the process may also have an additional mechanical function, possibly as a buttress, related to those compressive loads in the joint which originated from leaping. Arboreal marsupials have an enormous peroneus longus (gleaned from field dissections of phalangerids and didelphids) but have only a relatively small peroneal tuberosity on the first metatarsal. This structure, unlike the first metatarsal of most primates, tends to be stubby and robust, like those in *Perodicticus* and *Nycticebus*. The marsupial condition is completely unlike the euprimate peroneal process of the Eocene adapids and omomyids or the one in extant lemuroids.

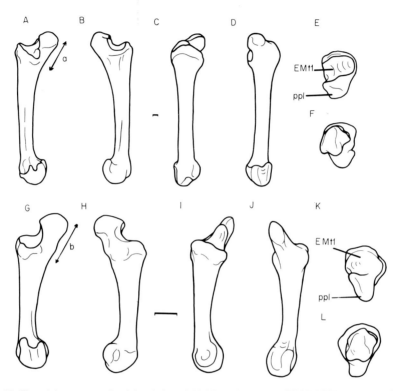

Figure 10. First right metatarsals of the cheirogaleid *Microcebus murinus*, USNM-VZ No. 83657 (A–F), and the indriine indriid *Avahi laniger*, USNM-VZ No. 83652 (G–L). Note the relatively more attenuated lever, associated with the peroneus longus, of the habitually vertically graspclinging and leaping *Avahi* (b) than that in *Microcebus* (a), compared to the length of the metatarsals. Both metatarsals were drawn the same length. From left to right views as in Figure 9 (abbreviations as in Table 1; scales represent 1 mm).

The possibility of a buttressing function in the joint by the peroneal process in the euprimates (except in metatarsifulcrumating anthropoids; see below) is supported by the observations of Gebo (1987), that *Tarsius* (discussed below), which has a reduced and relatively weak peroneus longus, retains a robust and hypertrophied tarsiiform peroneal process, albeit a relatively smaller one than is found in galagos. What may be significant is that in spite of the reduction of the peroneus longus, tarsiers still continue to land and grasp at great speeds, and therefore load the joint as do other relevant euprimates, like graspleaping lemuriforms (see Figures 10, 11, and 14). This fact may well explain the retention of a large peroneal process as a buttress.

Adapinae. The first metatarsal of *Adapis* and *Leptadapis* (Figure 9) contrasts with that of notharctines in the relatively less robust peroneal process (especially in *Adapis*). Both adapine taxa display the reduced depth of the process for the peroneus longus. The great proximal width of the bones is suggestive of a lorisine-like grasping complex, unlike the constellation of characters seen in notharctines. The first metatarsal evidence, coupled with the information gleaned from the proximal tarsals, is concordant with the interpretation of *Adapis* as a slow climber, somewhat lorisine-like, whereas *Leptadapis* displays a somewhat more conservative morphology, judged at least from its astragalus (Dagosto, 1983).

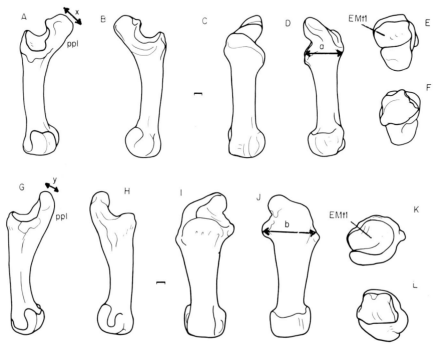

Figure 11. First right metatarsals of the galagine *Galago crassicaudatus*, USNM-VZ No. 267588 (A–F), and the lorisine *Nycticebus coucang*, USNM-VZ No. 30000 (G–L). Note the relatively greater depth of the joint suggesting greater resistance to dislocation in galagines and other lemuriforms (x), compared to the shallower joint of lorisines (y). The latter have relatively much wider metatarsals proximally (b) than other lemuriforms (a). Both metatarsals were drawn the same length. From left to right views as in Figure 9 (abbreviations as in Table 1; scales represent 1 mm).

Lemuriformes

The living and subfossil lemuriforms, while they retain an EMtlJ similar to that described for notharctines, display a modest range of grasping modifications in the entocuneiform and first metatarsal. A comparison of the first metatarsals of *Microcebus* and *Avahi* (Figure 10), in which both bones were drawn to the same absolute length, suggests that the process for the peroneus longus is a relatively longer lever in the habitually vertically graspclinging and leaping (Szalay & Dagosto, 1980) indriid than in the cheirogaleid, a less habitual leaper. Another graphic comparison of the first metatarsals of a species of *Galago* and *Nycticebus*, both drawn to the same length (Figure 11), suggests that in graspleapers like galagines (and all nonlorisine lemuriforms) the relatively greater depth of the metatarsal articulation compared to lorises might be causally related to greater resistance to dislocation. This in turn may reflect the loading due to the leaping component of the grasp. The joint in lorisines is shallower, with relatively wider metatarsals proximally than other lemuriforms, suggesting very different loading of this area. The broadly distributed stress resistance in the lorisine EMtlJ correlates well with the deliberate, powerful grasp and tensile loading (when hanging).

The lorisines show a characteristic shortening of the entocuneiform as one expects from their secondarily shortened climbing (as opposed to climbing and leaping) adapted tarsals (Figures 8D, E, F). In spite of the change in shape and resultant differences of mechanical

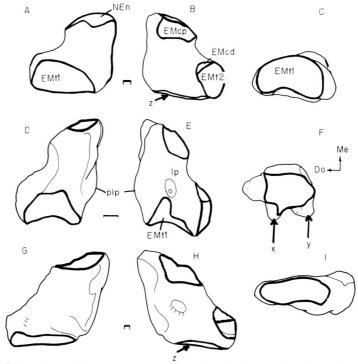

Figure 12. Right entocuneiform of the tarsiiform omomyid *Hemiacodon gracilis*, AMNH 12613, middle Eocene (D, E, F), compared to its homologs in the cercopithecid *Cercopithecus aethiops*, personal collection (A, B, C), and the subfossil archaeolemurid *Archaeolemur* sp., Academie Malgache, unnumbered (G, H, I). The complete lack of medial extension of the EMt1 facet is shown by "z" (B, H). The medially rotated plantar process, "y", and the hypertrophied lateral process "x" are shown in *Hemiacodon*. Medial (A, D, G), lateral (B, E, H), and distal (C, F, I) views, respectively (abbreviations as in Table 1; scales represent 1 mm).

function, the EMtlJ facets have retained an essentially primitive euprimate configuration. The lorisine EMt2 facet is retracted from the vicinity of the EMt1 facet, the latter the condition found in primitive euprimates and most other lemuriforms. The potentially most interesting taxa, such as *Megaladapis* and the palaeopropithecines, which could reveal aspects of constraints which influence the direction of morphological change in this joint, are unknown in this regard.

The entocuneiform of *Archaeolemur* (Figure 12) is probably the most distinct (among those known) in displaying a drastically reduced, mediolaterally narrowed EMt1 facet which, however, retains a basically sellar configuration. This is undoubtedly related to the reduced role of the hallux in locomotion and it is equally likely that the hallux was not habitually abducted. There is nothing in the joint, however, which would have prevented the hallux from performing grasping while climbing. Whether the pattern is due to the hallux having been reduced as a result of some ancestral adaptation related to pedal hanging, or to the cursorial habitus of the genus (as gleaned from other evidence) is difficult to decide. What is clear is that this bone was not bearing the kind of compressive loads we can infer in other lemuriforms.

The shared similarity of the form-function complex between the lemuroids (and to a large extent that seen in the cheirogaleids and galagines) with the notharctines is compelling evidence for interpreting this pattern to be the primitive lemuriform pattern. The most significant departure from the notharctines is in the offset of the EMt2 and EMcd facets.

Tarsiiformes
The entocuneiform and first metatarsal are known in *Hemiacodon* (Simpson, 1940; Szalay, 1976), and in addition the first metatarsals are shown here for cf. *Arapahovius* (Savage & Waters, 1978) and *Necrolemur* (see Figures 12–13).

The meager sample of omomyid entocuneiforms (Figure 12D, E, F) and first metatarsals displays the following characteristics not seen in notharctines or lemuriforms: (a) a hypertrophied tuberosity (the lateral process of the entocuneiform), probably for the entocuneiform–second metatarsal ligament; (b) laterally deflected large plantar process of the entocuneiform; (c) hypertrophied peroneal process on the first metatarsal compared to known contemporary adapids; (d) less elongated naviculoentocuneiform (NEn) articulation than in notharctines and lemuroids; (e) mediodorsal segment of the EMt1 facets is relatively longer compared to the plantar side than in the adapid or lemurid condition; (f) slightly less prominent EMcd and EMt2 facets, like in tarsiers, than those seen in notharctines and lemuriforms.

The hypertrophied lateral process of the entocuneiform seen in *Hemiacodon* (Figure 12F) also occurs in *Tarsius* (Figures 17A, B, C), and it is prominent in such platyrrhines as *Saimiri* (Figures 17D, E, F). This may represent a haplorhine synapomorphy, as this does not occur with any degree of prominence in lemuriforms known to us. An argument against this hypothesis, however, is the large, albeit differently sculpted, lateral process noted in *Plesiadapis*. The hypertrophied lateral process in *Hemiacodon*, together with the remaining tarsal evidence, supports the notion of habitual and explosive graspleaping. Loads generated as a result of such activity would result in increased need for keeping the tarsus firmly bound.

The large plantar process of the entocuneiform seen in notharctines and plesiadapids is present but it is characteristically deflected so its leading edge points laterally (see Figures 12D, E, F).

The proximal end of the entocuneiform of *Hemiacodon* and the corresponding male conarticular surface of the navicular is distinct from the notharctine and lemuriform conditions. Unlike in the latter, in *Hemiacodon* and cf. *Arapahovius* the much shortened ovoid articulation approaches the condition found in tarsiers in its brevity, albeit not in the ball-like shape of that articulation. Galagos also have small, shortened entocuneiform-navicular facets, but not the rotation displayed by tarsiers. According to the characters listed by Gebo (1987), omomyids share none of the features indicating the novel foot rotation of tarsiers. One of us (FSS), however, still believes that omomyids may have shared an incipient stage of this foot rotation mechanism. The small, unelongated ovoid NEn facet of cf. *Arapahovius* is particularly striking when compared to the invariably elongated ovoid homolog in all notharctines and lemuriforms. This may be highly significant and may point to an early functional stage from which the tarsiids pursued their extreme pedal modification, a topic recently analyzed by Gebo (1987). Furthermore, one of us (FSS) finds it significant that in both omomyids *Hemiacodon* and cf. *Arapahovius*, and therefore probably in the common ancestry of the Omomyidae, the naviculocuboid contact is well removed from the naviculomesocuneiform (NMc) articulation. This condition is similar to tarsiers, but unlike in notharctines and lemuriforms and is also distinct from the putative primitive eutherian condition. According to his interpretation the strepsirhines

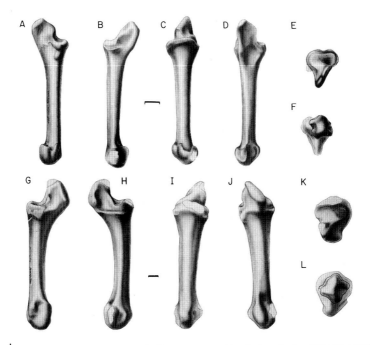

Figure 13. First left metatarsal of an early Eocene omomyid, cf. *Arapahovius*, UCMP 113310, Bitter Creek, loc. 70214 (A–F)), and first right metatarsal of the late Eocene omomyid, probably *Necrolemur*, Phosphorites of Quercy, Montauban collection (G–L). Views as in Figure 9 (scales represent 1 mm).

may also display an equally derived euprimate condition. MD, on the other hand, believes that the haplorhine condition is retained unchanged from the eutherian condition, and that it is the strepsirhines which exhibit a unique derived euprimate condition.

Perhaps the most striking special similarity of known omomyid distal tarsal features, shared also with *Tarsius*, to the primitive anthropoid condition is the enlargement of the dorsal half of the EMt1 articular surfaces compared to the reduced plantar portion (character listed above e). If our interpretation of the homology of the anthropoid joint morphology is correct (in that we believe that the ovoid articulation of these is the dorsal part of the joint of strepsirhines and tarsiiforms; see below), then it is likely that the omomyid condition is antecedent to the one seen in platyrrhines. This is a critical connection in the understanding of the direction of transformation because it represents an intermediate condition between the notharctine-primitive strepsirhine condition and the primitive anthropoid one.

The difference between adapid and tarsiiform EMt1 facets must be briefly explained here. When looking at the metatarsal EMt1 facet, the line that can be drawn along the highest point of the ridge of that saddle is the line of demarcation of what we call the dorsal and plantar portions of that facet, respectively. This is most obvious on the metatarsal facet (see Figures 15 and 16), but it can be discerned on the entocuneiform as well. This is a subtle but important difference, as it suggests a directional change in the tarsiiform feature

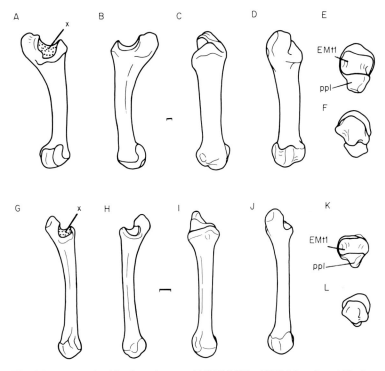

Figure 14. First left metatarsals of *Lepilemur leucopus*, AMNH-M No. 49668 (above), and *Tarsius* sp. (pers. coll.; below). "X" marks the articular area with the entocuneiform. Views as in Figure 12 (abbreviations as in Table 1; scales represent 1 mm).

towards the protoanthropoid emphasis on the dorsal half of the joint. The full mechanical significance of this difference from adapids is not clear, and the associated biorole cannot be deduced. Consequently, the potential phylogenetic significance of this character is difficult to assess.

The hypertrophied peroneal process of the Mt1 of *Hemiacodon* has been commented on by all students who studied these remains (Simpson, 1940; Szalay, 1976). The consensus has been that it is related to the powerful grasp through the tensile loads placed on the bone by the peroneus longus. But as we have noted above, the size of the process may not be causally related to the size of the peroneus longus alone, but may also be a buttress, related to the loading of the joint while leaping and landing. It appears to us that the originally well-developed process for the peroneus longus (this area is as yet unknown in archaic

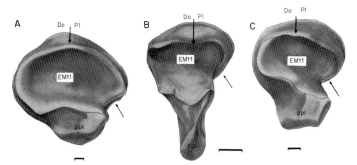

Figure 15. Comparison of the left metatarsal EMt1 facets, in proximal view, of the adapid *Leptadapis magnus* (A), the omomyid *Hemiacodon gracilis* (B), and the lemurid *Lemur catta* (C). The arrow pointing downward indicates the approximate subdivision of the joint into a dorsal (to the left) and a plantar (or ventral; to the right) half, demarcated by the most proximal line of the saddle joint on Mt1. The arrow pointing towards the plantar (or ventral) portion of the facet draws attention to that part of the joint which is plantar of the facet extending onto the dorsal surface of the peroneal process (abbreviations as in Table 1; scales represent 1 mm).

primates), a response for tensile loading by the tendon, was further transformed. This transformation, driven by selection resulting from graspleaping, manifested itself both by bone hypertrophy related to the compressive loads generated on the joint surface which is located on the articular side of the process, and by the increased lever advantage that a powerful joint conferred on animals which habitually employed this locomotor behavior.

The significance of relatively small distal mesocuneiform and narrow second metatarsal contact of the entocuneiform in omomyids, as in tarsiers, is difficult to evaluate. It may well represent a homologous tarsiiform condition or merely a convergence. Until foot function is better understood in omomyids, this remains a rather moot point.

As described above, the remarkably underdeveloped joint contacts of the entocuneiform –hallucial joint of *Tarsius* supports Gebo's (1987) account that grasping in that genus is accomplished mainly through the action of intrinsic muscles, rather than the peroneus longus. In comparing the relative proportions of the entocuneiform–hallucial complex to various strepsirhines, in particular to such leapers as *Lepilemur*, one is struck by the relatively smaller articular surfaces of the EMt1J of *Tarsius* (Figure 14). Although we did not attempt to quantify the difficult joint surface area to bone size relationships, we found the graphic comparison of the first metatarsals, drawn to the same size, highly instructive

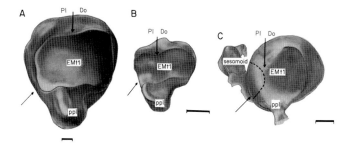

Figure 16. Comparison of the right metatarsal EMt1 facets, in proximal view, of the adapid *Northarctus* sp., AMNH 13050 (A), *Tarsius* sp., personal collection (B), and *Saimiri sciureus*, personal collection (C). The arrows, as in Figure 15, delimit the plantar portion of the facet (abbreviations as in Table 1; scales represent 1 mm).

(Figure 14). We suspect that the relative size of the articular areas in joints, as in *Lepilemur* and *Tarsius*, probably reflects the habitual loading to which these joints are subjected. Given this assumption, it is intriguing to note that the shape of the proximal first metatarsal in *Tarsius* is comparable in every detail to that of known omomyids, but to cf. *Arapahovius* in particular (Figures 13G–L). Our comparison shows that this genus, like *Tarsius*, seems to have not only a relatively small EMtlJ surface area, but also a reduced proximal ridge to which the peroneus longus probably attached. We have already commented on the persistence of a large process for the peroneus longus in *Tarsius*, in spite of the reduction of the tendon itself. Whether this reflects joint loading and/or a persistence of an ancient tarsiiform character predating the reduction of the tendon and switching to grasping primarily by the intrinsic musculature of the foot is a moot point at present.

From the various modifications (such as the *combination* of the large lateral process of the entocuneiform and the orientation and size of the entocuneiform plantar process, the proportion of the various facets, etc.) it appears, at least as a working hypothesis, that the grasping mechanism in omomyids has diverged and has undergone some reorganization from the putative ancestral euprimate pattern probably well mirrored in notharctines. The relatively great length of the navicular and the hypertrophied peroneal process of the first metatarsal are simply indicative of the modification of general tarsal elongation and maximization of hallucial grasp associated with specialized leaping behavior. In our view some early haplorhine (which could be called an omomyid) might well have been ancestral to the younger tarsiids. Gebo's (1987) account of the leaping related foot function in *Tarsius*, however, makes it clear that the euprimate grasp, which is expressed to an extreme in omomyids like *Hemiacodon*, has been drastically reorganized in *Tarsius*. The peroneus longus is dramatically reduced, and this lineage evolved a unique method of foot rotation to allow inversion through the rotation of the navicular at both ends. Interestingly enough, however, the laterally pointing leading edge of the plantar process on the entocuneiform in *Hemiacodon* still persists in *Tarsius*, although in a reduced form. If the functions of this bony crest are as suggested, then its presence in an omomyid-like form in the living tarsiids might indicate a heritage morphology unaffected by the highly evolved alternate grasping system of tarsiids.

In the following discussion of anthropoids it is important to remember that the proportion of the EMT1J in the haplorhine *Tarsius*, as reflected on the proximal metatarsal, is similar to that of omomyids: the plantar component of the joint is distinctly relatively smaller than we see it in adapids and lemurids. There is no doubt that the hypertrophy of the dorsal portion of the joint is a necessary precondition to the increase and transformation to the ovoid anthropoid condition discussed below. The dorsal part of the joint, however, is not ovoid in either omomyids or *Tarsius*. This critical transformation likely occurred in a lineage in which the selective advantage of the extremely stable EMtlJ was of lesser importance than the more mobile anthropoid derivative condition.

Anthropoidea

The most superficial and most striking observation, yet phylogenetically perhaps the most important one, is that when one contrasts the habitually abducted hallucial position of strepsirhines and nonanthropoid haplorhines with that of the non-hominoid anthropoids, the latter clearly show less habitual abduction. Anthropoids, particularly platyrrhines and cercopithecoids, have a hallux that is relatively smaller than the adapid or lemuriform one. Yet, in spite of the less-abducted hallux, anthropoids appear to be able to move the big toe in a greater range of positions than lemuroids. This inference is based on the simple manipulation of the first metatarsal on the entocuneiform. The non-sellar and less concave anthropoid EMt1 facet is much less constrained in its swings and rotations on the entocuneiform than the nonanthropoid euprimate equivalent permits.

The small peroneal process of the first metatarsal and the reduced angular value of the EMt1 facets support the behavioral observation, and furthermore the morphology of the EMt1 facets clearly points to the direction of change. In spite of the lack of habitual

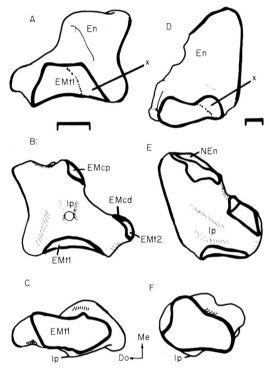

Figure 17. Comparison of the right entocuneiforms of *Tarsius* (A, B, C) and *Saimiri* (D, E, F), both personal collection. Note the area designated as "X" which corresponds to the plantar portion of the metatarsal EMt1 facet delimited on Figures 15 and 16. Medial (A, D), lateral (B, E), and distal (C, F) views, respectively. Abbreviations as in Figure 1 (scales represent 1 mm, and arrows depict anatomical directions, for a quadrupedal primate, for the distal views on bottom).

abduction of the hallux, the joint contact on the entocuneiform still extends to the medial side of the bone, attesting to the fully and habitually abducted ancestral position.

The anthropoid morphocline displays a varied enough pattern to assure a reasonable understanding of the transformations which occurred. This is being pursued in far greater detail by Strasser (in prep.) than we deal with here. As perhaps expected, the non-callitrichine platyrrhines display patterns which appear to be the appropriate ancestral stages for the grasping related morphology of the living catarrhines (see especially Figures 18–20). The joint in platyrrhines has been discussed in detail by Lewis (1972) and Wikander *et al.* (1986), and both have provided ample illustrations and examples. The salient points of the joint found in a platyrrhine like *Cebus* are the modified ovoid on the dorsal part of the entocuneiform EMt1 facet, and the plantar facet for the sesamoid separated by a slight furrow from the former. The conarticular metatarsal facet is a trough-like surface with virtually no restrictions for translation in a dorsoplantar direction.

We have, we believe, a strong indication that the platyrrhines derived their joint complex from a condition similar to that described for the Paleogene euprimates. A telling minor morphological peculiarity of the entocuneiform of *Alouatta* is that a raised bony surface of the lateral distal end of the bone lacks the lateral extension of the EMt1 facet seen in Paleogene forms and lemuriforms. This is analogous to the situation seen on the

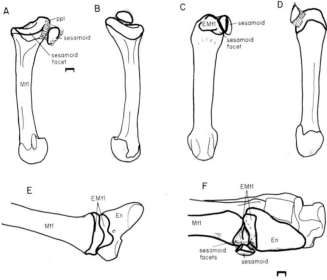

Figure 18. Position and articular relationships of the sesamoid associated with the platyrrhine EMt1J. Right metatarsal (A, B, C, D), left metatarsal and entocuneiform (E), and medial side of partial right foot (F) of *Saimiri sciureus*, personal collection (abbreviations as in Table 1; scale represents 1 mm).

sustentaculum of marmots (*Marmota*) which have the highly derived raised bony surface of tree squirrels but lack the actual joint contacts of the latter (Szalay, 1985, Figure 9). This condition in *Alouatta* is one of the clues suggesting that the platyrrhine condition is derived from the Paleogene euprimate pattern.

In his contribution on the EMt1J in anthropoids Lewis (1972, Figure 2) has given an excellent descriptive illustration of the joint in *Cebus*, showing the range of hallucial movement while the latter is abducted into the close-packed position. The position and articulation of the opposing surfaces and of the sesamoid are clearly depicted. In Figure 18 and in the various figures of Lewis (1972) and Wikander *et al.* (1986) the articular facets for the sesamoid are shown in various platyrrhines and hylobatids. Given the nature of contact on bone by synovial membrane and ligaments, we are not confident (nor are Wikander *et al.*, 1986) that the presence of the homologous sesamoid can be unequivocally established from articular facets on the entocuneiform or the first metatarsal. Nevertheless, Conroy's (1976) suggestion (see also Conroy & Rose, 1983) that a Mt1 attributed to *Propliopithecus zeuxis* has the appropriate faceting is very much expected in light of early catarrhine similarities to platyrrhines (Fleagle, 1983).

In callitrichines we see a sharp departure from the condition displayed by the remaining platyrrhines (Figure 19). In contrast to the latter, marmosets eliminate the medial extension of the EMt1 facet of other anthropoids so that the joint surface is largely restricted to the distal end of the entocuneiform. Although the joint is wide (not unlike in plesiadapids), the arc is drastically reduced compared to either the protoplatyrrhine or the euprimate condition. There can be little doubt that the reason for the change in this joint is historically tied to the suspensory loading of the cheiridia (i.e., claw reacquisition) compared to the compressive loading which operates in other euprimates. Although grasping can be performed by the reduced hallux of marmosets, we believe that it performs a less significant role.

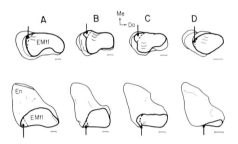

Figure 19. Comparison of the left entocuneiform in distal (above) and lateral (below) views, respectively. From left to right: A, *Pithecia pithecia*, USNM-VZ 241007; B, *Aotus trivirgatus*, USNM-VZ 396155; C, *Saimiri sciureus*, USNM-VZ 396197; D, *Saguinus geoffroyi*, USNM-VZ 257345. The arrows indicate the approximate portions of the EMt1 facet which articulates with the sesamoid of the joint (see especially Figure 18) (abbreviations as in table 1; scales represent 1 mm).

What the exact nature of foot mechanics is in marmosets compared to other platyrrhines, and how the biorole and mechanics of the foot and joints is integrated with the reacquisition of claws (but the retention of hallucial tegulae), is an intriguing and complex problem yet to be tackled. This is an area of future research where field studies should be at least as important as any further laboratory investigation of this and other similar problems.

In the extant gibbons and cercopithecids the joint for hallucial grasping is an elongated modified ovoid. The plantar "furrow" on the medioplantar side of the entocuneiform facet, so prominent in platyrrhines, is much less prominent in most catarrhines (detailed investigation of these is in preparation by Strasser). As noted above, this latter condition is a convincing connection between the more primitive euprimate joint of a basically sellar configuration and the well developed, but more primitive, anthropoid condition displayed by the last common ancestor of known platyrrhines.

Among the extant catarrhines (which we only mention briefly) the cercopithecids stand out in two important characteristics related to the EMt1J. The entocuneiform articular facet is largely on the medial side of the bone with no lateral extension (in effect the facet faces distally, see Figure 15), and the arc of that facet appears to be the smallest angular distance within the catarrhines. There is certainly no comparison among platyrrhine vs. catarrhine "monkeys" in regards to this character complex. Unlike Lewis (1972), we see no "monkey grade" similarities between platyrrhines and cercopithecids in the hallucial joint. Whatever similarities persist are unequivocally primitive anthropoid and euprimate retentions.

In hylobatids and *Pan* the angular distance of the joint surface on the entocuneiform is large (at least 90°), as one expects from such habitual climbers and graspers, but the peroneal process of the first metatarsal is not prominent, not more so than one observes in cercopithecids. Both in hylobatids and *Pan* there is an expansion of the distal facet on the entocuneiform to the lateral side of the bone both dorsally and plantarly.

The EMtlJ facets of the Miocene hominoids from Africa (Langdon, 1986) and of the recently studied remains of *Oreopithecus* (Szalay & Langdon, 1986) yield important information about these species. In the powerful grasper and climber *Pan* the facet on the entocuneiform is primarily a male ovoid, and the plantar and medial side displays a faint furrow, homologous to the pronounced one of KNM RU 5872 and 2036, specimens of large and small *Proconsul* species. In the Miocene forms the morphology of the entocuneiform facet is platyrrhine-like with a large plantar concavity on the medioplantar side of the joint

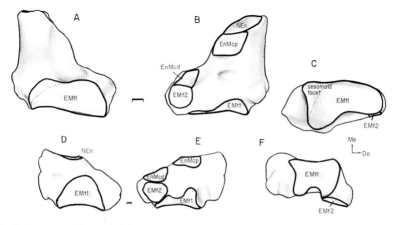

Figure 20. Left entocuneiform of *Cebus* sp., personal collection (A, B, C) and *Indri indri*, personal collection (D, E, F). Abbreviations as in Table 1 (scales represent 1 mm; arrows depict anatomical directions, for a quadrupedal primate, for the distal views on the right).

and a pronounced ovoid on the mediodorsal side. This is essentially similar in *Oreopithecus* also. In the Miocene catarrhines, as in several platyrrhines, the lateral extension of the entocuneiform EMt1 facet on the distal surface is pronounced. In *Pan* the mediodorsal ovoid of the entocuneiform is hypertrophied compared to the platyrrhine and Miocene catarrhine conditions, so that in effect the medioplantar concavity of the entocuneiform facets is all but eliminated. Figures in Szalay & Langdon (1986) clearly show that in *Pan* the ovoid configuration of the joint is the dominant one on the distal contour of the entocuneiform.

The EMtlJ morphology of the Miocene forms (*Oreopithecus* included) is decidedly more primitive than that found in hylobatids and cercopithecids, in that these latter families have lost all traces of the ancient euprimate sellar articulation.

Discussion and Conclusions

In the foregoing section we have concentrated on the comparative and functional morphology of selected living and fossil taxa and the available relevant Tertiary fossil material of primates. In the following section of the paper we will attempt to synthesize the comparative and functional information we were able to glean into a paleobiological account, discuss particularly important homologies, and attempt to reconstruct an evolutionary account (a phylogeny) of the hallucial grasping complex.

The earliest primates

The EMtlJ of the earliest eutherians has not been analyzed as carefully as the subject warrants. The first known EMtlJ facets in the archaic primates, as described for *Plesiadapis*, display a relatively large and broad sellar articulation at the distal end of the entocuneiform. This complex allowed relatively much less motion of the hallux on the entocuneiform than we see in the earliest euprimates. Although this is somewhat similar to that seen in *Ptilocercus*, and thus may point to an ancient archontan tie, it differs in its details from any known marsupial to such a degree that any hypothesis linking primate (or archontan) and marsupial arboreality as homologous seems unacceptable to us. Added to the evidence marshalled here, is of course, that discussed from the more proximal tarsals (Szalay, 1984, 1985).

The archaic primate-euprimate transition

Based on what we can discern from early Eocene adapids and omomyids, the contrast between the euprimate grasping complex and that seen in plesiadapids is considerable. The apparent magnitude of these differences, however, may be exacerbated by the lack of well known archaic primate feet. The shift from the ancestral archaic primate to the first euprimate probably involved: (1) the medial shift, from a largely distal position, of the entocuneiform EMtl facet, and consequently an increase in the arc of the joint; and (2) a change in form from a less restrictive, low arc, low saddle joint, to a more restrictive one. Although there is nothing we can say as yet of the morphology of the first metatarsal, it is likely that this became hypertrophied in the transition. These features strongly suggest a lemur-like habitual abduction of the powerful opposable hallux of the first euprimates. A female facet on the proximal end of the first metatarsal spanning nearly 180° of arc ensures a joint capable of withstanding enormous loads (unlikely to occur only during grasping and climbing) while the hallux and the opposing cheiridia essentially encircled the branch.

The habitual abduction of the hallux, by itself, is not causally correlated with this enhanced load-bearing ability. This is a unique euprimate solution requiring special explanation. In australidelphian microbiotheres and phalangeriforms a habitually extremely abducted hallux is based on a saddle functionally similar to the articulation seen between the human trapezium and first metacarpal. In the didelphid condition, from which the australidelphian one almost certainly evolved (Szalay, in prep.), the peculiar, hinge-like sellar joint is again accompanied by habitual hallucial abduction. In neither ancestral nor descendant metatherian conditions is there a large peroneal process on the first metatarsal, although these marsupials have an enormous peroneus longus. Euprimates, on the other hand, have a diagnostically hypertrophied peroneal process on the first metatarsal, and this process also bears on its dorsal surface, significantly, a large portion of the EMtl facet, in effect buttressing the joint against large compressive loads. It follows from the foregoing that a biorole explanation couched solely in terms of enhanced climbing ability in the protoeuprimate is not a satisfactory answer.

In addition to these features of the EMtlJ, the diagnostically elongated tarsals and deep astragali of Eocene euprimates reflect large compressive loading, and their stable upper ankle joint is capable of rapid extension and flexion along a large angular distance. They are also distinguished by such diagnostic osteological attributes which we find in the hip (expanded area for the gluteus medius) and femur (specifically the long and deep patellar groove).

Biorole explanations for the origin of euprimates must account for all of the unique features of the last common ancestor. We maintain, as we have before (Szalay & Delson, 1979; Szalay & Dagosto, 1980) that rapid, successive, leaping and landing with a habitual grasp (i.e., graspleaping), a mode of travel seen in many lemuriforms, accounts best for the form-function complex probably present in the protoeuprimate. Napier & Walker's (1967) view of the evolution of euprimate locomotion also stresses the importance of grasping and leaping in the common ancestor of the euprimates. The primary difference between their concept of the ancestral euprimate locomotor mode and ours is in their emphasis on vertical clinging and leaping, a habitually practiced behavior in some primates which we interpret to have been derived several times independently (see Szalay & Dagosto, 1980, for a discussion of various primate morphotype locomotor modes).

We do not see a single "grade" feature (an attribute attained independently in several lineages) in either the skull or skeleton in subsequent primate clades, with the exception of the ubiquitous encephalization occurring in most eutherian lineages. We maintain that characters often cited as representing "trends", implying multiple pathways, are primitive euprimate features, inherited with sundry modifications in the descendants. By a trend we understand a unilinear character transformation.

The graspleaping locomotion of the first euprimates is possibly most significant in its far reaching adaptive consequences on the senses, on the association areas of the brain, and on reproductive strategy. The increased speed of locomotion and the precariousness of landings dictated the selective forces which adaptively constrained the euprimates towards reducing the number of young per pregnancy, and shaping a nervous system capable of coping with information reaching its association areas much faster than before. Euprimate stereoscopy coupled with the orbital ring is probably also a direct consequence of the risks imposed on an animal as a result of high speed arboreal graspleaping (see also Szalay, 1975). We are aware that increase in body size, increase in life span, and living on islands have been also correlated with the reduction of the number of young per pregnancy. Similarly, the functional or biorole reasons for orbital rings are many, judged by the widely different distribution of this character outside of the euprimates.

Haplorhines

The role of the omomyid evidence is critical in our evaluation of the broad outlines of the history of primate pedal grasping and correlated locomotor behaviors. Omomyids not only retain the fundamental solutions of the euprimates to pedal grasping (as do the notharctines) but also document some changes early in that clade which indicate their close ties to the tarsiers. Thus, these ties are suggested not only from cranial characters (Cartmill, 1972; Szalay, 1976; Rosenberger & Szalay, 1980), but also from other areas of the skeleton (Szalay & Dagosto, 1980; Dagosto, 1985). Our present assessment of the EMt1 articulations also points to a subtle resemblance between omomyids and anthropoids, suggesting that omomyid morphology has the necessary preconditions to evolve into the distinct patterns seen in both tarsiers and protoanthropoids.

We find a number of similarities in the form of the naviculo-entocuneiform joint in omomyids and haplorhines. In cf. *Arapahovius* (as gleaned from the distal navicular), and in

Hemiacodon, but not in known adapids and lemuroids, the contact is much shortened along the transverse dimensions of the navicular. The articulation is ovoid, as it is in both tarsiers and strepsirhines, but much shortened as in tarsiers and galagos. Omomyids appear to be more similar in the pattern of distal navicular facets to tarsiers and platyrrhines than they are to adapids or lemurids. It is particularly interesting that the shortened naviculo–entocuneiform joint as well as the exclusion of the EMc facet on the navicular from the naviculo–cuboid contact is also present in such diversely adapted platyrrhines as *Cebus* or *Alouatta*. To one of us (FSS) this appears to be a haplorhine modification carried to the extreme in tarsiers. The few species of omomyids we know are therefore like primitive euprimates in retaining (and somewhat augmenting) an emphasis on a powerful hallucial grasp, but show a characteristic combination of haplorhine features. These are the hypertrophied lateral process of the entocuneiform for the EMt2 ligament (also present in plesiadapids), a larger mediodorsal surface on the EMt1 facets than on the plantar part of the joint, and perhaps the shortened navicular facet for the entocuneiform described above.

Anthropoidea

Anthropoid origins. The morphology and inferred function of omomyids, when compared to platyrrhines, is indicative of a possible evolutionary link between the former and the protoanthropoids. In the omomyid EMt1J there is a clearcut subdivision in the joint (see especially Figures 15 and 16). The proximal joint surface of the hallux clearly reflects the functional division in that the dorsal surface is large in contrast with the plantar one. This offset cylindrical plantar side of the joint appears to be a homolog of the area in platyrrhines which is both reduced and is in articulation with the proximal hallucial sesamoid. The subdivision of this joint in the manner described and the subsequent possibility of a homologous similarity between the dorsal components of the joint in omomyids and platyrrhines suggests an omomyid–protoanthropoid transformation. In other words, the transformation of the medial and dorsal articular area of the EMt1J into a modified ovoid from a fully sellar joint of a euprimate may represent a recognizable and therefore useful primitive anthropoid character. It appears that the first anthropoid built on this condition and reduced the plantar half of the articulation by expanding the dorsal one into the modified ovoid shape characteristic of anthropoids. Judged from the implications of this hypothesis the extreme, and therefore probably habitual, abduction in the haplorhine ancestor was not eliminated while changes for increased circumduction of the hallux were under way.

The anthropoid "prehallux". The homology of the bone found in the EMt1J of platyrrhines and hylobatids has received detailed attention from Lewis (1972). He argued that this element is a prehallux (what we call a sesamoid; see Figure 18), homolog of an archaic amniote tarsal bone medially articulating with the proximal part of the entocuneiform in metatherians and relatively primitive eutherians. Lewis (particularly 1972, but see previous references cited therein) argued essentially as follows. The flexor tibialis of early therians (i.e., what he finds in living metatherians), with its triple tendinous insertions is attached to (1) the flexor fibularis tendon, (2) the plantar aponeurosis, and (3) the base of Mt1, containing the alleged prehallux. The last branch overlapped the insertion of the tendon of the tibialis anterior onto the entocuneiform. From this ancestral condition he postulated, Lewis assumed that in primates the connection to the flexor fibularis tendon was lost and that this segment became the naviculometatarsal ligament. It is here that

Lewis presents the first homology hypothesis between the hallucial continuation of the flexor tibialis and the naviculometatarsal ligament of primates, the acceptance of which is necessary for the remaining argument. He considered the bone in the naviculometatarsal ligament which is attached proximally to the entocuneiform in many nonprimates (but independent from the tendon of tibialis anterior) to be the strict homolog of the sesamoid found in platyrrhines. This bone in platyrrhines (and hylobatids) receives one of the two branches of the tibialis anterior (see Lewis, 1972, Figure 1, for excellent illustrations and clear descriptions).

It is imperative that we now take careful stock of the necessary preconditions and consequences of the homology hypothesis of the anthropoid "prehallux": (1) the naviculometatarsal ligament of primates has to be a strict homolog of the third tendinous branch of a postulated primitive therian flexor tibialis condition (based on its presence in *Didelphis*, which is thus assumed to be a representative of the primitive therian condition from which eutherians and metatherians evolved their attributes); (2) the insertion of the tendon of the tibialis anterior onto the prehallux as this bone descended into the EMtlJ; (3) the lack of traces of the protoanthropoid condition either in any living lemuriform, fossil strepsirhine, living tarsiiform, or omomyid must be due to the complete loss of any indication of this transformation.

The latter must be true in this argument in spite of the strong probability, deduced from morphological considerations and not contradicted by paleontological evidence, that the joint at the base of the hallux in nonanthropoids more closely approximates the primitive euprimate condition than the one in platyrrhines. In light of the additional fact gleaned from rodents, like the living squirrels, that there can be at least four sesamoids in the medial ligamentous binding of the entocuneiform (Szalay, 1985), we find the arguments leading to the primate "prehallux" hypothesis unconvincing. The fact is that the prehallux (and prepollex) of various amniotes, didelphids and phalangeroids (among others) appear as ossifications early in ontogeny (along with the other ancient tarsal and carpal elements). In fact the prepollex/prehallux are bones which are derivatives of the early segmentation of the cartilaginous blastema of the radiale/tibiale in all amniotes. The absence of an early formation of a "prehallux" is paramount in attempts to explain its homology. In contrast with marsupials in which the prehallucial cartilaginous precursor is very early in ontogeny, there is strong indication that in platyrrhines the presence of the "prehallux" sesamoid is age related (see Wikander *et al.*, 1986, for platyrrhine data). We believe, therefore, that the admittedly inadequate developmental information is unsupportive of the homology hypothesis.

Among living catarrhines, with the exception of the hylobatids (Lewis, 1972; Wikander *et al.*, 1986), there is no sign of a medial sesamoid. It would not be surprising, however, as Conroy (1976) already reported, if the earliest catarrhines had an EMtlJ quite similar to forms like *Cebus* (Figure 20). In spite of such a likelihood both in the Fayum catarrhines and the Miocene forms (e.g., *Proconsul*), as this has been asserted by Lewis (1972), considerable empirical problems exist in the recognition of unequivocal sesamoid facets both on the entocuneiform and first metatarsal (Wikander *et al.* 1986). We have had the identical problems reported by these authors when examining various living catarrhines which we know lack an EMtlJ sesamoid. For this reason, as stated by Wikander *et al.* (1986), we cannot discern the condition of this joint in the Oligocene and Miocene catarrhines. Unlike Wikander *et al.* (1986), however, we are convinced that the platyrrhine and hylobatid sesamoids are homologous as they have the same exact topographical and developmental relationship, and because in no group of primates besides the anthropoids is there any indication for such a hallucial joint supporting sesamoid.

Other anthropoid distinctions. Aside from the sesamoid in the joint, which alone, and somewhat out of its historical and mechanical context, received attention in the past, the differences of the EMtlJ in regard to this element between lemuriforms and haplorhines have not been appraised. Two degrees of freedom in rotation of a sellar articulation will permit considerable stability under the apparent great loads of the omomyid grasp if the articulation is close packed. Any attempt to loosen the articulation (as in the human trapezium-pollex joint; see Napier, 1955) into a less restrictive saddle to gain the ability for some adjunct rotation, however, is likely to decrease the stability of the joint. In becoming anthropoid and gaining the increased motion of an ovoid articulation, the protoanthropoid reduced load bearing in the EMtlJ compared to its ancestry when it reduced the habitual hallucial abduction displayed in non-anthropoids.

In the protoanthropoids, possessing pedal morphology perhaps not unlike the extant *Cebus*, for whatever bioroles, the habitually abducted powerful (more stable) grasp of the hallux may have been modified to a less powerful (less stable) grasp, a hallux more aligned with the other cheiridia. This arrangement permitted a greater range of adjustments through a greater range of adjunct rotation. It is possible that the reduction of the extremely abducted hallux to one which is more nearly aligned with the rest of the foot has been the consequence of a large branch environment which favored a more plantigrade locomotion (Rollinson & Martin, 1981). Such an assessment is based on concrete morphology which suggests both transformational history and mechanical latitudes for hallucial function. It is extremely important to keep in mind that in the living platyrrhines, e.g., *Cebus* and *Alouatta*, the plantar and medioproximal portion of the EMtl facet of the entocuneiform and the reduced plantar component of the Mtl articulation uncannily reflect the ancestral saddle joint. The "new" modified ovoid of that joint on the distal and medial entocuneiform faithfully mirrors the ability of the hallux to align with the rest of the foot.

This particular protoanthropoid solution (the reduction and modification of the euprimate hallucial power grasp) may have set the stage for the exploitation in some lineages, such as atelines and hominoids, of more slow-climbing and less leaping-oriented pedal bioroles, behaviors clearly demanded by larger body size in some circumstances. Needless to say, the protoanthropoid alignment of the hallux did not in any way compromise the ability to climb using a grasp, or the speed of travel of small arboreal anthropoids. But the often reported rapid travel in the canopy by a variety of small and medium sized platyrrhines is more of a gallop (Kinzey, pers. comm.) than the bold and rapid leaping of the lemuroids. The lemurs characteristically employ an habitually abducted hallux even when slowly progressing on the ground. The emphasis on increased range of motion of the hallux in the protoanthropoid probably allowed a more flexible alignment of the first ray with the rest of the foot when rapidly travelling on large branches without sacrificing the hallucial grasp during climbing. To our knowledge no meaningfully comparative study exists which evaluates the speed, efficiency, or any other attribute of locomotor behaviors between similar sized lemuriforms, platyrrhines, and cercopithecids to allow us to test these ideas.

What we do see in the morphology of the living catarrhines is the near complete abandonment of any remnant of the archaic euprimate sellar articulation, still so obvious on the medial side of the entocuneiform EMtl facet in the platyrrhines. The entocuneiform part of the joint becomes an attenuated ovoid, essentially a half cylinder, on which the proximal hallux moves. Many details of the differences among living species of catarrhines were previously described (Lewis, 1972), and their variations examined and discussed (Wikander *et al.*, 1986). We feel confident to state, however, that at least in the living catarrhines the largely hinge-like ovoid articulation (considering the important caveat of the MacConaill dictum of conjunct rotation in reaching the close packed position) is a nearly universal pattern. The functional joint in catarrhines is, we believe, the strict homolog of the dorsal half of the primitive euprimate saddle joint.

The enormously stress resistant saddle joint of the early euprimates was modified by some tarsiiform into the modified ovoid of the earliest anthropoids. We believe this involved the acquisition of a new ossification (the sesamoid in the EMtlJ) to continue to brace the first metatarsal medioplantarly. The reduction of the process for the peroneus longus on the Mt1 (and consequently the plantar component of the metatarsal EMt1 facet) were probably correlated events in the protoanthropoids. The slight reduction and realignment of the hallux from a habitually abducted position to one more parallel with the other cheiridia was also one of the corollaries of this almost certainly adaptive change. What, then, were the behaviors of the protoanthropoids which dictated the adaptive responses? The graspleaping locomotion of the antecedent tarsiiform was probably channelled into more large branch running and climbing. But we certainly hope that more explicit and satisfactory answers will result from other research programs, using all the available behavioral and ecological observations, when the morphological evidence is re-examined.

Summary

The characteristic habitual abduction of the hallux in living marsupials and primates is accomplished by unequivocally distinct, nonhomologous solutions of joint mechanics and other bone related adaptations. The medial migration of the entocuneiform–metatarsal 1 joint in euprimates can be traced to an antecedent primate or archontan condition in which, as in other eutherians, the metatarsal articulates with the distal end of the entocuneiform. In all living marsupials the hallux articulates on the distomedial side of the entocuneiform. In addition to the evidence from the hallucial articular areas the morphology of the entire tarsus of the relevant marsupials and eutherians strongly corroborates this view.

The plesiadapid *Plesiadapis* is the only taxon so far which provides evidence for a mobile or abductable hallux in archaic primates. This genus did not have the habitually abducted big toe of the first euprimates, a condition reflected by the highly modified joint in representatives of the latter. The living lemurids preserve the postulated primitive euprimate condition, present also in the notharctines. The modified grasp and tarsus of tarsiers is specially similar to unique features of omomyids. Fossil and living tarsiiforms share the emphasis of the dorsal half of the entocuneiform–hallucial joint with the primitive anthropoid condition.

We see a probable transformation of the archaic primate saddle joint on the distal end of the entocuneiform into the characteristically medially expanded one of the first euprimates. The euprimate condition is further modified in the tarsiiforms into proportions more similar to anthropoids, although the essentially sellar articulation and the large process for the peroneus longus of the first metatarsal is retained. The protoanthropoids, whose condition is probably approached by some living platyrrhines, realigned the habitually abducted euprimate (and haplorhine) big toe with the remaining cheiridia, reduced the peroneal process of the first metatarsal, reduced the plantar half of the joint and acquired a sesamoid in its place while converting the sellar joint into a modified ovoid. The emphasis and transformation of the dorsal half of the joint and the appearance of the sesamoid are probably developmentally related to the timing of the change in shape and function.

There is no indication that the developmentally tibiale-related prehallux (or the radiale-related prepollex) of marsupials and some eutherians was present in known euprimates. It is unlikely that the sesamoid in the entocuneiform-hallucial joint of some anthropoids, unknown in strepsirhines or tarsiiforms, is the homolog of the therian prehallux.

Acknowledgements

We are grateful for permission to study specimens in the care of Drs G. G. Musser and M. J. Novacek of the American Museum of Natural History, New York; Dr D. E. Savage of the University of California, Berkeley; and Drs F. K. Jouffroy and D. E. Russell of the Museum National d'Histoire Naturelle, Paris.

We thank Dr Mike Rose for his generous and detailed review of the paper. We also thank another anonymous reviewer for helpful and constructive comments. Dr Eric Delson's careful editing of the manuscript is much appreciated. The study was supported by grants from the PSC-CUNY Research Award Program of the City University of New York to the senior author. Drawings were prepared by Anita Cleary and Patricia Van Tassel.

References

Bock, W. J. & von Wahlert, G. (1965). Adaptation and the form function complex. *Evolution* **19,** 269–299.

Cartmill, M. (1972). Arboreal adaptation and the origin of the Order Primates. In (R. Tuttle, Ed.) *The functional and evolutionary biology of primates*, pp. 97–122. Chicago, New York: Aldine-Atherton.

Cartmill, M. (1986). Climbing. In (M. Hildebrand, D. M. Bramble, K. F. Liem, & D. B. Wake, Eds) *Functional vertebrate morphology*, pp. 73–88. Cambridge: The Belknap Press of Harvard University Press.

Clark,, W. E. Le Gros (1926). On the anatomy of the pentailed tree shrew (*Ptilocercus lowii*). *Proc. Zool. Soc.* 1179–1309.

Conroy, G. C. (1976). Hallucial tarsometatarsal joint in an Oligocene anthropoid *Aegyptopithecus zeuxis. Nature* **262,** 684–686.

Conroy, G. C. & Rose, M. D. (1983). The evolution of the primate foot from the earliest primates to the Miocene hominoids. *Foot & Ankle* **3,** 342–364.

Dagosto, M. (1983). Postcranium of *Adapis parisiensis* and *Leptadapis magnus* (Adapiformes, Primates). *Folia primatol.* **41,** 49–101.

Dagosto, M. (1985). The distal tibia of primates with special reference to the Omomyidae. *Int. J. Primatol.* **6,** 45–75.

Fleagle, J. G. (1983). Locomotor adaptation of Oligocene and Miocene homonoids and their phyletic implications. In (R. L. Ciochon, & R. S. Corruccini, Eds) *New Interpretations of Ape and Human Ancestry*, pp. 301–324. New York: Plenum Press.

Gebo, D. L. (1987). Functional anatomy of the tarsier foot. *Am. J. phys. Anthrop.* **73,** 9–31.

Gregory, W. K. (1920). On the structure and relations of *Notharctus*, an American Eocene primate. *Mem. Am. Mus. nat. Hist.* **3,** 51–243.

Haines, R. W. (1958). Arboreal or terrestrial ancestry of placental mammals. *Quart. Rev. Bio.* **33,** 1–23.

Koenigswald, W. V. (1979). Ein lemurenrest aus dem Eozänen Ölschiefer der Grube Messel bei Darmstadt. *Paläont. Z.* **53,** 63–76.

Krause, D. W. & Jenkins, Jr., F. A. (1983). The postcranial skeleton of North American multituberculates. *Bull. Mus. Comp. Zool.* **150,** 199–246.

Langdon, J. H. (1986). Functional morphology of the Miocene hominoid foot. *Contrib. Primatol.* **22,** 1–225.

Lewis, O. J. (1972). The evolution of the hallucial tarsometatarsal joint in the Anthropoidea. *Am. J. phys. Anthrop.* **37,** 13–34.

Lewis, O. J. (1980*a*). The joints of the evolving foot. Part I. The ankle joint. *J. Anat.* **130,** 527–543.

Lewis, O. J. (1980*b*). The joints of the evolving foot. Part II. The intrinsic joints. *J. Anat.* **130,** 833–857.

Lewis, O. J. (1980*c*). The joints of the evolving foot. Part III. The fossil evidence. *J. Anat.* **131,** 275–298.

MacConaill, M. A. (1946). Studies in the mechanics of synovial joints. III. Hinge joints and the nature of intraarticular displacements. *Ir. J. Me. Sci.* **6,** 620–626.

MacConaill, M. A. (1953). The movements of bones and joints. V. The significance of shape. *J. Bone Joint Surg.* **30-B,** 322–326.

MacConaill, M. A. (1966). The geometry and algebra of articular kinematics. *Biomed. Eng.* **1,** 205–212.

MacConaill, M. A. (1973). A structuro-functional classification of synovial articular units. *Ir. J. Me. Sci.* **142,** 19–26.

Matthew, W. D. (1904). The arboreal ancestry of the Mammalia. *Am. Nat.* **38,** 811–818.

Matthew, W. D. (1937). Paleocene faunas of the San Juan basin, New Mexico. *Trans. Am. Phil. Soc.* **30,** 1–510.

Napier, J. R. (1955). The form and function of the carpo-metacarpal joint of the thumb. *J. Anat. Lond.* **89,** 362–369.

Napier, J. R. & Napier, P. H. (1967). *A Handbook of Living Primates.* London: Academic Press.

Napier, J. R. & Walker, A. C. (1967). Vertical clinging and leaping—a newly recognized category of primate locomotion. *Folia primatol.* **6,** 204–219.

Rollinson, J. & Martin, R. D. (1981). Comparative aspects of primate locomotion, with some special reference to arboreal cercopithecines. *Symp. Zool. Soc. Lond.* **48,** 377–426.

Rosenberger, A. L. & Szalay, F. S. (1981). On the tarsiform origins of Anthropoidea. In (R. L. Ciochon & B. Chiarelli, Eds) *Evolutionary Biology of the New World Monkeys and Continental Drift,* pp. 139–157. New York: Plenum Press.

Savage, D. E. & Waters, B. T. (1978). A new omomyid primate from the Wasatch Formation of southern Wyoming. *Folia primatol.* **30,** 1–29.

Simpson, G. G. (1940). Studies on the earliest primates. *Bull. Am. Mus. nat. Hist.* **77,** 185–212.

Szalay, F. S. (1975). Where to draw the nonprimate–primate taxonomic boundary. *Folia primatol.* **23,** 158–163.

Szalay, F. S. (1976). Systematics of the Omomyidae (Tarsiiformes, Primates): taxonomy, phylogeny, and adaptations. *Bull. Am. Mus. nat. Hist.* **156,** 157–450.

Szalay, F. S. (1982). A new appraisal of marsupial phylogeny and classification. In (M. Archer, Ed.) *Carnivorous Marsupials,* pp. 621–640. Sidney: Roy. Zool. Soc. New South Wales.

Szalay, F. S. (1984). Arboreality: is it homologous in metatherian and eutherian mammals? *Evol. Biol.* **18,** 215–258.

Szalay, F. S. (1985). Rodent and lagomorph morphotype adaptations, origins, and relationships: some postcranial attributes analyzed. In (W. P. Luckett & J-L. Hartenberger, Eds) *Evolutionary Relationships Among Rodents: A Multidisciplinary Analysis,* pp. 83–157 (NATO ASI series, series A, Life Sciences) New York: Plenum Press.

Szalay, F. S. & Dagosto, M. (1980). Locomotor adaptations as reflected on the humerus of Paleogene primates. *Folia primatol.* **34,** 1–45.

Szalay, F. S. & Delson, E. (1979). *Evolutionary History of the Primates.* New York: Academic Press.

Szalay, F. S. & Drawhorn, G. (1980). Evolution and diversification of the Archonta in an arboreal milieu. In (W. P. Luckett, Ed) *Comparative Biology and Evolutionary Relationships of Tree Shrews,* pp. 133–169. New York: Plenum Press.

Szalay, F. S. & Langdon, J. H. (1986). The foot of *Oreopithecus*: an evolutionary assessment. *J. hum. Evol.* **15,** 585–621.

Wikander, R., Covert, H. H., & DeBlieux, D. D. (1986). Ontogenetic, intraspecific, and interspecific variations on the prehallux in primates: implications for its utility in the assessment of phylogeny. *Am. J. phys. Anthrop.* **70,** 513–524.

Marian Dagosto

Department of Cell Biology and Anatomy, Johns Hopkins University School of Medicine, Baltimore MD (Current address: Departments of Anthropology and Cell Biology and Anatomy, Northwestern University, Evanston, IL 60208, U.S.A.)

Received 15 June 1987
Revision received 27 December 1987
and accepted 6 January 1988

Publication date June 1988

Keywords: euprimates, postcranium, foot, tarsus, Strepsirhini, adapids, omomyids.

Implications of postcranial evidence for the origin of euprimates

Lemuriforms, *Tarsius*, anthropoids, adapids, and omomyids share numerous derived postcranial features which support monophyly of the Euprimates and indicate that the archaic primate to euprimate transition was marked by a striking reorganization of the skeleton presumably related to a transformation in locomotor behavior. Analysis of the tarsus indicates that the ancestral euprimate differed from its plesiadapiform progenitor in features related to its increased ability to leap and climb using a grasp based on an opposable hallux. Adapids share derived pedal features with the extant lemuriforms, supporting the monophyly of the Strepsirhini. Omomyids, *Tarsius*, and anthropoids have the primitive condition of these traits.

Journal of Human Evolution (1988) **17**, 35–56

Introduction

Since the time of Mivart's (1873) often-quoted definition of the Order Primates, it has been recognized that this group of mammals is distinguished by several traits of the postcranial skeleton, most notably the opposable big toe and the possession of nails instead of claws, and that these traits are somehow related to arboreality. Nevertheless, there has also been a contradictory, but equally persistent, notion that primate limbs are generalized, or at least less specialized than those of other mammals. Although acknowledging nails, opposable big toes, and several other features (increased range of motion at the shoulder joint, increased range of pronation and supination at the elbow joint, more flexible ankle joint, increased mobility of the digits, development of friction pads) as distinctive traits of primates, Le Gros Clark (1959) was impressed by their retention of a clavicle, five digits, an unfused radius and ulna, and an unfused tibia and fibula, and concluded that primates had preserved "a generalized structure of the limbs" (p. 43). This was partly conditioned by his beliefs that primitive therians were arboreal (and thus primates simply retained "an ancient simplicity of structure and function", p. 172); that arboreality itself necessitated a lack of limb specialization; and his wide definition of Primates which included tree shrews, *Anagale*, and plesiadapids, which lacked some or all of these features.

There has been a near consensus in recent years that tree shrews should not be included in the Order Primates (Van Valen, 1965; Luckett, 1980), that ancestral eutherians were likely terrestrial (Szalay, 1984; but see Lewis, 1980) so that primate arboreality and structures associated with it are derived conditions; and that Plesiadapiformes, whether included within the Order Primates or not, are best regarded as a separate archaic group (e.g. Van Valen, 1969; Cartmill, 1972; Martin, 1972, 1986; Simons, 1972; Szalay & Delson, 1979; MacPhee *et al.*, 1983; Gingerich, 1986*a*). It has thus become clearer that: (1) the living primates plus adapids and omomyids form a natural group within the Order informally referred to as the euprimates (Hoffstetter, 1977; Cartmill & Kay, 1978; Szalay & Delson, 1979; Gingerich, 1981; but see Schwartz *et al.*, 1978); (2) although euprimate limbs do indeed retain some primitive mammalian features, this group can be

0047–2484/88/01/20035 + 22 $03.00/0

distinguished from all other mammals by a suite of derived cranial and postcranial features (Cartmill, 1972, 1974; Szalay, 1977); and that (3) arboreality *per se* cannot explain the origin of these features (Cartmill, 1972, 1974).

The purpose of this paper is to review the postcranial evidence for euprimate monophyly and for its major subdivisions (Strepsirhini and Haplorhini) and to attempt an adaptive explanation for the origin of these derived features. Only a few areas of the skeleton have been analyzed with the detail necessary to reach reasonably secure conclusions regarding polarity and functional significance. In this paper I will restrict my discussion to the tarsus. Detailed arguments on the criteria used to determine polarity and function of these features is found in Decker & Szalay (1974) and Dagosto (1985, 1986). Only the most pertinent points are repeated here. Rather than trying to interpret the entire euprimate morphotype, this discussion focuses on those features which have been determined to be derived (i.e., newly acquired by euprimates) and which thus distinguish the ancestral euprimate from its archaic ancestor (see Szalay, 1981). However, primitive retentions which might alter adaptive hypotheses based on derived features have been taken into account, even if not explicitly stated.

Even with only a few areas of the skeleton studied, it is possible to identify a number of derived postcranial features which are shared by lemuriforms, *Tarsius*, and anthropoids (Table 1), and thus clearly indicate monophyly of the living primates. Most of these features can be identified in omomyids and adapids as well (Table 2), including *Cantius*, an early Eocene adapid and *Teilhardina*, the oldest recognized omomyid, and were thus likely to have been present in the common ancestor of all of these groups. It is thus apparent that the primate–euprimate transition is marked by a dramatic reorganization of the postcranium. Although it is likely that at least some of these traits evolved together as a complex, changes affecting the limb skeleton are more numerous and as extensive as those affecting the skull and dentition. This implies that changes in locomotor behavior were of particular importance in the origin of euprimates.

Euprimate morphology

The upper ankle joint of euprimates differs from that of archaic primates (and archaic and modern eutherians) in several ways. The body of the astragalus is very high (Figure 1) with medial and lateral trochlear crests of approximately equal height; in some strepsirhines the medial crest is slightly higher than the lateral. This is a great departure from the primitive low bodied astragalus of *Protungulatum* or of many other mammals where the lateral crest exceeds the medial in height (e.g., *Plesiadapis*). The increased height of the body probably reflects an increase in the radius of curvature of the upper ankle joint relative to body size. Although these parameters cannot be directly compared in plesiadapiforms, fossil euprimates and living euprimates, a similar difference exists between leaping (lemurs, galagines) and non-leaping (lorisine) strepsirhines (Dagosto, 1986; Gebo, 1986a).

The equally high trochlear crests serve to reduce concomitant pronation and supination (definitions of these terms as in Dagosto, 1986) which accompanies dorsiflexion and plantarflexion at the upper ankle joint in mammals with high lateral crests (Jenkins & McClearn, 1984). This may seem contradictory to the primate trend of increasing the ability of the foot to invert and evert, but in euprimates this ability is concentrated in the subtalar joint and the transverse tarsal joint.

Table 1

Synapomorphies of Euprimates. Cranial and dental traits after Szalay & Delson (1979), Martin (1986), and Kay & Cartmill (1977). An asterisk in the second column indicates that the trait is unique to euprimates. G, grasping; C, climbing; L, leaping.

Trait		Behavioural significance
Cranial		
Postorbital bar		
Increased convergence and frontation of orbits		
Enlarged eyes		
Enlarged brains		
Dental		
Protocone on P3		
Mesiodistally compressed trigonid on M3		
Trigonid widely open lingually on M1		
Postcranial		
1. Elongate, cylindrically shaped humeroulnar joint	*	G/C
2. Deep groove separating the humeroradial and humeroulnar joints	*	G/C
3. Medial and lateral crests of astragalar trochlea equal in height		L
4. Posterior trochlear shelf moderately developed	*	L
5. Increase depth of trochlea		L
6. Elongation of tarsus	*	L
7. Components of upper ankle joint moderately medially rotated	*	G/C
8. More spherical astragalonavicular joint		G/C
9. Sellar-shaped calcaneocuboid joint	*	G/C
10. Reduction and more proximal placement of peroneal process		L
11. Sellar entocuneiform-first metatarsal joint	†	G/C
12. Enlarged peroneal process on first metatarsal	*	G/C
13. Enlarged hallux		G/C
14. Claws replaced by nails	*	G/C

† Marsupials have a sellar-shaped joint, but the anatomical details are very different (Szalay & Dagosto, 1988).

Euprimates increase the range of plantarflexion possible at the ankle by lengthening the degree of faceting for the upper ankle joint on the posterior part of the astragalus. The depth of the trochlea in euprimates (except for *Daubentonia* and *Avahi*) is increased over that in archaic primates, thus adding an increased degree of stability to the upper ankle joint. The astragalus is also lengthened by the addition of the posterior trochlear shelf (Figure 1), a feature which is unique to euprimates (Decker & Szalay, 1974) although it is modified or lost in many euprimate taxa. The mechanical significance of this feature is unclear.

Most euprimates are distinguished from archaic primates and other mammals by elongation of the tarsal elements, especially the talar neck, the navicular, and the part of the calcaneus anterior to the posterior astragalocalcaneal joint; but also of the cuboid and cuneiforms (Figure 2). Although some euprimates have relatively short tarsal bones (adapines, lorisines, apes), the euprimate ancestor was characterized by a tarsus more elongated than that of any known plesiadapiform or other eutherian. It has long been accepted that the lengthening of the tarsus is an adaptation to increase the effective length of the hindlimb so that a more powerful leap can be produced (e.g., Hall-Craggs, 1965). In other specialized mammalian jumpers (e.g., rabbits, elephant shrews, dipodid rodents) it is the metatarsal region, not the tarsus, which is elongated. Morton (1924) has explained the basis for this distinction. The primate metatarsus is involved in grasping, and cannot elongate for obvious mechanical reasons, thus the tarsus is the only segment of the foot available for this purpose.

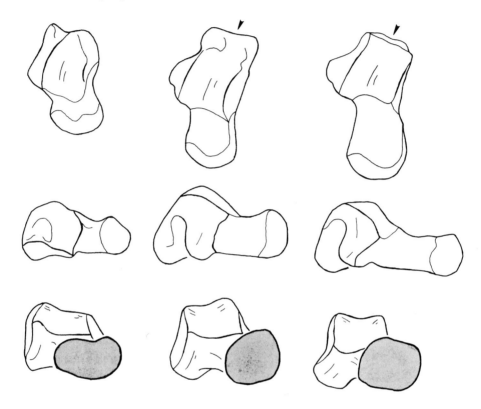

Figure 1. Right astragali of *Plesiadapis tricuspidens* (left); *Notharctus* sp. (center); and *Teilhardina belgica* (right). From top to bottom, dorsal, lateral, and anterior views. In dorsal view, arrow points to the posterior trochlear shelf. In the anterior view, the astragalonavicular facet is shaded. Astragali are drawn to the same trochlear width, which is a reasonably good estimator of body size in strepsirhines.

In plesiadapiforms, like in the majority of primitive mammals, there is a tubercle on the lateral side of the calcaneus which is very distally placed (at the level of the calcaneocuboid joint). In euprimates a smaller tubercle or swelling is more proximally placed (at the level of the posterior astragalocalcaneal joint) (Figure 3). These are probably homologous structures serving to channel the peroneus longus and brevis tendons and/or to anchor the retinaculum which binds these tendons. The change in placement of this tubercle has been related to the elongation of the tarsus (Decker & Szalay, 1974), and the shift in the primary function of this muscle from a foot evertor to an hallucal adductor (Gebo, 1986a), but there is as yet no satisfactory explanation for its reduction in size.

The features listed above are reflections of the importance of leaping in the earliest euprimates. Increasing the range of upper ankle joint motion, especially for plantarflexion,

Table 2 **Eocene primates for which data is available for the determination of euprimate postcranial traits. Numbers of traits refer to Table 1. One example specimen is listed for each taxon; in most cases, the trait is known in more than one individual of the species. Except where noted, all taxa conform to the euprimate morphology described in Table 1.**

Trait	Taxon	Specimen	Reference
1, 2	Adapids:		
	Cantius mckennai	USGS 6759	1
	Cantius trigonodus	USGS 5900	2
	Smilodectes gracilis	AMNH 11484	3, 4
	Notharctus sp.	AMNH 11478	4, 5
	Adapis parisiensis[1]	NMB QW 1482	3
	Leptadapis magnus	NMB QD 664	3
	Omomyids:		
	(Bitter Creek, Wyoming)	UCMP 113301	3
	Hemiacodon gracilis	AMNH 29126	3
	Microchoerus sp.	NMB QJ 620	3
	Necrolemur sp.	NMB QK 989	3
3–5, 7, 8	Adapids:		
	Cantius ralstoni	UW uncat.	22
	Cantius mckennai[2]	UMMP 81819	4
	Cantius trigonodus	AMNH 16852	6
	Cantius frugivorous	YPM 28492	4
	Cantius abditus	USGS 6784	7
	Smilodectes gracilis	AMNH 11484	4
	Notharctus sp.	AMNH 11478	5
	Messel adapid	HLD ME7430	8
	Adapis parisiensis	Montauban, uncat.	9
	Leptadapis magnus	NMB QF 496	9, 10
	Caenopithecus lemuroides[2]	NMB, no uncat.	11
	Omomyids:		
	(Bitter Creek)	UCMP uncat.	12
	Tetonius sp.	AMNH 88818	13
	Arapahovius gazini	UCMP 118293	14
	Hemiacodon gracilis	AMNH 12613	15
	Washakius insignis	AMNH 97288	13
	(Bridger Basin) 2 species	AMNH no, uncat.	12
	Teilhardina belgica	IRNSB 4386	13, 16
	Necrolemur antiquus	BFI 811	17
6[3]	Adapids:		
	Cantius ralstoni		22
	Cantius mckennai		4
	Cantius trigonodus		6
	Smilodectes gracilis		4
	Notharctus sp.		5
	Messel adapid		8
	Omomyids:		
	Tetonius sp.		
	Arapahovius gazini		14
	Hemiacodon gracilis		15
	(Bridger Basin)		12
9, 10	Adapids:		
	Cantius ralstoni	UW uncat.	22
	Cantius mckennai	UMMP 81825	4

Table 2—*continued*

Trait	Taxon	Specimen	Reference
	Cantius trigonodus	AMNH 16582	6
	Cantius abditus	DPC 1245	4
	Cantius frugivorous	YPM 28492	4
	Cantius venticolis	AMNH 117081	11
	Smilodectes gracilis	AMNH 11484	4
	Notharctus sp.	AMNH 11478	5
	Messel adapid	HLD ME 7430	8
	Adapis parisiensis	NMB QE 553	9
	Leptadapis magnus	NMB QF 491	9
	Caenopithecus lemuroides	NMB uncat.	11
	Omomyids:		
	(Bitter Creek)	UCMP uncat.	12
	Tetonius sp.	AMNH 88821	13
	Arapahovius gazini	UCMP 118498	14
	Hemiacodon gracilis	AMNH 12613	15
	Washakius insignis	AMNH 88886	13
	(early Bridger) 2 species	AMNH uncat.	12
	Teilhardina belgica	IRNSB 4385	13, 16
	Necrolemur sp.[4]	AIZ 637	18, 19
	Microchoerus sp.[4]	AIZ 8763	18
11–13	Adapids:		
	Cantius ralstoni[5]	AMNH 29144	20
	Cantius mckennai	UMMP 81820	4
	Cantius frugivorous	YPM 28492	4
	Cantius trigonodus[5]	AMNH 16582	6
	Smilodectes gracilis	USNM 25686	4
	Notharctus sp.	AMNH 12570	5
	Messel adapid	HLD ME 7430	8
	Adapis parisiensis[6]	Montauban uncat.	21, 23
	Leptadapis magnus[6]	NMB QL 787	21, 23
	Omomyids:		
	(Bitter Creek)[6]	UCMP uncat.	12
	Hemiacodon gracilis	AMNH 12613	15
	Necrolemur sp.[6]	Montauban uncat.	23
14	Adapids:		
	Cantius sp.	UGS 18338	4
	Smilodectes gracilis	USNM 25686	4
	Notharctus sp.	AMNH 11474	5
	Messel adapid	HLD ME 7430	8
	Omomyids:		
	(Bitter Creek)	UCMP uncat.	20
	(early and late Bridger)	AMNH uncat.	20

[1] *Adapis* has conditions more similar to the inferred primitive primate or eutherian condition. These are interpreted as secondarily derived (Dagosto, 1983).

[2] Head of talus is missing, therefore the morphology of feature 8 cannot be determined.

[3] This grouping only lists taxa where the navicular, cuboid or cuneiforms are known. All taxa known by calcanea (except *Adapis*, *Leptadapis* and *Caenopithecus*) are characterized by elongation of the anterior calcaneus.

[4] Distal end of calcaneus is missing, therefore the morphology of feature 10 cannot be determined.

[5] Only the entocuneiform is known.

[6] Only the first metatarsal is known.

Table 2—*continued*

Abbreviations

AIZ, Anthropologisches Institut, Zurich.
AMNH, American Museum of Natural History, New York.
BFI, Universite de Montpellier.
DPC, Duke Primate Center.
HLD, Hessesiches Landesmuseum, Darmstadt.
IRNSB, Institut Royale de Science Naturelle, Belgium.
NMB, Naturhistorishes Museum, Basel.
UCMP, University of California at Berkeley Museum of Paleontology.
UMMP, University of Michigan Museum of Paleontology.
USGS, United States Geological Survey, Denver.
USNM, United States National Museum, Washington.
UW, University of Wyoming Museum of Paleontology.
YPM, Yale Peabody Museum.

References

 1. Gebo in press, *b*.
 2. Rose & Walker, 1985.
 3. Szalay & Dagosto, 1980.
 4. Covert, 1985, 1988.
 5. Gregory, 1920.
 6. Matthew, 1915.
 7. Gebo, in press, *a*.
 8. von Koenigswald, 1979.
 9. Decker & Szalay, 1974.
10. Stehlin, 1912.
11. Dagosto, 1986.
12. Dagosto, in prep.
13. Szalay, 1976.
14. Savage & Waters, 1978.
15. Simpson, 1940.
16. Teilhard de Chardin, 1927.
17. Godinot & Dagosto, 1983.
18. Schmid, 1979.
19. Schlosser, 1907.
20. This paper.
21. Dagosto, 1983.
22. Dagosto & Gebo, in prep.
23. Szalay & Dagosto, 1988.

restricting conjunct pronation–supination motions at the upper ankle joint, increasing its stability, and lengthening the foot are all features mechanically advantageous for leaping. The relationship between these features and leaping can also be appreciated by comparing a specialized leaper (e.g., a galago or *Lepilemur*) with a non-leaper (e.g., a loris). Galagos show the greatest development of these features, lorises the least. It is also noteworthy that in other mammals where quick, stable motion at the upper ankle joint is desirable (other leapers like rabbits, cursors like artiodactyls) similar mechanical, but not always anatomical, adaptations occur (see for example, Schaeffer, 1947; Szalay, 1985; Bleefeld, in press).

Another set of euprimate pedal features is most likely associated with climbing and/or grasping. In almost all extant lemuriforms the articular surface of the medial malleolus of the tibia is rotated to face 30–40° from the sagittal plane, the distal part of the malleolus is highly convex to articulate with a concave astragalar facet which ends in a marked cup on the astragalar neck (Figure 4, top row). The dorsal tibioastragalar articulation is curved medially as it progresses distally. These curvatures dictate that as the astragalus moves from a plantarflexed to a dorsiflexed position it also abducts relative to the tibia (Hafferl, 1932; Lewis, 1980; Dagosto, 1985). These features are most exaggerated in climbers like lorisines, and least expressed in leapers suggesting a possible mechanical relationship to climbing. Lewis (1980) however, has related these features specifically to grasping.

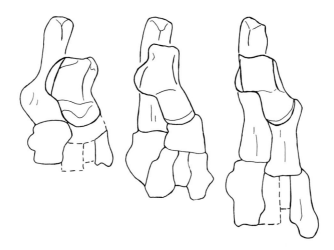

Figure 2. Right tarsal bones of *Pleisadapis tricuspidens* (left), *Notharctus* sp. (center), and *Hemiacodon gracilis* (right). *Plesiadapis tricuspidens* is a composite based on elements which may not be from the same individual. Dashed lines represent unknown elements. Tarsal length is determined by adding the lengths of the calcaneus and cuboid. Feet are drawn to the same trochlear width.

In other euprimates these features are not as well expressed as in lemuriforms. They are lemuriform-like in *Adapis* and *Leptadapis*, but notharctines have less talar cupping and curvature. In tarsiiforms and anthropoids medial malleolar rotation and medial cupping and twisting of the tibioastragalar joints is present, but slight. This can be explained either as an indication of more specialized leaping in these forms (Gregory, 1920; Covert, 1985; Dagosto, 1985) or that the euprimate ancestor was more conservative in these features than are extant lemuriforms. The more extreme condition of these features (especially the rotation of the medial malleolus) is perhaps better regarded as a strepsirhine synapomorphy (Covert, 1988).

The subtalar and transverse tarsal joints are the sites at which inversion and eversion (which are important in the ability of the foot to adopt variable orientations) take place. Thus, the functional significance of modifications at these joints is most likely related to climbing and grasping. The subtalar joint exhibits relatively few changes from the primitive primate or archontan condition. The posterior astragalocalcaneal facets may be slightly more elongated, related to the development of the posterior trochlear shelf (Decker & Szalay, 1974). The confluence of the posterior astragalocalcaneal, astragalonavicular, and spring ligament facets is more definite and regular than in plesiadapiforms. The anterior astragalocalcaneal facet is much longer than in *Plesiadapis*. This may simply be the result of a relatively longer astragalar neck and distal calcaneus. Elongation of the tarsal elements is usually ascribed to changes in foot lever mechanics advantageous for leaping.

However, the relatively long astragalar necks of *Potos*, *Felis wiedii*, and *Ptilocercus*, mammals which show increased ranges of subtalar motion (Szalay & Drawhorn, 1980; Jenkins & McClearn, 1984) suggest that some lengthening of the astragalar neck may instead be the result of elaboration of subtalar motion.

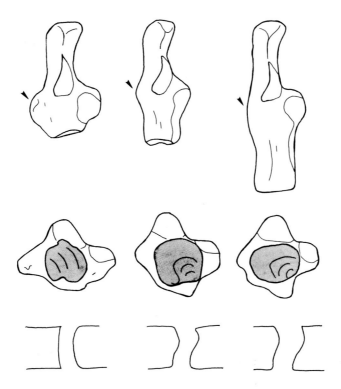

Figure 3. Right calcanea of *Pleisiadapis tricuspidens* (left); *Notharctus* sp. (center); and *Teilhardina belgica* (right). Top, dorsal view; middle, anterior view; bottom schematic view of an anterior–posterior section through the middle of the calcaneocuboid joint. In dorsal view the arrow points to the peroneal tubercle. In anterior view the calcaneocuboid facet is shaded. In bottom figure note the calcaneocuboid "pivot" of euprimates formed by the depression on the plantar half of the calcaneus (left of each pair) and the protuberance on the plantar half of the cuboid (right of each pair). All specimens are drawn to the same mediolateral width (excluding the peroneal tubercle in *Plesiadapis*) which is a reasonable estimator of body size in strepsirhines.

The transverse tarsal joint of euprimates is more advanced than in the archaic primates (Decker & Szalay, 1974). The astragalonavicular joint is more spherical, and thus allows axial rotation (Figure 1). The calcaneocuboid joint is sellar rather than ovoid in shape (Figure 3). The joint takes the from of a pivot articulation; there is a protuberance of moderate size on the cuboid and a corresponding depression on the calcaneus. These modifications increase the degree to which the calcaneus and foot distal to it inverts and everts by increasing the degree to which adjunct supination and pronation are allowed at the transverse tarsal joint. In other mammals (Jenkins & McClearn, 1984), and probably in plesiadapiforms, motion at the transverse tarsal joint is a combination of pronation–supination and medio-lateral translation. In euprimates the pronation–supination component is enhanced and the translational component is reduced. Jenkins & McClearn

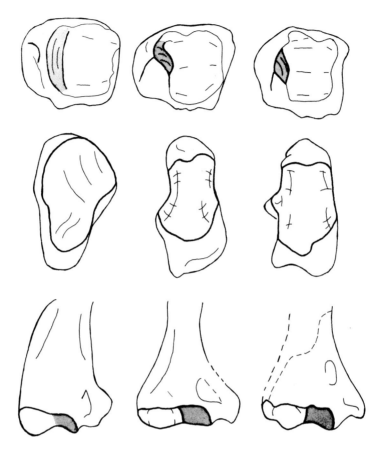

Figure 4. Morphology of left distal tibia (distal view) (top row); right entocuneiform (distal view) (middle row); and right distal humerus (anterior view) (bottom row) in *Plesiadapis tricuspidens* (left), *Cantius* sp. (middle) and a Bridgerian omomyid, probably *Hemiacodon* (right). In the top row, the medial malleolar surface of the distal tibia is shaded. In the bottom row the trochlea (ulnar articulation) is shaded.

(1984) found that in *Potos* and *Sciurus* inversion (supination) is associated with a lateral translation of the navicular relative to the astragalus. Dagosto (1986) found that in euprimates, on the other hand, supination is associated with little or a slight medial translation of the navicular. This difference in the details of transverse tarsal joint motion between euprimates and other mammals is reflected in the morphological differences in the astragalonavicular and calcaneocuboid joints. It probably also reflects a shift in the motion axis of the transverse tarsal joint away from the calcaneocuboid joint towards the astragalonavicular joint. This in turn may be related to the functional independence of the hallux.

The effect of adoption of an opposable hallux on the form of the foot is evidenced primarily in the shape of the entocuneiform-first metatarsal joint, and the increased size of the hallux. As noted above, some of the changes in the midtarsal region and the upper ankle joint which were hypothesized to be related to climbing, may also be correlated. This is, of course, because the ability to grasp is one of the adaptations that clawless mammals (like euprimates) develop in order to climb (Cartmill, 1985). Thus, the behaviors of climbing and grasping and the morphological adaptations related to them are inextricably linked in euprimates.

The construction of the euprimate entocuneiform-first metartarsal joint is unique among mammals, differing substantially even from other animals with prehensile feet such as didelphid and phalangerid marsupials (Szalay & Dagosto, 1988). In prosimian primates (strepsirhines and tarsiiformes) the joint is constructed differently than in anthropoids, but the evidence suggests that the prosimian type joint is the morphology most likely to have been present in the ancestral euprimate (Szalay & Dagosto, 1988), and it is that morphology which will be briefly described here. More detailed treatment is presented in Szalay & Dagosto (1988).

The entocuneiform and first metatarsal are known in all notharctines (except *Copelemur*) and in *Hemiacodon*. In addition, the first metatarsal is known in *Adapis*, *Leptadapis*, *Necrolemur*, and in an omomyid from the middle Wasatchian of Wyoming. In comparison to all other known mammals, including plesiadapiforms, euprimates have an entocuneiform-first metatarsal articulation which, on the entocuneiform, is strongly saddle shaped being convex mediolaterally and concave dorsoplantarly (Figure 4, middle row). The joint surface wraps around the medial side of the bone, covering approximately 180° of arc. The facet on the first metatarsal is similarly saddle shaped and the bone bears a hypertrophied tubercle for the peroneus longus, an important adductor of the hallux. This morphology is an efficient and stable system for promoting the swing and conjunct rotation required to oppose the hallux to the remaining digits. The hallux of the ancestral euprimate was thus large, set off medially from the other digits, and was capable of strongly opposing them.

Living primates are the only group of mammals in which claws on all of the digits are replaced by nails. Some primates have claw-like unguals (*Daubentonia*, callitrichids, *Tarsius pumilus*), but the reduction of the important deep matrix layer (Le Gros Clark, 1936; Thorndike, 1968) and the modifications of the terminal phalangeal joint from the primitive mammalian condition along the lines followed by other primates (Le Gros Clark, 1936; Rosenberger, 1977; Musser & Dagosto, 1987) in these taxa, plus the fact that Eocene euprimates had nails, strongly suggests that these "tegulae" are redeveloped from a more nail-like condition.

Nailed digits (hallucal, pollicial, and lateral digits) are known in *Notharctus* (Gregory, 1920), *Smilodectes* (Covert, 1985), a middle Eocene adapid from Germany (von Koenigswald, 1979), and *Cantius* (Covert, 1988), and it is likely that this was the typical condition for adapids. Previous to this report, evidence concerning omomyid unguals has been lacking. West (1976) has collected a large number of mammalian postcranial bones from sites in the Bridger Basin. Among these fossils, which are dated to the early middle Eocene (Bridgerian), are numerous remains of omomyids (currently under study by Dagosto & Szalay). Seventy-three nail bearing terminal phalanges are in the collection (Figure 5). They greatly resemble the terminal phalanges of living lemuriforms and tarsiers, and differ from the phalanges of those rodents or marsupials which have nailed or ungual-less big toes or thumbs. There would seem to be little doubt that they belong to the omomyid primates known from these sites. These terminal phalanges can be allocated to the hallux, pollex and lateral digits, although for the lateral digits, it is impossible to sort hand from foot, or to know if all lateral digits bore nails. The dental material from these sites has not yet been studied, but based on the postcrania, there are three omomyid taxa present. The majority of the phalanges apparently belong to the largest of these taxa, which

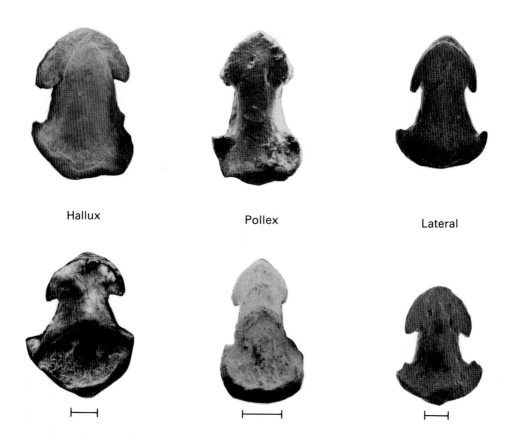

Hallux   Pollex   Lateral

Figure 5. Omomyid terminal phalanges from the Bridger Basin. Top row, dorsal view; bottom row, ventral view. Scale is 1 mm.

is most likely *Hemiacodon gracilis* or an allied species. An hallucal and lateral digit terminal phalanx are also known from middle Wasatchian Bitter Creek sites. The fauna from here is also yet to be analyzed, but a preliminary assessment suggests that *Tetonius* and *Anemorhysis* are present at this locality (Savage, personal communication).

Tarsius pumilus has long and laterally compressed claw-like nails, but its terminal phalanges are not different from those of other tarsiers (Musser & Dagosto, 1987). Thus, it is impossible to deduce the actual shape of the ungual in any fossil taxon. But it is clear that omomyids at least had a nail-bearing ancestry, and that the taxa for which terminal phalanges are known did not have tegulae like *Daubentonia* or callitrichids. The fact that both adapids and omomyids had similar nail-bearing phalanges is good evidence that this trait was present in their common ancestor.

Although the reason for replacing a hallucal claw with a nail in a grasping foot is clear (Cartmill, 1974), the reason for the replacement of all claws in euprimates is still obscure. Cartmill (1974) has thoroughly reviewed the issue and concludes that the most likely explanation is the exploitation of a relatively small branch environment by the ancestral euprimate.

Strepsirhini and Haplorhini

Among the most hotly debated topics in primate phylogeny is the relationship of the Eocene families Adapidae and Omomyidae to the living primates. There are two major schools of thought: (1) the only living descendant of the Omomyidae is the living genus *Tarsius*; lemuriforms and anthropoids are both descended from adapids (Schwartz *et al.*, 1978; Gingerich, 1980; Rasmussen, 1986) and (2) omomyids gave rise to tarsiers and anthropoids; adapids are ancestral to, or the sister group of, only the tooth combed lemurs (Szalay, 1977; Cartmill & Kay, 1978; Szalay & Delson, 1979; Rosenberger & Szalay, 1980; Rosenberger *et al.*, 1985).

Postcranial features have not figured prominently in this debate, but have been cited to support both positions. Only one feature has been used to support hypothesis 2: the presence of a downturned humeral trochlea in omomyids and anthropoids (Szalay & Dagosto, 1980; Rosenberger & Szalay, 1980). Gingerich (1980) does not list any shared derived features uniting adapids and anthropoids but cites two traits, a fused tibiofibula and anterior calcaneal lengthening which he assumes are characteristic of omomyids and derived for primates, logically concluding that omomyids are too derived postcranially to have given rise to anthropoids. Since the publication of Gingerich's work it has been shown that tibiofibular fusion is only known in one omomyid (*Necrolemur*, a very late Eocene form) although it may also be present in *Nannopithex* (Weigelt, 1933). *Hemiacodon* shows no evidence of fusion (Dagosto, 1985). In distal tibial morphology omomyids are in fact more similar to anthropoids than are adapids (Dagosto, 1985). All known omomyids are indeed characterized by some lengthening of the anterior calcaneus and navicular, but except for *Necrolemur*, this lengthening is not as extensive as in *Tarsius*, *Galago* or even *Microcebus*, but is more typical of that seen in the other extant cheirogaleids (Schmid, 1979; Gebo, in press, *a*). In extant lemuriforms (with the exception of indriids) an increase in body size is accompanied by a decrease in relative anterior calcaneal lengthening (Gebo & Dagosto, 1988). It thus seems likely that a transition from a small omomyid ancestor to a larger bodied anthropoid would have been accompanied by a relative reduction in the length of the tarsus.

Two recent studies of lower primate foot anatomy (Dagosto, 1986; Gebo, 1986*a*) have discovered additional features relevant to this issue. Two are features of the astragalus and one concerns the details of the naviculocuboid articulation. In extant lemuriforms the astragalofibular joint facet on the astragalus slopes gently laterally for its entire extent generally making an angle of more than 100° with the dorsal edge of the astragalus (Figure 6). A cline exists in lemuriforms: the flaring is most developed in the slow climbing lorisines (and *Adapis*) and least developed in the specialized leapers like galagines, *Lepilemur*, indriids, and notharctines. In *Tarsius* and anthropoids the astragalofibular facet is vertical (it makes an angle of 90° with the dorsal edge of the astragalus) except at its plantar surface where it makes an abrupt flare, resulting in a small pointed process. From its proximal to its distal end the facet is flat, whereas in lemuriforms, it is convex proximally and becomes concave distally resulting in sinuous astragalofibular movement (Dagosto, 1986). All adapids for which the astragalus is known (seven genera representing both subfamilies) have the lemuriform condition; all omomyids for which the astragalus is known (nine genera representing all three subfamilies) have the *Tarsius*-anthropoid condition. The astragalofibular joint of adapids and lemuriforms is unique among eutherians (although some marsupials exhibit an even more extreme flare [Szalay, 1984]); the haplorhine

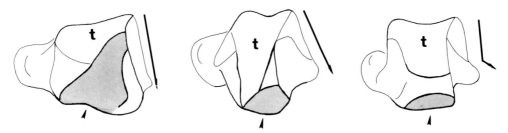

Figure 6. Posterior view of the right astragalus in *Plesiadapis gidleyi* (left), *Notharctus* (center), and *Teilhardina* (right). Arrows on lateral side indicate the slope of the fibular facet. *Notharctus* exhibits one of the least sloping facets of strepsirhines. Also note the position of the groove for the flexor fibularis tendon (bottom arrow) relative to the posterior part of the joint surface (t).

condition is similar to that exhibited by plesiadapiforms and by other eutherians (although the plantar process is normally absent).

In the extant lemuriforms the groove for the flexor fibularis (flexor hallucis longus) tendon lies lateral to the posterior part of the tibioastragalar joint. In *Tarsius* and anthropoids the groove is plantad and central to the facet (Figure 6) (Gebo, 1986a,b). Again, all known adapids have the morphology typical of lemuriforms; all known omomyids have the morphology typical of *Tarsius* and anthropoids. Again, the strepsirhine condition appears to be unique among mammals, while the haplorhine condition is typically mammalian. Plesiadapiforms exhibit a range of morphologies for this trait. In *P. tricuspidens* the enormous flexor fibularis groove occupies almost the entire posterior surface of the astragalus, and may be described as both lateral and plantad to the facet. *P. gidleyi*, although most similar to *P. tricuspidens*, is more moderate; the facet is largely plantad.

The naviculocuboid articulation differs in strepsirhines and haplorhines (Dagosto, 1986). In the living lemuriforms and notharctines (the navicular is unknown in adapines) the navicular and cuboid have a broad area of articulation. As a result, the contact facet on the navicular is contiguous with (and plantad to) both the ectocuneiform and mesocuneiform facets (Figure 7). These features are less well developed in *Cantius*, *Daubentonia*, and galagines in which the cuboid facet just approaches the lateral edge of the mesocuneiform facet rather than extending completely under it. In contrast, in *Tarsius*, anthropoids, omomyids (this region is known in *Hemiacodon* and *Arapahovius*) and in all other mammals examined, the cuboid facet is related only to the ectocuneiform facet (usually lateral to it) and never extends under the mesocuneiform facet or even approaches its lateral edge. Lemuriforms and adapids also have navicular-cuneiform facets which are more separate and have greater curvatures than the very flat, confluent facets of haplorhines (Dagosto, 1986; Gebo, 1986a,b).

Two additional strepsirhine synapomorphies (more extreme medial malleolar rotation and an enlargement of the posterior trochlear shelf) are recognized by Covert (1988). Beard & Godinot (1988) describe morphology of the ulnocarpal articulation which links *Adapis* (but not *Smilodectes*) to lemuriforms.

Although evidence comes from a somewhat limited number of fossil taxa, it is not insignificant that in all cases the distribution of pedal features is unequivocal: adapids are similar to the living lemuriforms while omomyids resemble *Tarsius* and anthropoids. The simplest phylogenetic interpretation of all three of these features is the same. Extant

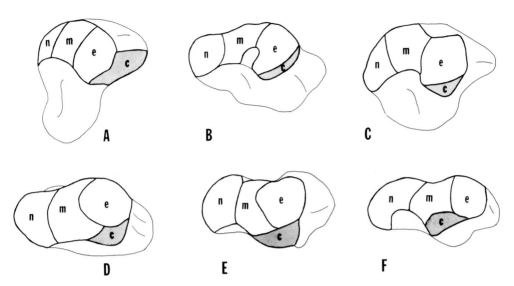

Figure 7. Distal view of navicular tilted slightly medially and dorsally in non-primate therians (A); *Hemiacodon* (B); *Saimiri* (C); *Cantius ralstoni* (D); *Notharctus* (E); and *Lemur* (F). C, cuboid facet; E, ectocuneiform facet; M, mesocuneiform facet; N, entocuneiform facet.

tooth-combed lemurs and adapids share traits of the foot which are derived for mammals (and primates), and thus they form a sister-group within the euprimates. These traits, along with incisor form (Rosenberger *et al.*, 1985), support the monophyly of the Strepsirhini, which has previously been widely assumed, but not clearly demonstrated by synapomorphies (Cartmill & Kay, 1978; Rasmussen, 1986). It is also obvious that anthropoids do not share derived strepsirhine pedal features with adapids, thus it is unlikely that they are descended from this group. Like *Tarsius* and omomyids, anthropoids retain the primitive eutherian condition for these traits. The fact that all haplorhines are similar in terms of these primitive features cannot bolster the cause of haplorhine monophyly, but of course, if adapids are ruled out as possible anthropoid ancestors, omomyids are the only group left which could have given rise to them, unless one postulates an independent origin from the euprimate stock. In any case it appears that at least in terms of these aspects of pedal morphology, it is the strepsirhines that are specialized while the haplorhines are persistently primitive.

The astragalofibular joint of strepsirhines indicates that more complex rotations take place at this joint than the more simple sagittal motion of haplorhines (Dagosto, 1986; Gebo, 1986a,b). Within strepsirhines these types of motions are more exaggerated in climbers than leapers. The laterally placed flexor hallucis longus groove indicates an habitually inverted foot (Gebo, 1986b). The reorientation of the navicular-cuboid articulation in strepsirhines results in a greater folding of the midtarsal region of the foot in which the mesocuneiform is approximated to the cuboid (Dagosto, 1986). The morphology of the navicular-cuneiform joints allows minor readjustments of the distal tarsus to increase conformation of the grasping foot on the support. Strepsirhines thus have modifications of the foot which are probably best interpreted as adaptations to facilitate grasping and climbing with a probable bias for vertical supports (Gebo, 1986b).

A very important determinant of locomotor and thus morphological differences between primitive strepsirhines and haplorhines is likely to have been body size. Although today the average haplorhine is larger than the average strepsirhine, at the time these groups diverged from each other the opposite was true (Fleagle, 1978). The great majority of early and middle Eocene adapids are estimated to have weighed more than 1000 g (Gingerich *et al.*, 1982). *Cantius torresi*, an earliest Wasatchian adapid, is estimated to have weighed as much as 1100 g (Gingerich, 1986*b*). In contrast, almost all known Eocene haplorhines are estimated to have weighed less than 500 g; *Teilhardina belgica*, the earliest known omomyid, probably weighed at most 90 g (Gingerich, 1981). Based on this dichotomy in body size alone, one would expect that adapids would have emphasized climbing in their locomotion, while omomyids were more likely to have been frequent leapers (Cartmill & Milton, 1977; Fleagle, 1978, 1985; Jungers, 1979). The morphological distinctions between strepsirhine and haplorhine pedal morphology may be a reflection of this. However, all prosimian primates, regardless of body size, climb (Gebo, 1987); it is more than likely that this was true of Eocene primates as well. One is left with several possibilities: (1) the actual mechanics of grasping and climbing differed in early strepsirhines and haplorhines; (2) there were major differences in support size or orientation choice; or (3) increases in body size alone required morphological readjustments for grasping/climbing, even though these behaviors were performed to the same extent by both groups (but anthropoids, which are also large, did not develop these traits). If the association between derived strepsirhine foot features and large body size is valid, it implies that the ancestral euprimate was small (i.e., less than 500 g) rather than large.

Discussion

This study and the independent work of Covert (1988) have shown that the ancestor of all living primates, adapids, and omomyids was distinguished from its archaic primate predecessor by a set of postcranial traits as well as dental and cranial characters. Indeed, the reorganization of the postcranium appears to be at least as extensive and profound as changes affecting the skull or dentition. The transition from a primitive primate tarsus to a euprimate one involves numerous changes in the upper ankle joint, the transverse tarsal joint, and other aspects of foot structure. The upper ankle joint loses most of its ability to pronate and supinate during dorsiflexion and plantarflexion, but its range of motion in dorsi- and plantar flexion is increased. Stability of the joint is increased by deepening the trochlear groove. These features, along with the elongation of the tarsal region, reflect the importance of leaping in the earliest euprimates. Other aspects of the upper ankle joint (medial malleolar morphology) reflect the conjunct abduction of the foot which accompanies dorsiflexion and is probably related to climbing and grasping. The subtalar joint of euprimates shows very little change from the primate condition, except for the greater elongation of the anterior astragalocalcaneal facets and the more regular confluence of the facets on the head of the astragalus. These features reflect the continued importance of elaborate subtalar motion in euprimates.

The transverse tarsal joint is distinguished by the change of the astragalonavicular joint to spheroidal form and the development of the calcaneocuboid pivot (actually a sellar joint). Together, these changes indicate an increased degree of axial rotation at the transverse tarsal joint at the expense of translational movements. The loss of concomitant pronation–supination abilities at the upper ankle joint as a result of modifying the joint for leaping forced euprimates to increase this ability at other joints. The morphological changes in the subtalar joint and the transverse tarsal joint are related to increasing the range of pronation and supination at these joints.

Other clearly euprimate hallmarks are the enlargement of the hallux and the specialization of the entocuneiform-first metatarsal joint into a sellar shaped unit. Claws are replaced by nails. These features reflect the importance of grasping in the earliest euprimates.

Mechanical interpretation of these features and analogy with living primates indicate that they are related to three main functions: leaping, climbing, and grasping. The mode of locomotion of the common ancestor of all known euprimates was thus a type of arboreal quadrupedalism in which leaping and climbing were the predominant methods of displacement; the grasping, clawless extremities indicate that relatively small supports were preferred to large trunks and branches (Cartmill, 1974). The euprimate ancestor is likely to have been relatively small (less than 500 g).

These are the same general descriptors of locomotion that, with qualifications, I would apply to the lemurids, cheirogaleids, and the smaller platyrrhines and catarrhines. This locomotor mode has been termed "grasp-leaping" (Szalay & Delson, 1979; Szalay & Dagosto, 1980) to emphasize the importance of both grasping (and thus climbing) and leaping in the emergence of euprimates. Martin (1972) reconstructed a similar ancestral locomotor mode for "primates" (= euprimates), but called it hindlimb dominated locomotion.

This notion is in conflict with other proposed ideas of the ancestral euprimate locomotor mode. Napier & Walker (1967) advanced the hypothesis that vertical clinging and leaping may have been the type of locomotor behavior used by the earliest primates. This idea has previously been criticized on several grounds (Szalay, 1972; Cartmill, 1972; Stern & Oxnard, 1973), and a study of the foot (Dagosto, 1986) yielded no evidence that the primitive euprimate was specialized for vertical clinging postures, although features related to this are extremely difficult to identify. Nor was the ancestral euprimate as specialized for leaping as are indriids, bushbabies, tarsiers, or *Lepilemur*.

I do not share the opinion of Godinot & Jouffroy (1984) that the anatomy of *Adapis* is a good model for the primitive euprimate condition, or that the ancestral euprimate was an arboreal, quadrupedal walker-runner and not a leaper. Although the hand of *Adapis* may preserve some primitive euprimate features (but see Beard & Godinot, 1988), an analysis of the rest of the postcranium shows that it was a specialized climbing, quadrupedal, non-leaping primate (Dagosto, 1983; Gebo, in press *a*). For reasons outlined above, mostly having to do with the morphology of the upper ankle joint and the proportions of the foot bones, both the primitive euprimate and the primitive strepsirhine were likely to have been accomplished leapers, unlike *Adapis*. The relative uniqueness of *Adapis*-like morphology among living and fossil euprimates, and the fact that it first appears well after the origin of euprimates are additional arguments which support the contention that *Adapis* is derived, not primitive, in its postcranial morphology (Dagosto, 1983). Ford (1988) also proposes the hypothesis that leaping was not a significant component of locomotion of the ancestral euprimate, but for similar reasons, I disagree.

Early in their history euprimates divided into two groups: a large bodied Adapidae and a small bodied Omomyidae. Current evidence suggests that, at least in terms of postcranial features, the Adapidae are more derived than the Omomyidae (see also Gebo, 1986*b*). The more extensive elongation of the tarsus, the lesser degree of medial malleolar rotation and associated astragalar features, and close apposition of the tibia and fibula, are the only pedal features in which omomyids might be considered more derived than the euprimate ancestor or adapids. The possibility exists that these are primitive euprimate features (e.g., see Dagosto, 1985). In this case, the alternate conditions would be additional strepsirhine synapomorphies. These features indicate that omomyids emphasized the leaping component of locomotion while adapids apparently emphasized climbing. All adapids for which the relevant antomy is known share derived pedal features with the extant Lemuriformes supporting monophyly of the clade Strepsirhini. Anthropoids and *Tarsius*, like omomyids, preserve the primitive form of these traits.

Several issues remain. This paper has only examined the morphology of one region of the body, but numerous features of phylogenetic and adaptive importance have been identified. Analyses of other regions of the postcranium will doubtless produce more information. For instance, a study of the elbow joint by Szalay & Dagosto (1980) revealed that all euprimates are characterized by a complex of traits including a very spherical humeroradial joint and radial head, an elongated, cylindrically shaped humeroulnar joint, and a salient ridge and deep groove demarcating the boundary between the two joints (Figure 4, bottom). This morphology increases the degree of pronation and supination at the humeroradial joint while shifting the primitive weight bearing/forearm stabilization role of this joint to the humeroulnar articulation which thus increases in relative size and attains a cylindrical shape recalling that of the primitive humeroradial articulation. It was hypothesized by Szalay & Dagosto (1980) that this suite of features implied well developed powers of grasping and complex forearm movements (and thus probably climbing) in the ancestral euprimate.

On preliminary examination, other derived features which are likely to have been part of the euprimate morphotype are general elongation of the limb bones, the reduction of the ischium, the anterior–posterior expansion of the femoral condyles and the enlargement and rounding of the lateral edge of the patellar groove, and retroversion of the tibial condyles. Szalay, Tattersall & Decker (1975) and Szalay (1977) have noted the dorsoventral expansion of the iliac blade, and Ford (1980a,b) has analyzed the polarity and distribution of numerous postcranial features.

This study identifies three locomotor behaviors (leaping, grasping, and climbing) which are likely to have had the most impact on the evolution of the euprimate skeleton. Did these behaviors evolve together or did some precede others (Cartmill, 1982)? Features associated with all three behaviors appear simultaneously in the fossil record as evidenced by the morphology of early Sparnacian *Teilhardina belgica* (Teilhard de Chardin, 1927; Szalay, 1976) and early Wasatchian *Cantius mckennai* (Covert, 1988). If Morton's (1924) analysis of the reasons behind tarsal lengthening is correct, then hallucal grasping must have preceded or at least evolved concurrently with leaping. On the other hand, if the novelties of the euprimate midtarsal region are a result of increasing rotation at the transverse tarsal joint because rotation was lost at the upper ankle joint as a result of leaping, then modifications for leaping must have preceded or at least evolved concurrently with these climbing/grasping adaptations.

A related issue concerns possible associations between derived postcranial features of euprimates and derived dental and cranial traits. Has one trait, set of traits, or the behaviors associated with these traits influenced the evolution of others? For example, did orbital convergence in euprimates result from a need to accurately judge distance when leaping (Collins, 1921); a need to accurately judge distance in order to capture small, mobile insect prey (Cartmill, 1972); or was it simply a passive result of facial shortening due to atrophy of the olfactory apparatus (Jones, 1917)?

These points are part of a larger question. Did all the features of the euprimate morphotype originate because of selective pressures arising from a single way of life? Cartmill (1972, 1974) has proposed such a scheme for euprimate origins which includes most of the features of the morphotype relating claw loss, grasping feet, and orbital frontation and convergence to a lifestyle of nocturnal visually directed predation in a fine branch arboreal environment. Leaping can be incorporated into this model if one assumes that: (1) the ancestral euprimate was small, and (2) the most likely way for a small primate to cross gaps between arboreal supports is by leaping (Fleagle, 1985).

As noted by Cartmill (1982), such scenarios are valid only if the elements of the complex evolved simultaneously rather than having been attained sequentially. We know that at some point prior to the early Eocene, a species with this complex of traits existed and that this species was ancestral to adapids, omomyids, and living primates. What we still do not know "is whether the suite of "euprimate" traits evolved as a complex, or, if not, in what order the various traits were added to the suite" (Cartmill, 1982, p. 168). The existence of such a complete (but not uncriticized; see Szalay & Delson, 1979) model for euprimate origins, and the fact that elements of the complex appear simultaneously (but also fully developed) in the fossil record provides some support for the idea that these traits developed concurrently.

Conclusion

The postcranium of euprimates is distinctively different from that of archaic primates and other mammals, indicating that changes in locomotor behavior were important in the origin of this group. Analysis of postcranial features strongly supports the hypothesis of euprimate monophyly. The ancestral euprimate had derived features of the tarsus which indicate that leaping, climbing, and grasping with a large, opposable hallux were important components of its locomotor repertoire, but it is at present impossible to judge if one behavior and its attendant set of traits preceded another in the primate–euprimate transition.

Tarsal anatomy provides evidence that adapids and lemuriforms form a natural group within the Euprimates. The derived features of this group (the Strepsirhini) are related to climbing and grasping, and may have evolved as a response to increased body size in the Adapidae.

Acknowledgements

I thank D. L. Gebo and reviewers for their comments on the manuscript; A. C. Walker for taking the SEM photographs for Figure 6; R. M. West for permission to describe the omomyid phalanges; and the many curators who allowed me to study specimens in their institutions. This work was partly supported by NSF BNS82-03982 and a postdoctoral fellowship from Johns Hopkins University Medical School.

References

Beard, K. C. & Godinot, M. (1988). Carpal anatomy of *Smilodectes gracilis* (Adapiformes, Notharctinae) and its significance for Lemuriform phylogeny. *J. hum. Evol.* **17,** 71–92.
Bleefeld, A. R. (in press). Lagomorph systematic relationships as shown by their tarsal morphology and function (Lagomorpha, Mammalia). *Biol. J. Linn. Soc.*

Cartmill, M. (1972). Arboreal adaptations and the origin of the Order Primates. In (R. Tuttle, Ed.) *The Functional and Evolutionary Biology of Primates*, pp. 97–122. Chicago: Aldine-Atherton.

Cartmill, M. (1974). Pads and claws in arboreal locomotion. In (F. A. Jenkins, Ed.) *Primate Locomotion*, pp. 45–83. New York: Plenum Press.

Cartmill, M. (1982). Basic primatology and prosimian evolution. In (F. Spencer, Ed.) *A History of American Physical Anthropology, 1930–1980*, pp. 147–186. New York: Academic Press.

Cartmill, M. (1985). Climbing. In (M. Hildebrand, D. M. Bramble, K. F. Liem, and D. B. Wake, Eds) *Functional Vertebrate Morphology*, pp. 73–88. Cambridge: Harvard University Press.

Cartmill, M. & Kay, R. F. (1978). Cranio-dental morphology, tarsier affinities and primate sub-orders. In (D. J. Chivers & K. A. Joysey, Eds) *Recent Advances in Primatology*, Vol. 3, pp. 205–213. New York: Academic Press.

Cartmill, M. & Milton, K. (1977). The lorisiform wrist joint and the evolution of "brachiating" adaptations in the Hominoidea. *Am. J. phys. Anthrop.* **47**, 249–272.

Clark, W. E. Le Gros (1936). The problem of the claw in primates. *Proc. Zool. Soc. London* **1936**, 1–25.

Clark, W. E. Le Gross (1959). *The Antecedents of Man*. Chicago: Quadrangle Books.

Collins, E. T. (1921). Changes in the visual organs correlated with the adoption of arboreality and with the assumption of the erect posture. *Trans. Opthalm. Soc. U.K.* **41**, 10–90.

Covert, H. H. (1985). Adaptation and evolutionary relationships of the Eocene primate family Notharctidae. Ph.D. Dissertation, Duke University.

Covert, H. H. (1988). Ankle and foot morphology of *Cantius mckennai*: adaptations and phylogenetic implications. *J. hum. Evol.* **17**, 57–70.

Dagosto, M. (1983). Postcranium of *Adapis parisiensis* and *Leptadapis magnus* (Adapiformes, Primates). *Folia primatol.* **41**, 49–101.

Dagosto, M. (1985). The distal tibia of primates with special reference to the Omomyidae. *Int. J. Primat.* **6**, 45–75.

Dagosto, M. (1986). The joints of the tarsus in the strepsirhine primates. Ph.D. Dissertation, City University of New York.

Decker, R. L. & Szalay, F. S. (1974). Origin and function of the pes in the Eocene Adapidae (Lemuriformes, Primates). In (F. A. Jenkins, Ed.) *Primate Locomotion*, pp. 261–291. New York: Academic Press.

Fleagle, J. G. (1978). Size distributions of living and fossil primate faunas. *Paleobiology* **4**, 67–76.

Fleagle, J. G. (1985). Size and adaptation in primates. In (W. L. Jungers, Ed.) *Size and Scaling in Primate Biology*, pp. 1–19. New York: Plenum Press.

Ford, S. M. (1980a). A systematic revision of the Platyrrhini based on features of the postcranium. Ph.D. Dissertation, University of Pittsburgh.

Ford, S. M. (1980b). Phylogenetic relationships of the Platyrrhini: the evidence from the femur. In (R. L. Ciochon & A. B. Chiarelli, Eds) *Evolutionary Biology of the New World Monkeys and Continental Drift*, pp. 317–329. New York: Plenum Press.

Ford, S. M. (1988). Postcranial adaptations of the earliest platyrrhine. *J. hum. Evol.* **17**, 155–192.

Gebo, D. L. (1986a). The anatomy of the prosimian foot and its application to the primate fossil record. Ph.D. Dissertation, Duke University.

Gebo, D. L. (1986b). Anthropoid origins—the foot evidence. *J. hum. Evol.* **15**, 421–430.

Gebo, D. L. (1987). Locomotor diversity in prosimian primates. *Am. J. Primat.* **13**, 271–281.

Gebo, D. L. (in press a). Foot morphology and locomotor adaptation in Eocene primates. *Folia primatol.*

Gebo, D. L. (in press b). Humeral morphology of *Cantius*, an early Eocene adapid. *Folia primatol.*

Gebo, D. L. & Dagosto, M. (1988). Foot anatomy, climbing, and the origin of the Indriidae. *J. hum. Evol.* **17**, 135–154.

Gingerich, P. D. (1980). Eocene Adapidae, paleobiogeography, and the origin of South American Platyrrhini. In (R. L. Ciochon & A. B. Chiarelli, Eds) *Evolutionary Biology of the New World Monkeys and Continental Drift*, pp. 123–138. New York: Plenum Press.

Gingerich, P. D. (1981). Early Cenozoic Omomyidae and the evolutionary history of tarsiiform primates. *J. hum. Evol.* **10**, 345–374.

Gingerich, P. D. (1986a). *Plesiadapis* and the delineation of the Order Primates. In (B. Wood, L. Martin & P. Andrews, Eds) *Major Topics in Primate and Human Evolution*, pp. 32–46. Cambridge: Cambridge University Press.

Gingerich, P. D. (1986b). Early Eocene *Cantius torresi*—oldest primate of modern aspect from North America. *Nature* **319**, 319–321.

Gingerich, P. D., Smith, H. & Rosenberg, K. (1982). Allometric scaling in the dentition of Primates and prediction of body weight from tooth size in fossils. *Am. J. phys. Anthrop.* **58**, 81–100.

Godinot, M. & Dagosto, M. (1983). The astragalus of *Necrolemur* (Primates, Microchoerinae). *J. Paleont.* **57**, 1321–1324.

Godinot, M. & Jouffroy, F. K. (1984). La main d'*Adapis* (Primates, Adapidae). In (E. Buffetaut, J. M. Mazin & Salmion, Eds) *Actes du symposium paleontologique G. Cuvier*, pp. 221–242. Montbeliard.

Gregory, W. K. (1920). On the structure and relations of *Notharctus*: an American Eocene primate. *Mem. Am. Mus. nat. Hist.* **3**, 51–243.

Hafferl, A. (1932). Bau und funktion des Affenfusses: Ein beitrag zur Gelenk und Muskelmechanic. II. Die Prosimier. *Z. Anat. Ent. Ges.* **99**, 63–112.

Hall-Craggs, E. C. B. (1965). An osteometric study of the hindlimb of the Galagidae. *J. Anat.* **99**, 119–126.

Hoffstetter, R. (1977). Phylogénie des Primates: confrontation des resultats obtenus par les diverses voies d'approache de probleme. *Bull. et Mém. de la Soc. Anth. de Paris, Série 13*, **4**, 327–346.

Jenkins, F. A. & McClearn, D. (1984). Mechanisms of hind foot reversal in climbing mammals. *J. Morph.* **182**, 197–219.

Jones, F. W. (1917). *Arboreal Man*. London: Arnold.

Jungers, W. L. (1979). Locomotion, limb proportions, and skeletal allometry in lemurs and lorises. *Fol. primat.* **32**, 8–28.

Kay, R. F. & Cartmill, M. (1977). Cranial morphology and adaptation of *Palecthon nacimienti* and other Paromomyidae (Plesiadapoidea, ?Primates), with description of a new genus and species. *J. hum. Evol.* **6**, 19–53.

Lewis, O. J. (1980). The joints of the evolving foot. Part I. The ankle joint. *J. Anat.* **130**, 527–543.

Luckett, W. P. (1980). The suggested evolutionary relationships and classification of tree shrews. In (W. P. Luckett, Ed.) *Comparative Biology and Evolutionary Relationships of Tree Shrews*, pp. 3–31. New York: Plenum Press.

MacPhee, R. D. E., Cartmill, M. & Gingerich, P. D. (1983). New Paleogene primate basicrania and the definition of the order Primates. *Nature* **301**, 509–511.

Martin, R. D. (1972). Adaptive radiation and behavior of the Malagasy lemurs. *Phil. Trans. Roy. Soc. London* **264**, 295–352.

Martin, R. D. (1986). Primates: A Definition. In (B. Wood, L. Martin & P. Andrews, Eds) *Major Topics in Primate and Human Evolution*, pp. 1–31. Cambridge: Cambridge University Press.

Matthew, W. D. (1915). A revision of the lower Eocene Wasatch and Wind River faunas. IV. Entelonychia, Primates, Insectivora (part). *Bull. Am. Mus. nat. Hist.* **34**, 429–483.

Mivart, St G. (1873). On *Lepilemur* and *Cheirogaleus*, and on the zoological rank of the Lemuroidea. *Proc. Zool. Soc. London* **1873**, 484–510.

Morton, D. (1924). Evolution of the human foot. *Am. J. phys. Anthrop.* **7**, 1–52.

Musser, G. G. & Dagosto, M. (1987). The identity of *Tarsius pumilus*, a pygmy species endemic to the montane mossy forests of central Sulawesi. *Am. Mus. Nov.* **2867**, pp. 1–53.

Napier, J. H. & Walker, A. (1967). Vertical clinging and leaping: a newly recognized category of locomotor behavior of primates. *Folia primatol.* **6**, 204–219.

Rasmussen, D. T. (1986). Anthropoid origins: a possible solution to the Adapidae–Omomyidae paradox. *J. hum. Evol.* **15**, 1–12.

Rose, K. D. & Walker, A. C. (1985). The skeleton of early Eocene *Cantius*, oldest lemuriform primate. *Am. J. phys. Anthrop.* **66**, 73–90.

Rosenberger, A. L. (1977). *Xenothrix* and ceboid phylogeny. *J. hum. Evol.* **6**, 461–481.

Rosenberger, A. L., Strasser, E. & Delson, E. (1985). Anterior dentition of *Notharctus* and the Adapid-Anthropoid hypothesis. *Fol. primat.* **44**, 15–39.

Rosenberger, A. L. & Szalay, F. S. (1980). On the tarsiiform origins of the Anthropoidea. In (R. L. Ciochan & A. B. Chiarelli, Eds) *Evolutionary Biology of the New World Monkeys and Continental Drift*, pp. 139–157. New York: Plenum Press.

Savage, D. E. & Waters, B. T. (1978). A new omomyid primate from the Wasatch formation of Southern Wyoming. *Fol. primat.* **30**, 1–29.

Schaeffer, B. (1947). Notes on the origin and function of the artiodactyl tarsus. *Am. Mus. Nov.* **1356**, pp. 1–23.

Schlosser, M. (1907). Beitrag zur Osteologie und systematischen Stellung der Gattung *Necrolemur*, sowie zur Stammesgeschichte der Primaten überhaupt. *Neuen Jahrb. Min. Geol. Paleont.*, pp. 197–226.

Schmid, P. (1979). Evidence of microchoerine evolution from Dielsdorf (Zurich region, Switzerland) a preliminary report. *Folia primatol.* **31**, 301–311.

Schwartz, J. H., Tattersall, I. & Eldredge, N. (1978). Phylogeny and classification of the Primates revisited. *Yrbk phys. Anthrop.* **21**, 95–133.

Simons, E. L. (1972). *Primate Evolution*. New York: MacMillan.

Simpson, G. G. (1940). Studies on the earliest primates. *Bull. Am. Mus. nat. Hist.* **77**, 185–212.

Sthlin, H. G. (1912). Die Säugetiere des scweizerischen Eocäens: *Adapis. Abh. schwiez. palaont. Gesch.* **38**, 1165–1298.;

Stern, J. T. & Oxnard, C. E. (1973). Primate locomotion: some links with evolution and morphology. *Primatologia* **4**, 1–93.

Szalay, F. S. (1972). Paleobiology of the earliest primates. In (R. Tuttle, Ed.) *The Functional and Evolutionary Biology of the Primates*, pp. 3–35. Chicago: Aldine–Atherton.

Szalay, F. S. (1976). Systematics of the Omomyidae (Tarsiiformes, Primates): taxonomy, phylogeny, and adaptations. *Bull. Am. Mus. nat. Hist.* **156**, 157–450.

Szalay, F. S. (1977). Constructing primate phylogenies: a search for testable hypotheses with maximum empirical content. *J. hum. Evol.* **6**, 3–18.

Szalay, F. S. (1981). Phylogeny and the problem of adaptive significance: the case of the earliest primates. *Folia primatol.* **36,** 157–182.

Szalay, F. S. (1984). Arboreality: is it homologous in metatherian and eutherian mammals? In (M. K. Hecht, B. Wallace & G. T. Prance, Eds) *Evolutionary Biology* **18,** 215–258. New York: Plenum Press.

Szalay, F. S. (1985). Rodent and lagomorph morphotype adaptations, origins, and relationships: some postcranial attributes analyzed. In (W. P. Luckett & J.-L. Hartenberger, Eds) *Evolutionary relationships among rodents—a multidisciplinary analysis.* NATO ASI series, series A, Life Sciences, Vol. 92, pp 83–132. New York: Plenum Press.

Szalay, F. S. & Dagosto, M. (1980). Locomotor adaptations as reflected on the humerus of Paleogene primates. *Fol. primat.* **34,** 1–45.

Szalay, F. S. & Dagosto, M. (1988). The evolution of hallucial grasping in primates. *J. hum. Evol.* **17,** 1–33.

Szalay, F. S., Tattersall, I. & Decker, R. L. (1975). Phylogenetic relationships of *Plesiadapis:* postcranial evidence. In (F. S. Szalay, Ed.) *Approaches to Primate Paleobiology.* Contributions to Primatology **5,** 136–166.

Szalay, F. S. & Delson, E. (1979). *Evolutionary History of the Primates.* New York: Academic Press.

Szalay, F. S. & Drawhorn, G. (1980). Evolution and diversification of the Archonta in an arboreal milieu. In (W. P. Luckett, Ed.) *Comparative Biology and Evolutionary relationships of Tree Shrews,* pp. 133–169. New York: Plenum Press.

Teilhard de Chardin, P. (1927). Les mammifères de l'Éocène inferieur de la Belgique. *Mem. Mus. Roy. Hist. nat. Belgique* **36,** 1–33.

Thorndike, E. E. (1968). A microscopic study of the marmoset claw and nail. *Am. J. phys. Anthrop.* **28,** 247–262.

Van Valen, L. (1965). Tree shrews, primates and fossils. *Evolution* **19,** 137–151.

Van Valen, L. (1969). A classification of the primates. *Am. J. phys. Anthrop.* **30,** 295–296.

von Koenigswald, W. (1979). Ein Lemurenrest aus dem eozänen Ölschiefer der Grube Messel bei Darmstadt. *Palaont. Z.* **53,** 63–76.

Weigelt, J. (1933). Neue Primaten aus mitteleozänen (oberlutetischen) Braunkhole des Geiseltals. *Nova acta Leopolda* **1,** 97–156.

West, R. M. (1976). Paleontology and geology of the Bridger formation, southern Green River basin, Southwestern Wyoming. Part 1. History of field work and geological setting. *Contrib. Biol. Geol. Milwaukee Public Mus.* **7,** 1–12.

Herbert H. Covert

Department of Anthropology, The
University of Colorado, Boulder,
CO 80309-0233, U.S.A.

Received 29 June 1987
Revision received 16 December
1987 and accepted 13 January
1988

Publication date June 1988

Keywords: Eocene Primates,
ankle, foot, adaptation,
phylogeny.

Ankle and foot morphology of *Cantius mckennai*: adaptations and phylogenetic implications

The morphology of the ankle and foot of the early North American adapid *Cantius mckennai* is described. Significant euprimate traits include a rotated medial malleolus on the tibia, a posterior trochlear shelf on the talus, distal elongation of the calcaneus, a prominent process on the calcaneal articular facet of the cuboid, a large concavoconvex facet for the hallucial metatarsal on the medial cuneiform, a large concavoconvex facet for the medial cuneiform on the hallucial metatarsal, and dorsoplantarly flattened distal phalanges. Functional analysis of the preserved elements suggest that this animal had a mobile ankle, a transverse tarsal joint that allowed a great deal of inversion, an opposable big toe, and grasping digits. Such functional attributes strongly suggest an arboreal lifestyle for *Cantius mckennai*. The presence of a large posterior trochlear shelf on the talus and a strongly rotated medial malleolus are argued to be important indicators of the strepsirhine affinities of adapid primates.

Journal of Human Evolution (1988) **17,** 57–70

Introduction

While early Eocene adapids are known from thousands of dental remains, only three skeletal elements of these animals had been figured as of the mid-1970s (Matthew, 1915 figured the talus, calcaneus, and medial cuneiform of *Cantius trigonodus*). During the past few years, however, interest has been focused on the skeleton of early primates, including adapids, and a number of reports on the morphology of these animals have appeared (e.g., Szalay & Decker, 1974; Decker & Szalay, 1974; Szalay & Dagosto, 1980; Dagosto, 1983, 1985; Covert, 1983, 1985; Rose & Walker, 1985; Gebo, 1985, 1986). Such information is of utmost importance for determining the adaptations of these early primates and the higher level relationships within this order. It has been noted for decades that adapid primates resemble extant lemurids and indriids and omomyid primates resemble tarsiers (e.g., Gregory, 1915, 1920; Simons, 1961). It has also often been argued that these resemblances indicate broad phylogenetic relationships. These notions, however, are questionable and in debate because careful character analyses of traits among these animals are lacking so that the polarity of specific traits are not well known. For example, Cartmill & Kay (1978) argued that omomyids could be linked to haplorhines (tarsiers and anthropoid primates) on the basis of derived carotid artery morphology but that adapids could not be linked specifically to either strepsirhines or haplorhine primates on the basis of derived morphology. In contrast, Gingerich & Schoeninger (1977) argued that, on the basis of phenetic similarities, adapid primates are linked phylogenetically with both anthropoid and strepsirhine primates. They also used phenetic characteristics to link omomyid primates with tarsiers. More recently, Rasmussen (1986) has suggested that there are no known derived characteristics uniquely linking adapids with strepsirhine primates. Based on cranial, dental, and skeletal evidence, he placed adapids in the ancestry of anthropoid primates (but not the tooth-combed primates) and omomyids in the ancestry of tarsiers.

What is needed at this time is careful consideration of which traits shared by the Eocene euprimates (adapid and omomyid primates) are derived from the plesiadapiform condition (Figure 1, traits A). We also need to evaluate how the early omomyids and adapids differ

0047–2484/88/01/20057 + 14 $03.00/0

from one another and then determine the polarity of such traits through outgroup
comparison with the plesiadapiforms (Figure 1, traits B and C). We can then go to later
occurring strepsirhine and haplorhine primates and see if diagnostic differences of these
groups can be traced back to the beginning of the euprimate radiation.

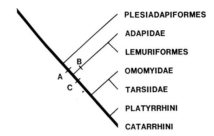

Figure 1. Cladogram of hypothesized relationships among primates. Traits A are derived characters that
define Euprimates and include: some distal elongation of the calcaneus, a posterior trochlear shelf, an
opposable big toe, and flattened nails. Traits B are derived characters that define Strepsirhini and
include: tibial medial malleolus that is oriented anterolaterally and a well developed posterior trochlear
shelf of talus. Traits C are derived characters that define Haplorhini and include: an enlarged
promontory artery and medial entry into the bulla by the internal carotid (Szalay & Delson, 1979).

   In this paper I attempt to do three things that will contribute to our understanding of
primate history. One, I describe the morphology of the ankle and foot of an early Eocene
adapid, *Cantius mckennai*. This is of interest because these specimens provide some of the
earliest known evidence of adapid skeletal morphology. The importance for both adaptive
and phylogenetic studies is obvious. Two, I briefly compare the morphology of these
specimens with that of early Eocene omomyids and with that of plesiadapiform primates
(the outgroup). Three, I discuss a small set of features that may indicate that adapids are
strepsirhines.
   The new specimens described here are from the collections of the University of Michigan
(UM) from the northwestern portion of the Bighorn Basin, Wyoming and the United
States Geological Survey (USGS) from the southern portion of the Bighorn Basin.
   According to Gingerich & Simons (1977) *Cantius mckennai* is an early member of the
North American adapid radiation dating to early lower Graybull deposits (Figure 2). This
taxon is dentally one of the most primitive members of the Adapidae, thus the morphology
of other anatomical regions are of great interest. None of the skeletal material described in
this paper was found in direct association with dental specimens. The skeletal material
described here is attributed to *C. mckennai* because (a) it is clearly euprimate material
(based on morphology); (b) it is too large to represent any known early Eocene omomyid;
and (c) *C. mckennai* is the only known adapid in the deposits where these specimens have
been collected.

Skeleton of *Cantius mckennai*

Covert (1983, 1985) described fragmentary skeletal remains of *Cantius frugivorus* and *Cantius
abditus*, and Rose & Walker (1985) described a well-preserved partial skeleton of *Cantius
trigonodus*. It is agreed by these authors that *Cantius* more closely resembles the middle
Eocene adapids *Smilodectes* and *Notharctus* in morphology than they do any extant primate.
The large-bodied Malagasy primates provide the best living analogs for these animals.

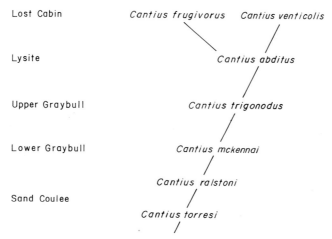

Lost Cabin   *Cantius frugivorus*   *Cantius venticolis*

Lysite   *Cantius abditus*

Upper Graybull   *Cantius trigonodus*

Lower Graybull   *Cantius mckennai*

Cantius ralstoni

Sand Coulee

Cantius torresi

Figure 2. Proposed phylogeny of the genus *Cantius* (Covert, 1985). North American mammal age zonations for the early Eocene are in left column. Note that *C. mckennai* is an early occurring member of this genus.

Anatomical detail and function

Tibia. UM 81820 is a distal tibial fragment (Figure 3). The tibia's talar articular surface has a shallow groove for the medial margin of the talar trochlea. The talar surface is not perpendicular to the tibial long axis, rather it is set obliquely so that its medial margin is more distal than its lateral margin. The medial malleolus is rotated such that its articular surface is facing anterolaterally. A small, somewhat triangular and slightly concave distal fibular facet is located on the anterior half of the lateral surface of the distal tibial epiphysis.

The distal tibia of *Cantius mckennai* differs little, if any, from those of later occurring members of *Cantius* and the middle Eocene notharctines *Notharctus* and *Smilodectes* (Covert, 1985). It is also quite similar to the tibia of the large extant Malagasy primates. The notharctines lack a posterior extension of the articular surface for the talus which is often seen on the extant forms' tibiae. This additional articular surface may make the tibiotalar joint slightly more stable for large Malagasy primates than that of the notharctines because it increases the articular surfaces habitually in contact between these two bones. The oblique cant of the talar facet on the distal tibia along with the anterolateral orientation of the articular surface of the medial malleolus of the notharctine tibia are distinct similarities to the tibia of extant strepsirhines. As Dagosto (1985) noted, these traits indicate that

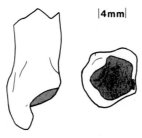

|4mm|

Figure 3. Distal right tibia fragment (UM 81823). Posterior view (left) and distal view (right). In distal view note that the medial malleolus is rotated to face anterolaterally. Articular surfaces for talus and fibula are shaded. Line drawings in this and following figures were made using a Wild M7S scope with camera lucida.

dorsiflexion of the foot is accompanied by abduction and some eversion. The small distal fibular facet is quite similar to that of extant Malagasy primates. This is representative of the "mobile fibula" articulation (Barnet & Napier, 1953), a joint that allows some axial rotation of the fibula on the tibia. This is an important capability that maintains ankle stability yet allows for foot mobility while moving on an uneven substrate.

Talus. UM 81819, 81827, and 81828 are damaged tali of *C. mckennai* (Figure 4). All preserve the trochlea and the latter two also preserve the posterior trochlear shelf. No distinct variations exist in the known parts. The posterior trochlear shelf is well-developed. A groove is present on the posterior end of this shelf for the tendon of flexor digitorum fibularis. Small tubercles are present on the medio- and lateroposterior extremes of the

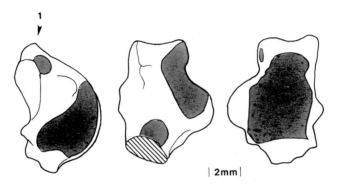

Figure 4. Right talar body (UM 81827). From left to right: lateral, plantar, and dorsal views. Note well-developed posterior trochlear shelf (1). Articular surfaces are shaded, damaged area is indicated by cross-hatching.

trochlear shelf. The posterior calcaneal facet (PCF) is quite concave and is oriented at approximately 45° to the neck of the talus. Distal and medial to the PCF is the deep interosseous sulcus. The most complete specimen, UM 81827, preserves a short portion of the talar neck. On its plantar surface, directly distal to the interosseous sulcus, is a part of the anterior calcaneal facet. Both medial and lateral sides of the trochlea are steep, their delimiting margins are sharp and only slightly higher than the trochlear floor. The facet for the medial malleolus of the tibia is semilunar in shape and slightly concave. It extends slightly onto the medial aspect of the talar neck. Plantoposterior to the facet is a depressed surface marking the insertion of the posterior talotibial ligament. The lateral facet for the fibular malleolus is triangular in shape with the inferior apex of the triangle extending laterally. This facet is convex proximodistally and concave dorsoplantarly. Posterior to this facet is a deep fossa, lying on the lateroposterior surface of the posterior trochlear shelf. There is also a small and indistinct second articular surface for the lateral malleolus. While this feature is unusual for primates, Decker & Szalay (1974) reported that it occurs in other notharctines and is occasionally present in later occurring strepsirhines.

    The talus of *Cantius mckennai* is nearly identical in form to those of later occurring *Cantius* and the middle Eocene notharctines and it most closely resembles that of *V. variegata* among the extant primates (Covert, 1985). Features shared by these animals and lacking in some other lemurids and most indriids include sharply defined medial and lateral trochlear borders, and a tall trochlear body. It differs from the talus of plesiadapids, most omomyids, platyrrhines, and catarrhines in having a well developed posterior trochlear shelf. It also differs from these animals in the lateral projection of the facet for the lateral malleolus (Gebo, 1986).

The tibial facet on the medial aspect of the trochlea and the squatting facet on the talar neck both suggest that the talus and hence the foot of *C. mckennai*, was capable of a great deal of dorsiflexion. The medial deviation of the talar neck is characteristic (but not uniquely so) of animals with an opposable big toe.

The talus of notharctines functioned with the tibial and fibular malleoli as a hinge. The primary movements were dorsi- and plantarflexion of the foot. In addition, considering the concave nature of the facet for the medial malleolus and the convex facet for the lateral malleolus, the foot of *C. mckennai* was probably habitually slightly inverted. This inversion would increase with plantarflexion and decrease with dorsiflexion. Additional information on the mechanics of these joints and others discussed in this paper are presented in Decker & Szalay (1974) and Szalay & Decker (1974).

Calcaneus. UM 81825 is a nearly complete calcaneus, lacking only the medial most portion of the sustentaculum and plantolateral portion of the cuboid facet (Figure 5). A small peroneal tubercle is present on the lateral surface of the calcaneus. This tubercle lies slightly nearer to the cuboid facet than to the heel. The sustentaculum is on the medial

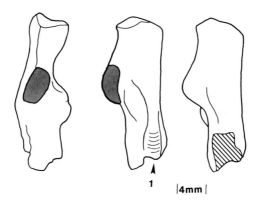

Figure 5. Right calcaneus (UM 81825). From left to right: dorsal, medial, and lateral views. Note well-developed groove for the tendon of flexor digitorum fibularis (1). Articular surfaces are indicated by shading (because of abrasion the articular surface on the sustentaculum cannot be delineated) and damaged area is indicated by cross-hatching.

surface of the calcaneal body directly across from the peroneal tubercle. The sustentaculum projects medially from the body and has a distinct groove for the tendon of the flexor digitorum fibularis running on its plantar surface parallel to the calcaneal long axis. This groove continues to the distal end of the calcaneus. The calcaneal body projects distally beyond the sustentaculum. Near the distal end of the plantar surface is a prominent tubercle marking the origin of the calcaneocuboid ligament. Because of damage, the morphology of the cuboid facet cannot be determined. The proximal talar facet is long (proximodistally) and narrow, with a proximodistal convexity. Its articular surface runs downward and only slightly laterally ending directly dorsal to the peroneal tubercle. A distal talar facet is present on the dorsal surface of the sustentaculum and runs distally on the medial aspect of the calcaneal dorsal surface.

The calcaneus of *Cantius* is quite similar to those of *Smilodectes* and *Notharctus*, and also quite similar to those of the large extant Malagasy primates (Covert, 1985). One consistent difference in the calcaneus of notharctines and living Malagasy primates is the location and size of the peroneal tubercle. In the extant forms the peroneal tubercle is reduced in size and more proximally positioned (Decker & Szalay, 1974). Aside from the morphology of the peroneal tubercle, the calcaneus of notharctines differs in no consistent fashion from those of *Lemur* and *Varecia* and only in the relative height of the heel from those of indriids (indriids have a slightly taller heel). The calcaneus of *C. mckennai* differs from that of plesiadapiforms in a number of features including the distal projection of the calcaneal body beyond the sustentaculum and in the small size of its peroneal tubercle.

Gregory (1920) suggested that the calcanei of *Notharctus* and *Cantius* (recognized as *Pelycodus* by Gregory) differed in that the latter had a shorter "lower end of calcaneum" (p. 95) than the former. As the sample of *Cantius* calcanei has grown it now appears that this element cannot be distinguished from that of *Notharctus* on the basis of the relative size of its lower or distal end (Table 1). Because the calcaneal heel index includes the proximal talar facet as part of the distal end of the calcaneus, data presented in Table 1 do not preclude that some variation in the relative proportions of this facet and the portion of the calcaneus distal to it may exist among notharctines.

Table 1.   Calcaneal heel index (heel to anterior end of posterior talar facet/calcaneal length × 100). The lower the index value, the longer the distal portion of the calcaneus. The *Cantius* sample includes specimens from species occurring later in the early Eocene than *C. mckennai*.

Taxon	n	Mean	Range	S.D.
Lemur fulvus	3	57·0	51·3–65·4	7·4
Varecia variegata	3	57·8	55·4–60·4	2·5
Hapalemur griseus	3	52·2	50·8–53·2	1·2
Lepilemur mustelinus	7	51·8	48·6–55·7	5·0
Avahi laniger	3	56·8	54·6–59·0	2·2
Propithecus verreauxi	3	56·8	56·7–56·9	0·1
Indri indri	3	56·3	55·3–58·2	1·6
Cantius	4	58·3	57·1–60·5	1·6
Notharctus tenebrosus	2	57·2	57·1–57·4	0·2
other *Notharctus*	5	59·0	57·9–60·4	1·0
Smilodectes	1	62·7		

Note that the values for *Cantius* overlap universally with those of the later-occurring notharctines and overlap extensively with those of the extant large-bodied Malagasy primates; n = sample size.

The distinct groove for the flexor digitorum fibularis on the calcaneus of *C. mckennai* is suggestive of at least fairly powerful grasping digits (Szalay & Decker, 1974).

Cuboid. Two cuboids are known for *Cantius mckennai* (UM 81823, right: complete; UM 81824, left: proximolateral two-thirds present) (Figure 6). The cuboid is a fairly long, narrow, and shallow bone. The calcaneal articular facet is oval; its dorsoplantar diameter

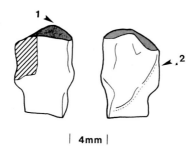

| 4mm |

Figure 6. Right cuboid (UM 81823). Dorsal view (left) and plantar view (right). Note the process on the calcaneal facet (1) and the groove for the tendon of peroneus longus (2). Articular surfaces are shaded, damaged area is indicated by cross-hatching.

is twice as wide as its mediolateral. The facet has a well-developed process in the middle of its plantar aspect. The cuboid's lateral surface is dominated by a distinct deep groove for the tendon of peroneus longus. This groove initially runs proximodistally and then curves medially across the distal portion of the plantar surface of the cuboid. A flange of bone extends from the lateral surface of the cuboid to cover the initial portion of this groove. The distal articular surface of the cuboid for the fourth and fifth metatarsals, is nearly rectangular, not particularly large, and slightly concave. A large facet for the lateral surface of the navicular and the proximolateral surface of the lateral cuneiform covers the proximal two-thirds of the medial side of the cuboid. There is another small facet for the lateral cuneiform on the distal aspect of this surface. The area in between these facets probably served for the attachment of the interosseous ligaments that provide foot stability. The plantar surface has a centrally located raised area for the insertion of the plantar calcaneocuboid ligament.

The cuboid varies little among notharctines and in all details it is similar to the cuboid of lemurids (Covert, 1985). The indriid cuboid differs from those of notharctines and most other strepsirhines in lacking the flange dorsal to the peroneus longus tendon groove on the lateral side of the cuboid.

Important characters for evaluating function include (1) the projecting process on the calcaneal articular facet; (2) the depth and distinctiveness of the groove for the tendon of peroneus longus; and (3) the concavity of the facet for the metatarsals. According to Szalay & Decker (1974), this first character is one of the hallmarks of being a primate and being arboreal: they view this as indicating that the foot could be freely inverted at the transverse tarsal joint. The second character implies a well developed peroneus longus and taken along with the morphology of the hallucial metatarsal, described below, a powerful grasping big toe. The third character indicates that metatarsals four and five could move somewhat freely on the cuboid.

Navicular. This element is represented by two specimens, UM 81827 (left and damaged) and UM 81831 (right and complete) (Figure 7). The following description is based on UM 81831. The proximal surface for the talar head is strongly concave dorsoplantarly and moderately concave mediolaterally. It is nearly twice as high as wide. The most plantomedial point of this surface is a distinct process (posteroinferior process of Decker & Szalay, 1974). The dorsal, medial, and plantar surfaces of the navicular have slightly roughened areas for attachment of stabilizing interosseous ligaments. The lateral surface is quite roughened. Dorsodistally there is a small facet for the cuboid and plantoproximally there is a small indistinct facet, possibly for the calcaneus. A number of ligaments insert and originate on the navicular's lateral surface including the plantar calcaneonavicular ligament (spring ligament) and the naviculocuboid interosseous ligaments. All three cuneiforms articulate with the distal surface of the navicular. Plantomedially there is a small oval convex facet for the medial cuneiform; lateral to that is a slightly larger and less convex facet for the intermediate cuneiform; and dorsolaterally, a large round concave facet for the lateral cuneiform.

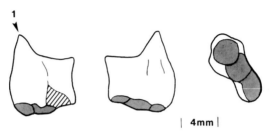

Figure 7. Right navicular (UM 81831). From left to right: plantolateral, dorsomedial, and distal views. Note well-developed posteroinferior process (1). Articular areas for cuneiforms are shaded and articular area for cuboid is cross-hatched.

Variation in the navicular among the notharctines resembles that seen in modern lemurids and indriids (Covert, 1985). *Lepilemur, Hapalemur,* and *Lemur* all have a navicular that is similar to that of *C. mckennai,* i.e., they are gracile, long elements with a pronounced posterioinferior process. *Varecia,* which is a larger animal, has a navicular that is more similar to that of *Notharctus* and *Smilodectes* in being robust and less elongate with a less well developed posteroinferior process. As Decker & Szalay (1974) noted, the navicular of *Indri* and *Propithecus* differ from those of notharctines in being even more robust, less elongated, and having a much reduced posteroinferior process. Finally, the navicular of notharctines differs slightly and consistently from that of the large extant Malagasy primates in the arrangments of the facets for the cuneiforms. When viewed distally, the right navicular of the extant forms is S-shaped; that is, the dorsal facet (for the lateral cuneiform) is turned slightly laterally and the plantar facet (for the medial cuneiform) is turned slightly medially. In notharctines, the navicular has these facets oriented in a more nearly straight line with that of *C. mckennai* curving the least. In addition, the facet for the intermediate cuneiform on the notharctine navicular does not project as far distally as do those of the extant lemuriforms examined.

The functional implications of the structural differences among and between the naviculars of notharctines and large extant Malagasy primates are not perfectly clear. The transverse arch in the notharctine foot may have been less high than in the foot of the extant forms (Decker & Szalay, 1974). This difference would seem to be trivial. Further, because of the similarity in morphology of the talonavicular, talocalcaneus, and calcaneocuboid articulations of *C. mckennai* and those of the extant Malagasy primates it seems reasonable to conclude that the former had a mobile transverse tarsal joint like the latter.

Medial cuneiform. A single, perfectly preserved, right medial cuneiform (UM 81820) is known for *C. mckennai* (Figure 8). Proximally, there is a small concave facet for the navicular. It is teardrop shaped, with the apex located dorsally. The medial surface is fairly smooth with its distal end over twice as tall as its proximal end. This increase in height occurs abruptly on the dorsal surface near its distal end. The distal surface is dominated by the large facet for the hallucial metatarsal. This facet is weakly concave dorsoplantarly and strongly convex mediolaterally. Medially it is most extensive at its plantar end. There are two articular facets present on the lateral surface of the medial cuneiform. One is located along the dorsal two-thirds of its proximal end and is for the intermediate cuneiform. The second is located on the dorsal portion of the lateral surface's distal end. This facet is for both the intermediate cuneiform (proximal) and second metatarsal (distal).

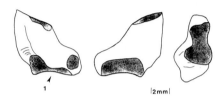

Figure 8. Right medial cuneiform (UM 81820). From left to right: lateral, medial, and distal views. Note large concavoconvex articular facet for the hallucial metatarsal (1). Articular areas are shaded.

The medial cuneiform of *C. mckennai* quite closely resembles those of later occurring *Cantius, Notharctus,* and *Smilodectes* (Covert, 1985). Overall differences between the morphology of notharctine and large extant Malagasy primates' medial cuneiforms are quite minor. There are, however, two fairly consistent differences between these groups. First, as noted by Decker & Szalay (1974), the well-developed plantar surface of the medial cuneiform of *S. gracilis* and *Notharctus* projects directly plantarly. On the other hand, this surface has a slightly lateral inflection in the large living Malagasy primates. The medial cuneiform of *C. mckennai* is nearly identical to those of the other notharctines. Decker & Szalay (1974) suggested that the medial cuneiform morphology of the extant forms creates "a more complete osseous containment" (p. 272) for digital flexor tendons. Second, the distal articulation for the intermediate cuneiform and articulation for the base of the second metatarsal is located relatively less dorsally on the notharctine medial cuneiform than they are on those of most large extant Malagasy primates. In combination with the navicular facets for the cuneiforms, this suggests that there was a slightly lower transverse arch present in the notharctine foot. Extensive variation is seen within large extant Malagasy primates ranging from *Lepilemur*—which overlaps in this respect with notharctine morphology—to indriids, where facets for the intermediate cuneiform and second metatarsal are almost completely dorsal to the hallucial metatarsal facet.

The great similarity in the concavoconvex hallucial metatarsal facet documents that all notharctines, for which medial cuneiform morphology is known, clearly had an opposable big toe.

Hallucial metatarsal. A portion of a right hallucial metatarsal (UM 81822) is known for *C. mckennai* (Figure 9). This specimen lacks its distal end and much of the lateral portion of the proximal epiphysis. The proximal articular facet for the medial cuneiform is long and slightly convex dorsoplantarly and concave mediolaterally. A large peroneal process is present on the lateroplantar surface of the metatarsal's proximal end. The shaft of this bone is round in cross-section.

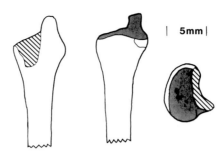

Figure 9. Right proximal hallucial metatarsal (UM 81822). From left to right: lateral, medial, and proximal views. Articular surfaces are shaded, damaged area is indicated by cross-hatching.

The hallucial metatarsal of *C. mckennai* is quite similar to those of later occurring notharctines and large extant Malagasy primates (Covert, 1985). Its articular facet for the medial cuneiform documents that the first pedal digit was opposable to the other digits (as does the morphology of the cuneiform). Also, the notharctine hallux appears to have been a relatively large digit in relationship to their other digits and in comparison with the hallux of similar sized extant strepsirhines (Gregory, 1920; Covert, 1985). This probably is indicative of a particularly powerful opposable big toe.

Distal phalanges. Two distal phalanges are known for *Cantius mckennai* (USGS 18338, 18339) (Figure 10). The proximal surface of this bone is dominated by the articular facet for the middle phalange's head. This surface has a medial and lateral concavity separated by a midline crest running from the plantar to the dorsal edge of this surface. The proximal end of the distal phalanx is significantly broader and deeper than its shaft. Midway between the proximal and distal ends of the distal phalanx the shaft becomes more compressed dorsoplantarly and wider mediolaterally yielding a broad-flattened end that is arrowhead-shaped. This, of course, is indicative of a flattened nail instead of a claw (Le Gros Clark, 1959). The flattened distal end is weakly curved plantarly (or palmarly).

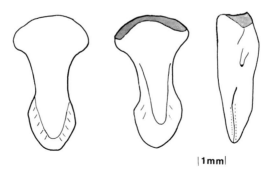

Figure 10. Distal phalanx (USGS 18338). From left to right: dorsal, plantar, and lateral views. Articular surface is shaded.

The morphology of these distal phalanges is quite similar to those of later occurring notharctines and the large extant Malagasy primates. This indicates that early *Cantius* was capable of grasping objects in a euprimate-like fashion.

Discussion

Adaptive importance

What do these specimens indicate about movement capabilities and habitat preference for *Cantius mckennai?*

The rotation of the medial malleolus of the tibia is such that when the foot is dorsiflexed it is also abducted. The talus of *Cantius* functioned with the tibial and fibular malleoli as a hinge with primary motions being dorsi- and plantarflexion of the foot. In addition, considering the concave nature of the facet for the medial malleolus and the convex facet for the lateral malleolus, the foot was probably habitually slightly inverted. This inversion would increase with plantarflexion and decrease with dorsiflexion. The distinct groove for the flexor digitorum fibularis on the calcaneus of *Cantius mckennai* is suggestive of at least fairly powerful grasping digits. The morphology of the calcaneal facet of the cuboid indicates that the foot could be freely inverted at the transverse tarsal joint. The deep and distinct groove for the tendon of peroneus longus on the cuboid suggests that peroneus longus was a strong muscle and, taken along with the morphology of the hallucial metatarsal, indicates a powerful grasping big toe. The concave nature of the metatarsal facet on the cuboid indicates that metatarsals four and five could be flexed and extended on the cuboid. The concavoconvex hallucial metatarsal facet documents that *Cantius mckennai* had an opposable big toe. The articular surface for the medial cuneiform indicates that this digit was capable of a wide range of movement towards and away from the other digits. The morphology of the distal phalanges indicate the presence of flattened nails which, in turn, indicate primate-like grasping abilities.

In sum, a number of features of the ankle and foot indicate that it was a mobile grasping organ. Thus, this animal was capable of moving about in an arboreal setting. Further, because detailed similarities exist between the ankle and foot of *C. mckennai* and the lemurids, I would suggest that this early notharctine probably lived in and moved about an arboreal setting much like these extant animals do today. This notion can be greatly strengthened, (or rejected!) with the discovery of more complete and associated skeletal material for *C. mckennai*.

It would be interesting to evaluate the grasp of *C. mckennai* in light of Gebo's (1985) recent statement that lemurids and indriids differ from all other primates (including adapids) in the way they grasp objects with their feet. Unfortunately, such an evaluation is not possible without the lateral cuneiform and third and fourth metatarsals of *C. mckennai*.

Phylogenetic importance

The skeletal material described in this article is among the earliest known for the family Adapidae. In that a number of skeletal elements of omomyid and adapid primates have been described previously and these animals share a number of traits, one would predict that *Cantius mckennai* should also show these traits. No surprises were found in the present analysis: the ankle and foot of *C. mckennai* shows no novel morphology when compared to those of other Eocene Euprimates.

Are adapids strepsirhine primates and are omomyids haplorhines?

Omomyids and adapids do share a large number of the features that are included in the above description and many of these appear to be derived over any known Paleocene mammal. Examples of such features include some tibial medial malleolus rotation (slight in omomyids, greater in adapids) (Dagosto, 1985); a posterior trochlear shelf on talus (small in omomyids, more developed in most adapids); tall, narrow talar trochlea (equally well-developed in most omomyids and adapids); a calcaneus with some anterior elongation (ranging from slight for *Adapis* and *Leptadapis* to great for microchoerine omomyids with notharctines and other omomyids intermediate); an opposable big toe, and flattened nails (Wible & Covert, 1987). I interpret these features as being characteristic of the last common ancestor of euprimates. Are there any features in which early omomyids and adapids differ and that the adapids share with modern strepsirhine primates (traits B, figure 1)? The answer is yes and more importantly, a number of these features are easily interpreted as being derived over the euprimate morphotype.

1. Distal tibia. Dagosto (1985) noted that primates differ from primitive mammals in having a tibial medial malleolus that is rotated so that its articular surface is oriented somewhat anterolaterally. Dagosto also reported that primates can be subdivided on the basis of this trait: strepsirhines usually show more rotation than do haplorhines in this feature. She further argued that it was not clear whether the strepsirhine or haplorhine condition was derived. As described above, the morphology of the tibial medial malleolus of *Cantius mckennai* is like that of extant strepsirhines. I agree with Dagosto's (1988) interpretation that the variation seen among primitive mammals, haplorhines, and strepsirhines represent a morphocline for which strepsirhines (including adapids) show the most derived condition.

2. Posterior trochlear shelf of talus. Plesiadapiforms lack a posterior trochlear shelf (Szalay & Decker, 1974). Tali of *Teilhardina belgica*, an early Eocene anatomorphine, and *Hemiacodon gracilis* have been figured by Szalay (1976); the talus of *Arapahovius gazini* has been figured by Savage & Waters (1978); and I have examined tali of two additional early and middle Eocene omomyids. All of these tali preserve very small posterior trochlear shelves. Only the talus of the late Eocene *Necrolemur antiquus* described by Godinot & Dagosto (1983) has a large posterior trochlear shelf. Because of its frequent occurrence among omomyids, I consider a weakly developed posterior trochlear shelf to be primitive for this group. All known notharctines (Covert, 1985), *Leptadapis magnus* (Dagosto, 1984), and the adapine described by von Koenigswald (1979) have a well developed posterior trochlear shelf (Figure 13). Of the adapids for which we have tali only *Adapis parisiensis* has a very small posterior trochlear shelf (Dagosto, 1983). Because of its frequent occurrence among

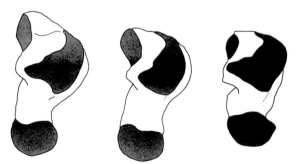

Figure 11. From left to right: medial views of the right talus of *Smilodectes gracilis*, *Hemiacodon gracilis* and *Plesiadapis tricuspidens* (modified from Szalay & Delson, 1979). Compare the posterior regions of the talus (top side of figure), *P. tricuspidens* lacks a posterior trochlear shelf (primitive), *H. gracilis* has a weakly developed posterior trochlear shelf (euprimate morphotype), and *S. gracilis* shows a well developed trochlear shelf (derived strepsirhine condition). Shaded areas illustrate fibular facet, posterior calcaneal facet, and talar head. All have been scaled to the length of the lateral malleolar articular surface.

adapids I consider a well-developed posterior trochlear shelf to be primitive for this group and derived over the euprimate morphotype (the common omomyid condition). Early anthropoids from the Fayum lack or have a weakly developed posterior shelf (Gebo & Simons, 1987) while many extant strepsirhine primates have a well developed posterior trochlear shelf (Decker & Szalay, 1974). I conclude, as does Dagosto (1988), that on the basis of the known distribution of this trait that the well developed posterior trochlear shelf characteristic of most adapids is a shared derived trait of strepsirhine primates.

Gebo (1986) has recently identified two unusual features of the adapid foot that links them specifically with later occurring strepsirhines. First, the lateral malleolar facet on the talus of extant lemuriforms and adapids extends laterally much further than it does in other euprimates and plesiadapiforms. Second, the flexor digitorum fibularis groove on the posterior talar trochlea of adapids and lemuriforms lies lateral to the posterior trochlear facet while in other euprimates and plesiadapiforms it lies in the midline of the posterior trochlear facet. Gebo argues that the adapid–lemuriform condition in both of these cases is derived over that of other primates.

Conclusions

1. A number of tarsals are now known for notharctine primates from the beginning of the early Eocene. In all details they closely resemble those of the middle Eocene notharctines and to only a slightly lesser extent those of large-bodied extant strepsirhines.

2. A number of features on these tarsals indicate that *Cantius mckennai* was arboreal.

3. These specimens corroborate hypotheses about the nature of the earliest euprimate feet. The presence of a large opposable big toe, flattened nails, distal elongation of the calcaneus, and a mobile transverse tarsal joint are among features present that would have been predicted for this animal.

4. The presence of a large posterior trochlear shelf on the talus and the presence of a large medial malleolus that is strongly rotated are quite likely important indicators of the strepsirhine affinities of these adapids.

Acknowledgements

I thank Richard F. Kay, Elywn L. Simons, and especially Elizabeth Strasser and Blythe A. Carlson for commenting on various drafts of this paper. I also thank D. Gebo, M. Dagosto, C. Beard and the *Journal of Human Evolution* reviewers for informative discussions and comments on the morphology of early primates. Special thanks go to Dr Philip Gingerich (UM) and Dr Thomas Bown (USGS) for permission to study and describe the new *Cantius mckennai* material. Finally, I thank the following persons and institutions for access to skeletal material of extant and extinct primates in their collections: the American Museum of Natural History, Drs G. Musser and R. H. Tedford; National Museum of Natural History (Smithsonian Institution), Drs R. Thorington and R. Emry; Yale Peabody Museum, Dr J. H. Ostrom; Museum of Paleontology at the University of Michigan, Dr P. D. Gingerich; the Museum of Comparative Zoology at Harvard University, Drs F. A. Jenkins, Jr. and M. Rutzmoser; U.S. Geological Survey (Denver), Dr T. M. Bown; The University of Wyoming Geological Museum, Dr J. A. Lillegraven; and the Duke University Primate Center, Dr E. L. Simons.

References

Barnett, C. H. & Napier, J. R. (1953). The rotary mobility of the fibula in eutherian mammals. *J. Anat. Lond.* **86,** 1–9.

Cartmill, M. & Kay, R. F. (1978). Craniodental morphology, *Tarsius* affinities, and primate suborders. In (D. J. Chivers & K. A. Joysey, Eds) *Recent Advances in Primatology.* Vol. 3. *Evolution,* pp. 205–214. London: Academic Press.

Covert, H. H. (1983). Newly discovered post-cranial remains of *Pelycodus. Am. J. phys. Anthrop.* **60,** 185–186.

Covert, H. H. (1985). Adaptations and evolutionary relationships of the Eocene primate family Notharctidae. Ph.D. Dissertation, Duke University.

Dagosto, M. (1983). Postcranium of *Adapis parisiensis* and *Leptadapis magnus* (Adapiformes, Primates): adaptations and phylogenetic significance. *Folia primatol.* **41,** 49–101.

Dagosto, M. (1985). The distal tibia of primates with special reference to the Omomyidae. *Int. J. Primat.* **6,** 45–75.

Dagosto, M. (1988). Implications of postcranial evidence for the origin of euprimates. *J. hum. Evol.,* **17,** 39–60.

Decker, R. L. & Szalay, F. S. (1974). Origins and function of the pes in the Eocene Adapidae (Lemuriformes, Primates). In (F. A. Jenkins, Jr., Ed) *Primate Locomotion,* pp. 261–291. New York: Academic Press.

Gebo, D. L. (1985). The nature of the Primate grasping foot. *Am. J. phys. Anthrop.* **67,** 269–277.

Gebo, D. L. (1986). Anthropoid origins—the foot evidence. *J. hum. Evol.* **15,** 421–430.

Gebo, D. L. & Simons, E. L. (1987). Morphology and locomotor adaptations of the foot of early Oligocene anthropoids. *Am. J. phys. Anthrop.* **74,** 83–102.

Gingerich, P. D. & Schoeninger, M. (1977). The fossil record and primate phylogeny. *J. hum. Evol.* **6,** 483–505.

Gingerich, P. D. & Simons, E. L. (1977). Systematics, phylogeny, and evolution of early Eocene Adapidae (Mammalia, Primates) in North America. *Contr. Mus. Paleon. Univ. Mich.* **24**(22), 245–279.

Godinot, M. & Dagosto, M. (1983). The astragalus of *Necrolemur* (Primates, Microchoerinae). *J. Paleont.* **57,** 1321–1324.

Gregory, W. K. (1915). I. On the relationships of the Eocene Lemur *Notharctus* to the Adapidae and to other primates. II. On the classification and phylogeny of the Lemuroidea. *Bull. Geol. Soc. Am.* **26,** 419–446.

Gregory, W. K. (1920). On the structure and relations of *Notharctus,* an American Eocene primate. *Mem. Am. Mus. nat. Hist.* n. s. **3**(2), 49–243.

Le Gros Clark, W. E. (1959). *The Antecedents of Man.* Edinburgh: University Press.

Matthew, W. D. (1915). A revision of the lower Eocene Wasatch and Wind River faunas. Part IV. Entelonychia, Primates, Insectivora (part). *Bull. Am. Mus. nat. Hist.* **34,** 429–483.

Rasmussen, D. T. (1986). Anthropoid origins: a possible solution to the Adapidae–Omomyidae paradox. *J. hum. Evol.* **15,** 1–12.

Rose, K. D. & Walker, A. C. (1985). The skeleton of early Eocene *Cantius,* oldest lemuriform primate. *Am. J. phys. Anthrop.* **66,** 73–89.

Savage, D. E. & Waters, B. T. (1978). A new omomyid primate from the Wasatch formation of Southern Wyoming. *Folia primatol.* **30,** 1–29.

Simons, E. L. (1961). Notes on Eocene Tarsioids and a revision of some Necrolemurinae. *Bull. Brit. Mus. (Nat. Hist.) Geol.* **5**(3), 43–69.

Szalay, F. S. (1976). Systematics of the Omomyidae (Tarsiiformes, Primates): taxonomy, phylogeny, and adaptations. *Bull. Am. Mus. nat. Hist.* **156,** 157–450.

Szalay, F. S. & Dagosto, M. (1980). Locomotor adaptations as reflected in the humerus of Paleogene primates. *Folia primat.* **34,** 1–45.

Szalay, F. S. & Decker, R. L. (1974). Origins, evolution, and function of the tarsus in Late Cretaceous Eutheria and Paleocene primates. In (F. A. Jenkins, Jr., Ed.) *Primate Locomotion,* pp. 223–259. New York: Academic Press.

Szalay, F. S. & Delson, E. (1979). *Evolutionary History of the Primates.* New York: Academic Press.

von Koenigswald, W. (1979). Ein Lemurenrest aus dem eozänen Ölschiefer der Grube Messel bei Darmstadt. *Paleont. Z.* **53,** 63–76.

Wible, J. R. & Covert, H. H. (1987). Primates: cladistic diagnosis and relationships. *J. hum. Evol.* **16,** 1–22.

K. Christopher Beard

Department of Cell Biology and
Anatomy, The Johns Hopkins
University, School of Medicine, 725
North Wolfe Street, Baltimore,
MD 21205, U.S.A.

Marc Godinot

Laboratoire de Paléontologie,
Université des Sciences et Techniques
du Languedoc, Place
Eugène-Bataillon, F-34060
Montpellier, Cedex, France

Received 5 June 1987
Revision received 23 November
1987 and accepted 9 December
1987

Publication date June 1988

Keywords: Adapiformes,
Lemuriformes, phylogeny, wrist
Eocene, Smilodectes gracilis, Adapi
parisiensis, Notharctus.

Carpal anatomy of *Smilodectes gracilis* (Adapiformes, Notharctinae) and its significance for lemuriform phylogeny

The oldest primate carpals collected to date are those of the middle Eocene notharctine *Smilodectes gracilis*. In general, carpal structure in *Smilodectes* appears to have been quite generalized compared to that of other living and fossil primates for which wrist anatomy is adequately known. This is reflected by the fact that *Smilodectes* exhibits several features of wrist anatomy that today are confined either to lemuriforms and *Tarsius* ("prosimians") or to haplorhines; in all cases the distribution of these traits among living and fossil primates and eutherians suggests that the character states present in *Smilodectes* are primitive for primates. Functionally, the wrist of *Smilodectes* appears to have been relatively less mobile than that of late Eocene *Adapis*, particularly in movements of pronation/supination at the midcarpal joint and abduction/adduction at the antebrachiocarpal and midcarpal joints. *Smilodectes*, *Notharctus*, and *Adapis* all lack the clearly derived midcarpal anatomy found in extant lemuriforms, strongly suggesting that lemuriforms are monophyletic with respect to these adapiform taxa. The ulnocarpal joint of *Smilodectes* is similar to that of extant haplorhines and other eutherians, while that of *Adapis* exhibits the presumably derived morphology that is otherwise restricted to extant lemuriforms. Together with dental anatomy and paleobiogeographic considerations, the ulnocarpal joint of *Adapis* suggests that this late Eocene adapiform shares more recent common ancestry with lemuriforms than either of these taxa does with *Smilodectes*.

Journal of Human Evolution (1988) **17**, 71–92

Introduction

In 1920 William King Gregory published an extremely thorough account of the postcranial anatomy of the middle Eocene adapiform primate *Notharctus*, a study whose ideas have stood essentially unchanged to the present day. Gregory's exemplary work, perhaps the least understood aspect of notharctine anatomy was the wrist, because only four of the ten osseous elements of the wrist of *Notharctus* were known in 1920. In the decades since Gregory's monograph, a number of skulls and partial skeletons of notharctine primates have been recovered from middle Eocene deposits of the Bridger Basin of southwestern Wyoming (e.g., Gazin, 1958; Covert, 1985). Several of the most complete of these specimens belong to the species *Smilodectes gracilis*. Perhaps the most important contribution that these specimens make to our knowledge of notharctine anatomy is that they jointly preserve the carpus in near entirety. Of the ten osseous elements of the wrist of *Smilodectes*, only the trapezoid and the prepollex remain unknown (Table 1). Moreover, most of the carpal elements of *Smilodectes* are represented by more than one specimen, so that at least some understanding of intraspecific variation is also possible.

Below, we describe the known carpals of *Smilodectes gracilis* and compare them with those of other fossil and Recent primates and eutherians. These observations are then used to assess several competing hypotheses of adapiform relationships and to draw some functional inferences about the possible adaptations of the wrist of *Smilodectes* and some other adapiforms.

Materials and methods

Carpal specimens representing a minimum of six individuals of *Smilodectes gracilis* and four individuals of *Notharctus* from the paleontological collections housed at the U.S. National

0047–2484/88/01/20071 + 22 $03.00/0

Museum of Natural History (USNM; Washington, D.C.) and the Carnegie Museum of Natural History (CM; Pittsburgh) form the basis for this study (Table 1). We have compared the proportions and anatomy of the carpus of *Smilodectes* to that of late Eocene *Adapis parisiensis* (Godinot & Jouffroy, 1984) and a wide variety of living primates including the following genera: *Microcebus, Cheirogaleus, Lemur, Varecia, Avahi, Daubentonia, Nycticebus, Perodicticus, Galago, Tarsius, Saimiri, Alouatta, Aotus, Chiropotes, Cacajao, Callicebus, Pithecia, Callimico, Callithrix, Leontopithecus, Ateles, Cercopithecus, Macaca, Presbytis,* and *Colobus.* In this paper we follow the classification of primates employed by Szalay & Delson (1979), except that we elevate their subfamilies Lorisinae and Galaginae to familial rank and that we use the taxon Anthropoidea in the conventional sense (i.e., as the combination of the infraorders Platyrrhini and Catarrhini).

Table 1   **List of notharctine wrist material utilized in this study**

Taxon	Specimen	Preserved carpal elements
Smilodectes gracilis	USNM 13251A	r. scaphoid, r. lunate, r. triquetrum, r. pisiform, r. centrale, r. capitate, r. trapezium, l. trapezium, l. pisiform
S. gracilis	USNM 13251B	r. scaphoid, r. lunate, r. capitate, l. scaphoid, l. lunate, l. triquetrum, l. pisiform, l. trapezium
S. gracilis	USNM 21815	r. triquetrum, r. hamate, l. capitate, l. hamate
S. gracilis	USNM 21954	r. hamate, l. scaphoid, l. lunate, l. capitate, l. hamate
S. gracilis	USNM 21964	r. hamate, l. triquetrum, l. capitate
S. gracilis	USNM 256745	l. lunate, l. pisiform
Northarctus tenebrosus	USNM 21932	r. capitate
N. tenebrosus	USNM 21951	r. hamate
N. tenebrosus	USNM 22009–11	r. scaphoid
N. robustior	CM 11910	l. scaphoid, l. capitate, l. hamate

Descriptive anatomy

The carpus of *Smilodectes*, like that of *Adapis* (Figure 1) and numerous other primates which lack fusion of the centrale and the scaphoid, is composed of ten separate osseous elements, including a prepollex. This was almost certainly the case for *Notharctus* as well, but the wrist of this genus is still not well known. Carpal anatomy can be difficult to describe because the carpus as a whole does not lie in a coronal plane (even when the forelimbs are taken to be in standard anatomical position, which clearly was not the habitual posture of *Smilodectes*).

For the sake of simplicity and consistency in this regard, we have followed as nearly as possible the terminology previously employed by Beard *et al.* (1986). Figure 1, a reconstruction of the distal radius and ulna, carpus and metacarpus of *Adapis parisiensis* (Godinot & Jouffroy, 1984), illustrates the relative positions and articular relationships of the ten carpal elements of a fairly generalized primate. It should be emphasized that we use "medial" as a synonym of "ulnar," and that "lateral" is equivalent to "radial."

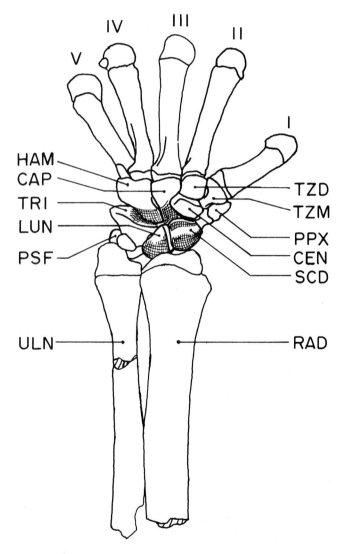

Figure 1. Distal left antebrachium, carpus, and metacarpus of *Adapis parisiensis* (RD 311) in dorsal view. The mediolateral curvature of the articulated carpus has been reduced in this reconstruction so that the articular relationships of the separate osseous elements of the wrist can be observed. Abbreviations are: hamate (ham); capitate (cap); triquetrum (tri); pisiform (psf); trapezoid (tzd); trapezium (tzm); prepollex (ppx); centrale (cen); scaphoid (scd); ulna (uln); and radius (rad). After Godinot & Jouffroy (1984: Figure 1).

Scaphoid

The proximal surface of the scaphoid of *Smilodectes* is dominated by the articular facet for the distal radius, which is more extensive dorsopalmarly than mediolaterally (Figure 2A). Overall, the articular surface is evenly convex but a slight ridge of bone defines the dorsal extent of the facet, making the curvature of the facet noticeably concave at its dorsal margin. The scaphoid tubercle protrudes laterally away from the facet for the radius. The proximolateral surface of the scaphoid tubercle bears an oval articular facet for the prepollex.

Distally, the scaphoid possesses two major articular facets (Figure 2C): a relatively large, flat facet for the proximal trapezium on the distal surface of the scaphoid tubercle and, immediately medial to this latter facet, a deeply concave facet for the centrale. The medial surface of the scaphoid bears a rather large, flat articular facet for the lunate (Figure 2B). Between the facets for the centrale and the lunate laterally there is a small, ellipitcal articular facet, which by analogy with several extant lemuriforms (e.g., *Lemur* and *Varecia*) must have articulated narrowly with the lateral part of the capitate head.

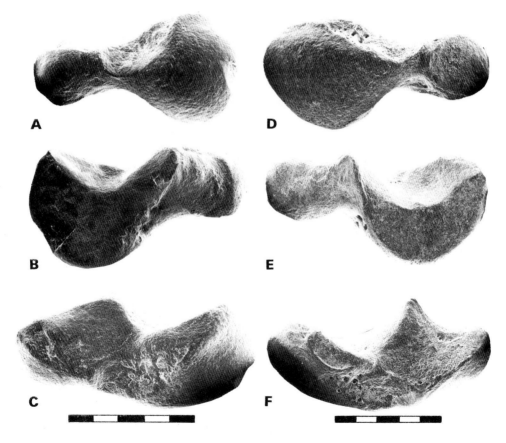

Figure 2. (A–C) USNM 13251A, right scaphoid of *S. gracilis* in proximal (A), medial (B), and distal (C) views. (D–F) CM 11910, left scaphoid of *N. robustior* in proximal (D), medial (E), and distal (F) views (both scales = 5 mm; left scale for *Smilodectes* specimen).

The scaphoids of *Notharctus robustior* (CM 11910; Figure 2D–F) and *N. tenebrosus* (USNM 22009-11) that we have studied are similar to each other in all respects except for size, yet differ morphologically from those of *Smilodectes* (see Figure 2). In *Notharctus* the proximal facet for the radius is more extensive mediolaterally than dorsopalmarly, the opposite condition from that found in *Smilodectes*. Distally, the concave facet for the centrale is deeper and mediolaterally wider in *Notharctus* than is the case in *Smilodectes*. Finally, the medial facet for the lunate in *Notharctus* is relatively smaller and different in shape from that of *Smilodectes*.

Centrale

Only a single centrale (USNM 13251A) of *Smilodectes* is contained in the present sample, and this specimen was preserved in articulation with its adjacent capitate (Figure 3F–H). Although this articulation obscures the anatomy of the medial side of the centrale, we did not attempt to separate the two bones for fear that the unique specimen might be damaged during preparation.

Most of the dorsal surface of the centrale is comprised of a convex articular facet for the overriding scaphoid. Only a restricted, distomedial corner of this surface is nonarticular. The distal surface of the centrale bears a large, flat facet for the trapezoid dorsally, and a much smaller facet for the trapezium palmarly. The facet for the trapezoid faces distally, so that the articulation between the trapezoid and the centrale occurs in an essentially transverse plane through the wrist.

As we noted above, the single known centrale of *Smilodectes* is preserved in articulation with its adjacent capitate. As a result, the anatomy of the medial side of the centrale cannot be directly observed. However, this surface must correspond to the curvature of the lateral part of the capitate head, with which it articulates. Thus we can infer from this region in USNM 21815, a well preserved left capitate of a different individual of *Smilodectes*, that the medial surface of the centrale of *Smilodectes* is only slightly concave dorsally, and that the concavity of this surface becomes most pronounced palmarly.

Lunate

Laterally, the lunate of *Smilodectes* possesses a poorly delimited region for articulation with the scaphoid (Figure 4B). There appears to have been little, if any, articular contact between the centrale and the lunate. The proximal surface of the lunate for articulation with the radius is mediolaterally very narrow, so that most of the radiocarpal articulation clearly occurred between the radius and scaphoid. The proximal surface of the lunate and the lateral surface of the bone lie in planes which are no more than about 30° removed from each other.

The medial side of the lunate bears a well defined facet for the triquetrum proximally (Figure 4A). Distally, the medial surface of the lunate forms a deep concavity for articulation with the proximolateral part of the capitate head. Overall, the lunate of *Smilodectes* is nearly symmetrical in medial view because of two well-developed dorsal and palmar processes which enclose the facet for the capitate head on either side.

Triquetrum

The lateral surface of the triquetrum of *Smilodectes* is chiefly devoted to an essentially flat articular facet for the lunate. On the proximodorsal surface of the bone (Figure 5A) is a large, concave facet for the ulnar styloid process, the dorsal extent of which is defined by a

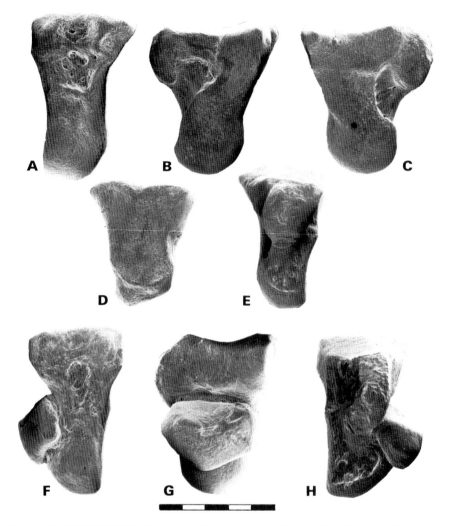

Figure 3. (A–E) USNM 21815, left capitate of *S. gracilis* in dorsal (A), medial (B), lateral (C), distal (D), and ventral (E) views. (F–H) USNM 13251A, articulated right centrale and capitate of *S. gracilis* in dorsal (F), lateral (G), and ventral (H) views (scale = 5 mm).

very prominent wing-like process. Palmar to the facet for the ulnar styloid process is a less extensive, flat facet for articulation with the pisiform. Both of these proximal triquetral facets are rather limited in their proximolateral-distomedial extent, the distomedial end of the triquetrum being nonarticular. The distal facet for articulation with the hamate (Figure 5B) is an irregular, concavoconvex surface. Laterally, the concave part of this facet would have articulated with the proximolateral, convex surface of the "spiral" facet of the hamate during relatively neutral wrist postures. Likewise, the convex part of the distal facet on the triquetrum, located dorsal and medial to the concave part of this facet, would have articulated with the more distomedial, concave surface of the "spiral" facet of the hamate during ulnar deviation and extension of the wrist.

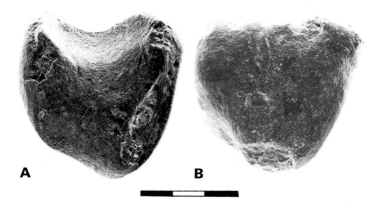

Figure 4. (A) USNM 256745, right lunate of *S. gracilis* in medial view. (B) USNM 13251A, left lunate of *S. gracilis* in lateral view (scale = 3 mm).

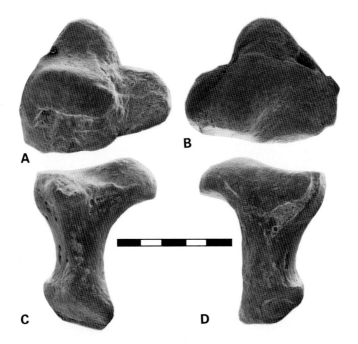

Figure 5. (A–B) USNM 13251A, right triquetrum of *S. gracilis* in proximal (A) and distal (B) views. (C–D) USNM 256745, left pisiform of *S. gracilis* in medial (C) and lateral (D) views (scale = 5 mm).

Pisiform

The proximal surface of the pisiform of *Smilodectes*, like that of most other primates, is comprised of a slightly convex lateral facet for articulation with the triquetrum and a concave medial facet for articulation with the ulnar styloid process. The facet for the triquetrum extends along the entire mediolateral surface of the proximal pisiform (Figure 5D). In contrast, the facet for articulation with the ulnar styloid process appears to be limited to the medial part of the proximal pisiform (Figure 5C).

The "shaft" of the pisiform is proximodistally quite long (relatively longer than any of the extant lemuriform taxa examined). The circumference of the shaft increases distally, eventually forming a bulbous enlargement for insertion of *m. flexor carpi ulnaris*.

Trapezium

In dorsal view the trapezium of *Smilodectes* is basically triangular (Figure 6D). Most of the dorsal surface of the trapezium is nonarticular, but at the proximomedial corner of this surface there is a small, slightly convex articular facet for the centrale.

The proximal surface of the trapezium is dominated by a well defined articular facet for the scaphoid tubercle (Figure 6D–E). This facet is essentially flat and is lateral to and contiguous with the articular facet for the centrale described previously. Ventral to the facet for the scaphoid tubercle is a relatively enormous palmar tubercle, medial to which the tendon of *m. flexor carpi radialis* passes on its way to insert into the ventral base of the second metacarpal in extant lemuriforms (Jouffroy, 1962). Lateral to the facet for the scaphoid tubercle on the proximal surface of the trapezium there is a well-defined bony process which bears a small articular facet for the prepollex (Figure 6E).

Distally, the articular facet for the first metacarpal is sellar-shaped, being mediolaterally concave and dorsoventrally convex (Figure 6A–B). Much of this facet extends onto the ventral surface of the bone, where it tapers gradually to a point at the base of the trapezial tuberosity. A small extension of the facet for the first metacarpal also lips onto the dorsomedial surface of the trapezium, where the articulation between trapezium and the first metacarpal must have occurred during extension of the pollical carpometacarpal joint.

The rectangular articular facet for the trapezoid, more extensive proximodistally than dorsoventrally, lies between the facets for the centrale and first metacarpal on the medial surface of the trapezium (Figure 6C). The facet is deeply concave proximodistally. There appears to have been little, if any, articular contact between the trapezium and the second metacarpal in *Smilodectes*.

Capitate

The distal surface of the capitate of *Smilodectes* possesses two contiguous, slightly concave facets for the base of the third metacarpal (Figure 3D). These facets are separated by a low ridge of bone which runs ventrolaterally from the center of the dorsal surface of the capitate, so that the more medial of the two distal facets for the third metacarpal is larger than the more lateral of these facets. Contrary to the condition in extant anthropoids, the distal capitate of *Smilodectes* does not bear an articular facet for the lateral base of the fourth metacarpal. The distomedial border of the capitate is interrupted about halfway between the dorsal and ventral margins of the bone by a depression for the medial ligament between the capitate and third metacarpal (Figure 3B, D). A corresponding depression for the lateral ligament does not occur in *Smilodectes* (for the distribution of these traits in other living and fossil primates, see Lewis, 1977; Beard *et al.*, 1986).

The articular facet for the hamate occupies most of the medial surface of the capitate in *Smilodectes* (Figure 3B). This facet extends the entire proximodistal length of the capitate dorsomedially, being dorsoventrally most extensive proximally. Ventral to the facet for the hamate about midway along the proximodistal axis of the bone is a well defined pit for the interosseous ligament between capitate and hamate.

The lateral surface of the body of the capitate of *Smilodectes* bears confluent articular facets for the medial part of the base of the second metacarpal distally and for the trapezoid proximally and dorsally (Figure 3C, G). The facet for the second metacarpal is slightly concave and faces laterally. The facet for the trapezoid is triangular and faces somewhat more proximally than does the facet for the second metacarpal.

The dorsal surface of the body of the capitate is nonarticular (Figure 3A, F). Unlike the condition in extant lemuriforms, the capitate body of *Smilodectes* does not exhibit lateral torsion with respect to the long axis of the bone, a feature which results in a mediolaterally wider capitate-trapezoid embrasure in the fossil form compared to extant lemuriforms. Likewise, the capitate body of *Smilodectes* is not expanded laterally, so that in dorsal view the capitate is essentially symmetrical. The lateral articular facet for the centrale extends only very slightly onto the dorsal surface of the capitate in *Smilodectes*.

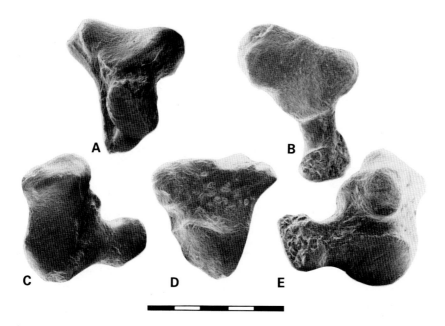

Figure 6. USNM 13251A, right trapezium of *S. gracilis* in ventral (A), distal (B), medial (C), dorsal (D), and lateral (E) views (scale = 5 mm).

The capitate head of *Smilodectes* bears articular facets for the lunate, scaphoid and centrale (Figure 3A, C, F–G). As noted previously, the facet for the centrale is almost completely limited to the lateral surface of the capitate head, where it is proximal to and confluent with the facet for the trapezoid described above. Also on the lateral surface of the capitate head is a small, poorly delimited region for articulation with the scaphoid. By manipulating the right scaphoid and articulated right centrale and capitate preserved in USNM 13251A, we were able to locate this region of articulation as proximal and ventral to the facet for the centrale. The medial part of the proximal surface of the capitate head, as well as all of the convex dorsal surface of this structure, is devoted to the facet for the lunate. The dorsal extremity of this facet is defined by a strong bony ridge, which must have served to limit midcarpal extension in *Smilodectes*.

Capitates of two different species of *Notharctus*, *N. tenebrosus* and *N. robustior*, were also available to us for this study. Although some very minor differences in capitate morphology appear to exist between these two species of *Notharctus*, the capitates of both species are easily distinguished from the capitate of *Smilodectes* on the basis of several traits (compare Figures 3 and 7). For example, in either medial or lateral view, the dorsodistal end of the capitate body in *Notharctus* forms a well-defined ridge of bone which protrudes distally well beyond the level of the rest of the capitate, so that the distal articular surface for the third metacarpal in *Notharctus* is much more concave than is the case in *Smilodectes*. Likewise, in dorsal view, this dorsodistal bony ridge protrudes farthest distally at a point slightly lateral to the center of the capitate body, whereas in *Smilodectes* there is much less distal relief on the capitate body and the dorsodistal end of the capitate forms a mediolaterally straight line. Also, the capitate body of *Notharctus* is noticeably expanded laterally and the lateral articular facet for the centrale extends well onto the dorsal surface of the bone (more so in *N. tenebrosus* than in *N. robustior*), while in *Smilodectes* this facet is almost completely restricted to the lateral surface of the bone. Despite the fact that the facet for the centrale extends well onto the dorsal surface of the capitate head in *Notharctus*, it is clear from the restricted medial extent of this facet that the centrale did not articulate with the hamate in *Notharctus* as it does in lemuriforms (see below).

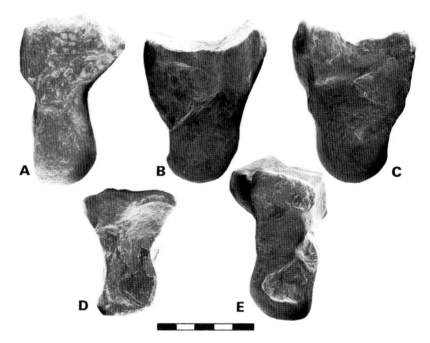

Figure 7. CM 11910, left capitate of *N. robustior* in dorsal (A), medial (B), lateral (C), distal (D), and ventral (E) views (scale = 5 mm).

Hamate

Medially, the hamate of *Smilodectes* bears a prominent "spiral" facet for articulation with the triquetrum (Figure 8E). Proximally, the rounded ventral part of this facet dominates the articular surface such that the proximal part of the facet is convex. Distomedially, the sharp dorsal margin of the facet becomes increasingly well defined and the facet as a whole becomes increasingly concave distally as a result. Overall, the facet for the triquetrum is oriented more proximodistally than mediolaterally.

The lateral surface of the hamate possesses a flat articular facet for the capitate, more extensive proximally than distally where it is limited to the dorsal part of the lateral hamate (Figure 8C). Ventral to the facet for the capitate lies a deeply excavated pit for attachment of the interosseous ligament between capitate and hamate. Distal to but continuous with the latter structure is a proximodistally oriented groove which, by analogy with extant lemuriforms, must have conveyed a carpometacarpal ligament binding the hamate to the lateral base of the fourth metacarpal.

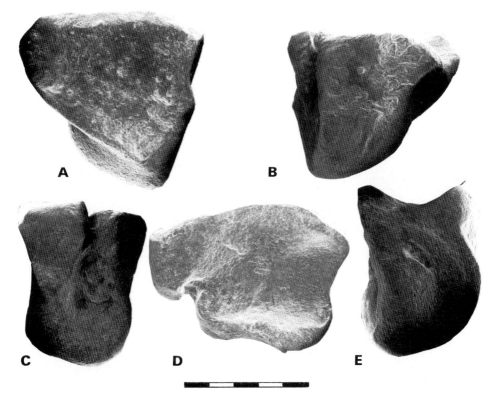

Figure 8. USNM 21825, left hamate of *S. gracilis* in dorsal (A), ventral (B), lateral (C), distal (D), and proximomedial (E) views (scale = 5 mm).

The dorsal surface of the hamate is largely nonarticular (Figure 8A). Proximolaterally, a small articular facet for the lunate is present which is confluent with the facet for the triquetrum medially.

Distally, the hamate of *Smilodectes* bears contiguous articular facets for the bases of the fourth and fifth metacarpals (Figure 8D). The distal surface of the hamate of *Smilodectes* as a whole is only modestly excavated. There is essentially no development of a hamulus ("hamate hook") ventromedially.

The hamate of *Notharctus robustior* (CM 11910) differs in several respects from those known for *Smilodectes* (compare Figures 8 and 9). The distal surface for articulation with the fourth and fifth metacarpals is much more deeply excavated and concave than is the case in *Smilodectes*. Laterally, the articular facet for the capitate extends onto the dorsodistal surface of the hamate, contrary to the condition in *Smilodectes*. Finally, the articular facet for the triquetrum is oriented more mediolaterally than proximodistally in *Notharctus*, and the dorsal border of this facet is better defined than is the case in *Smilodectes*.

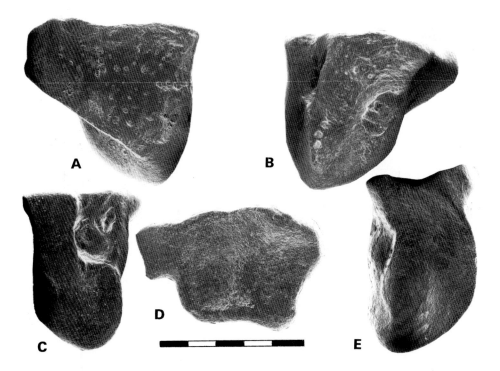

Figure 9. CM 11910, left hamate of *N. robustior*: same views as in Figure 8 (scale = 5 mm).

Comparative anatomy

Extant primates

Aside from the Lorisidae, which possess some structural convergences in wrist anatomy with the Hominoidea (Cartmill & Milton, 1977; Etter, 1978; Lewis, 1985) and are clearly derived in this respect among lemuriforms, the carpal anatomy of lemuriforms has been very poorly studied. Unfortunately, this paucity of data about wrist anatomy and adaptations among lemuriforms severely limits the functional inferences that can be drawn from the wrist of adapiforms.

Our examination of carpal structure among lemuriforms indicates that *Smilodectes* is similar in several aspects of wrist anatomy to the extant Malagasy primates, which is not surprising given the overall similarity in postcranial anatomy previously demonstrated between notharctines and certain lemuriforms (Gregory, 1920; Rose & Walker, 1985; Covert, 1985). Among the most obvious of such traits is the relatively small size of the lunate and capitate of *Smilodectes* compared to neighboring elements of the carpus, such as the scaphoid and the hamate (Table 2). Like *Smilodectes*, all lemuriforms except lorisids and galagids possess lunates which are very narrow mediolaterally (Gregory, 1920), as does *Tarsius* as well. The lunates of anthropoids are generally wider mediolaterally in such a way that the radiocarpal articulation is more evenly divided between the scaphoid and the lunate. Likewise, the capitate is relatively larger and articulates with the bases of both the third and the fourth metacarpals in anthropoids, whereas in lemuriforms, *Tarsius*, and *Smilodectes*, it articulates solely with the third metacarpal. The form of the trapezium is also very similar in *Smilodectes* and extant lemurids, indriids, and cheirogaleids, in which the bone is basically triangular in dorsal view. The trapezia of anthropoids bear relatively expanded lateral borders compared to the condition in most lemuriforms and *Smilodectes*, and the bone is much more square in dorsal view as a result (e.g., see Beard *et al.*, 1986: Figure 9).

While there are similarities in wrist anatomy between lemuriforms and *Smilodectes*, lemuriforms share at least two important features of carpal anatomy in common which are found in neither *Smilodectes* nor any other living primate or eutherian known to us. (Eutherian taxa examined in this regard include tupaiids, *Cynocephalus*, erinaceids, tenrecids, solenodontids, macroscelidids, and chrysochlorids). The first of these differences concerns the midcarpal joint as a whole and the form and articular relations of the centrale in particular (Figure 10). In both *Smilodectes* and extant haplorhines, the centrale is mediolaterally relatively narrow, and almost all of its dorsal surface is devoted to a facet for the scaphoid (Figures 3F–G, 10). Distally, the embrasure between the capitate and the trapezoid for the centrale (see Jenkins, 1981) is mediolaterally wide, and the articulation between the centrale and the trapezoid occurs in a transverse plane through the wrist. In contrast, among lemuriforms the centrale is a relatively larger bone (especially in the mediolateral dimension), and about half of the dorsal surface of the centrale lies medial to its articular facet for the scaphoid, where it is subcutaneous (Figure 10). The embrasure between the capitate and the trapezoid for the centrale is very narrow and notch-like in dorsal view in lemuriforms, as opposed to the relatively much wider capitate-trapezoid embrasure in extant haplorhines. As a result, the articulation between the centrale and the trapezoid occurs in an oblique plane through the wrist in lemuriforms, in such a way that the facet on the centrale for the trapezoid faces distoradially, rather than distally as it does in *Smilodectes* and haplorhines. Owing partly to the relatively great mediolateral extent of the centrale, this bone articulates slightly with the hamate in lemuriforms. This last articulation never occurs among extant haplorhines (Nayak, 1933; Jouffroy, 1975), and it clearly did not occur in *Smilodectes* or *Notharctus* either.

The second major way in which the wrists of extant lemuriforms differ from those of *Smilodectes* and extant haplorhines is in details of the ulnocarpal articulation. Among most primates (all except Lorisidae and Hominoidea, which have independently reduced or lost articular contact between the ulnar styloid process and the proximal carpal row; see Lewis, 1972, 1985; Cartmill & Milton, 1977) the ulnar styloid process articulates with both the triquetrum and the pisiform. In general, this articulation appears to be an adaptive compromise which limits capabilities of ulnar deviation, but enhances stability, at the ulnocarpal joint (Beard *et al.*, 1986). Although an ulnocarpal articulation occurs in the vast majority of primates, the detailed anatomy of this articulation differs between lemuriforms

Table 2 **Measurements (in mm) of notharctine carpals (see Table 1 for taxonomic allocations of specimens)**

Specimen	Dorsopalmar	Mediolateral	Proximodistal
Scaphoids			
USNM 13251A	5·0	8·7	4·5
USNM 13251B (right)	4·7	8·9	4·8
USNM 13251B (left)	4·7	9·0	4·4
USNM 21954	4·5	—	4·7*
USNM 22009-11	4·2	8·9	4·4
CM 11910	5·0	10·4	5·2
Centrale			
USNM 13251A	4·1	2·2	2·3
Lunates			
USNM 13251A	5·0	2·7	4·6
USNM 13251B (right)	4·9	2·6	4·4
USNM 13251B (left)	4·9	2·6	4·6
USNM 256745	5·1	2·9	5·1
Triquetra			
USNM 13251A	6·0	6·8	4·0
USNM 13251B	5·9	6·2	4·3
USNM 21815	5·7	6·4	4·3
USNM 21964	5·9*	6·6*	4·5
Pisiforms			
USNM 13251A (right)	2·9	—	6·4
USNM 13251A (left)	2·8	—	6·8
USNM 13251B	3·0	5·3	—
USNM 256745	2·9	5·2	7·2
Trapezia			
USNM 13251A (right)	4·9	4·6	4·4
USNM 13251A (left)	—	4·7	4·6
USNM 13251B	4·5	—	4·7
Capitates			
USNM 13251A	5·3*	4·0	7·1
USNM 13251B	5·5	4·1	7·4
USNM 21815	5·2	4·2	7·0
USNM 21954	5·1	4·1	7·5
USNM 21964	5·3	4·0	7·0
USNM 21932	5·4	4·8	7·8
CM 11910	6·6	5·3	9·1
Hamates			
USNM 21815 (right)	4·3	6·6	6·4
USNM 21815 (left)	4·4	6·7	6·4
USNM 21954 (right)	4·5	6·5	6·9
USNM 21954 (left)	4·3*	6·1	6·7
USNM 21964	4·5	—	—
USNM 21951	—	—	7·2
CM 11910	5·9	8·1	8·1

* Estimated measurement.

on the one hand and haplorhines and *Smilodectes* on the other. These differing morphologies are best illustrated by the proximal pisiforms of the two groups (Figure 11). In non-hominoid living haplorhines and *Smilodectes*, two facets, for the triquetrum and the ulnar styloid process respectively, occupy the proximal pisiform, and these facets are either approximately equal in mediolateral extent, or, more commonly, the mediolateral extent of the facet for the triquetrum exceeds that of the facet for the ulnar styloid process (Figure 11; see also Beard *et al.*, 1986: Figure 8). In lemuriforms the same two facets are present, but the mediolateral extent of the facet for the ulnar styloid process is typically twice that of the facet for the triquetrum, which is limited to the lateral part of the proximal pisiform (Figure 11). Also, the facet for the ulnar styloid is very deeply concave in lemuriforms, and a strong bony ridge defines the medial border of the facet. While the homologous facet is also concave in extant haplorhines, it is qualitatively more deeply excavated in lemuriforms, and the lemuriform-type of ulnocarpal morphology never occurs in haplorhines. *Smilodectes* resembles living haplorhines in all of these aspects of the ulnocarpal articulation.

Figure 10. Dorsal view of articulated left carpus of (left) the lemuriform *Lemur macaco* and (right) the anthropoid *Saimiri sciureus*. Note contrasting anatomy and articular relationships of the centrale (C) in these taxa (see text for details).

Adapis parisiensis

The anatomy of the wrist of a recently collected specimen of *Adapis parisiensis* from the late Eocene of France was described in detail by Godinot & Jouffroy (1984). As recently discussed by Dagosto (1983), there are numerous features of the postcranium of *Adapis* which differ significantly from the likely more generalized condition found in notharctines, and we find that such differences are found in the wrist as well. *Adapis* differs from *Smilodectes* in possessing a more mediolaterally expanded radial facet on the lunate, lateral expansion of the capitate body in the region for articulation with the trapezoid, slight extension of the facet for the centrale on the capitate onto the dorsal surface of the bone (it is limited to the lateral surface in *Smilodectes*), a more proximodistally oriented facet for the triquetrum on the hamate, and a relatively flat, rather than sellar-shaped, pollical carpometacarpal articulation. The functional significance of these features is discussed below. Like *Smilodectes* and all living primates except lemuriforms, the centrale of *Adapis* is mediolaterally narrow, most of its dorsal articular surface is devoted to a facet for the scaphoid, the articular facet for the trapezoid is oriented distally rather than distoradially, and there is no articular contact between the centrale and the hamate (Figure 1; see also Godinot & Jouffroy, 1984). In contrast, the ulnocarpal articulation in *Adapis* is like that of extant lemuriforms, and unlike that of *Smilodectes* and extant haplorhine primates (Figure 11; Godinot & Jouffroy, 1984: Plate 1, Figures m–n). *Adapis* resembles only extant lemuriforms in possessing a very deeply excavated facet for the ulnar styloid process which is mediolaterally wider than the more laterally positioned facet for the triquetrum on that bone.

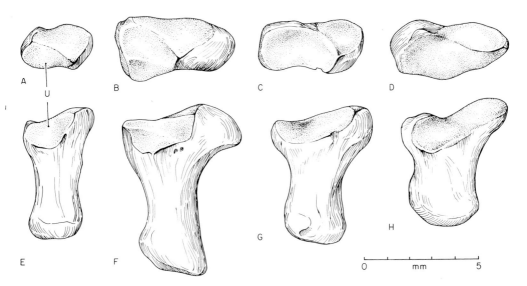

Figure 11. Proximal (A–D) and proximomedial (E–H) views of selected primate left pisiforms. Taxa illustrated are, from left to right: *Saimiri sciureus* (A, E); *Smilodectes gracilis* (USNM 256745; B, F); *Adapis parisiensis* (RD 311; C, G); *Lemur catta* (reversed from right side; D, H). Note relatively great concavity and mediolateral extent of facet for ulnar styloid process (U) in *Adapis* and *Lemur* (see text for details).

Discussion

Functional considerations

Although the wrist of *Smilodectes* is almost completely known, a coherent hypothesis relating carpal structure to function and support usage in *Smilodectes* remains difficult to formulate. In part this difficulty is a result of the lack of studies of carpal anatomy and adaptation among extant lemuriforms (to which *Smilodectes* bears a considerable overall phenetic similarity), as noted above. Perhaps the most that can be safely inferred concerning the functional anatomy of the wrist in *Smilodectes* is that there were no striking adaptations for extreme carpal mobility in the extinct form, as there are in extant lorisids and hominoids (e.g., Cartmill & Milton, 1977; Etter, 1978; Lewis, 1972, 1985; Jenkins, 1981). Contrary to both lorisids and hominoids, *Smilodectes* lacks mediolateral expansion of the facet for the radius on the proximal lunate and possesses a typically primate form of articulation between the ulnar styloid process and the pisiform and triquetrum (i.e., there is no reduction or elimination of contact between the ulnar styloid process and the proximal carpal row as occurs in lorisids and hominoids (cf. Cartmill & Milton, 1977; Lewis, 1985). Likewise, there appears to have been very little relative movement between the centrale and the capitate in *Smilodectes*, contrary to the condition in some of the more acrobatic extant primates such as *Hylobates* and *Ateles* (Jenkins, 1981). Conversely, *Smilodectes* also lacks apparent adaptations for carpal stability during parasagittal movement found in some extant ceboids and especially cercopithecoids. Most notable among these is the relatively proximodistal orientation of the "spiral" facet on the hamate in *Smilodectes*, a trait which would have allowed at least moderate capabilities of ulnar deviation at the midcarpal joint in the fossil form. Among most cercopithecoids the "spiral" facet of the hamate is oriented quite mediolaterally (O'Connor, 1975), which seems to be an adaptation for midcarpal stability at the expense of deviational capability in these quadrupedal primates. The triangular shape (in dorsal view; see Figure 6D) of the trapezium of *Smilodectes* is also typical of extant *Tarsius* and lemuriforms, and is not characteristic of anthropoids (Robertson, 1984; Beard *et al.*, 1986, Figure 9). While the lack of a trapezoid of *Smilodectes* prevents direct knowledge of the precise position of the trapezium within the distal carpal row, the triangular shape of the trapezium of *Smilodectes* strongly suggests that the pollex was highly divergent from the rest of the manual digits, as is the case in extant "prosimian" primates (Robertson, 1984). This character may be an adaptation for the use of relatively large diameter (? vertical) supports, but further study of hand use in extant *Tarsius* and lemuriforms is required before this hypothesis can be accepted with confidence.

Covert (1985) recently assessed the functional anatomy of most of the postcranium (but not the carpus) of *Smilodectes gracilis* and concluded that:

> In sum it appears that *S. gracilis* was probably an acrobatic leaper like indriids and *Lepilemur* . . . and a capable forelimb clinger like these animals. However, . . . it [*S. gracilis*] may have been a more capable orthograde climber and/or forelimb hanger than large extant Malagasy primates. The inferred enhanced mobility of the wrist may also be related to these behaviors (Covert, 1985, p. 474).

Our examination of carpal anatomy in *S. gracilis* is therefore in general agreement with Covert's analysis of the rest of the postcranium of this fossil primate, except that we find no morphologic evidence for the enhanced carpal mobility inferred by him for *Smilodectes*. None of the clear adaptations for enhanced carpal mobility found in such extant orthograde climbing and/or suspensory primates as lorisids, atelines, and hominoids (Lewis, 1972, 1985; Cartmill & Milton, 1977; Jenkins, 1981; Beard *et al.*, 1986) occur in *Smilodectes*.

As is generally the case for *Smilodectes*, the functional or adaptive reasons for the differences in wrist morphology observed above between lemuriforms and most other primates are not completely apparent and deserve further study. This is particularly the case with respect to the morphology of the midcarpal joint in lemuriforms. The peculiar anatomy of the ulnocarpal joint of lemuriforms, and particularly the very deeply excavated, mediolaterally extensive articular facet for the ulnar styloid process on the pisiform, may merely reflect the relatively great length of the ulnar styloid process itself among lemuriforms (e.g., Covert, 1985), rather than being an osteological adaptation for different ulnocarpal movement capabilities *per se*. This is so because the relatively great length of the ulnar styloid process among lemuriforms requires a mediolaterally wider carpal articular surface in lemuriforms than in other primates in order to achieve a similar degree of antebrachiocarpal ulnar deviation. The deep excavation of the facet for the ulnar styloid on the pisiform in lemuriforms may be a bony adaptation for increased stability of the ulnocarpal joint during ulnar deviation, and is the opposite type of configuration from that of primates that have enhanced capabilities of ulnar deviation (Lewis, 1972, 1985; Cartmill & Milton, 1977; Beard *et al.*, 1986). A possible biological role for such an ulnocarpal articulation in lemuriforms would be to stabilize the ulnocarpal joint during prolonged clinging to vertical supports (during which the antebrachiocarpal joint is typically deviated ulnarly), but this hypothesis requires further testing and other adaptive explanations are also possible.

While the functional significance of the differences in carpal anatomy between extant lemuriforms and haplorhines is not completely understood, at least some of the qualitative differences in carpal morphology between *Adapis* and *Smilodectes* noted above clearly do reflect different movement capabilities in these two adapiform genera. In the proximal carpal row, the most important difference between *Adapis* and *Smilodectes* is the relatively great mediolateral width of the facet for the radius on the lunate of *Adapis*. Among living primates, mediolateral expansion of this facet also occurs in Lorisidae and Hominoidea (even more so than in *Adapis*; see Cartmill & Milton, 1977; Etter, 1978; Lewis, 1985; Beard *et al.*, 1986), in which it is correlated with enhanced ranges of abduction/adduction at the antebrachiocarpal joint. *Adapis* also differs significantly from *Smilodectes* in details of capitate morphology which are likely related to midcarpal pronation/supination mobility. Most important among these features are the lateral expansion of the capitate body and the extension of the lateral facet for the centrale onto the dorsal surface of the bone in *Adapis*, both of which would have allowed greater relative movement between the centrale and capitate in *Adapis* than in *Smilodectes* (see discussion of these and other features by Jenkins, 1981; Rose, 1984). Midcarpal ulnar deviation likewise appears to have been enhanced in *Adapis*, as is evident from the relatively proximodistal rather than mediolateral orientation of the "spiral" facet for the triquetrum on the hamate in this form compared to *Smilodectes*. All of the preceding features point to enhanced carpal mobility in *Adapis* compared to *Smilodectes* and are therefore in general agreement with the rest of the postcranium (Dagosto, 1983), which suggests greater ranges of joint mobility in *Adapis*, possibly associated with a greater emphasis on climbing and palmigrade quadrupedalism and a decreased emphasis on leaping compared to *Smilodectes* (Dagosto, 1983; Godinot & Jouffroy, 1984). While the wrist of *Adapis* was clearly adapted for a greater range of mobility than that of *Smilodectes*, extreme adaptations for carpal mobility and pollical prehensility such as those found in extant lorisids (Cartmill & Milton, 1977; Etter, 1978; Lewis, 1985) are not present in *Adapis*. Curiously, the sellar-shaped pollical carpometacarpal joint of *Smilodectes* appears to have allowed a greater range of motion than that of *Adapis*, which is essentially flat (Godinot & Jouffroy, 1984).

Only three of the carpal elements of the various species of *Notharctus* were available for the present study (Table 1), thereby severely limiting any detailed functional comparisons between the wrist of *Notharctus* and those of either *Smilodectes* or *Adapis*. Nevertheless, the three carpals of *Notharctus* that we studied do exhibit some features worthy of discussion in a functional context. Pending recovery of more nearly complete carpal remains of *Notharctus*, we make some preliminary observations in this regard below.

Compared to both *Smilodectes* and *Adapis*, *Notharctus* exhibits mediolateral expansion of the facet for the radius on the scaphoid, possibly indicating a greater range of abduction/adduction movement at the antebrachiocarpal joint. The lack of a lunate of *Notharctus* makes this inference very tentative however, because the increased capability of antebrachiocarpal abduction/adduction inferred for *Adapis* relative to *Smilodectes* is entirely due to mediolateral expansion of the facet for the radius on the lunate rather than the scaphoid in the former genus. Similarly, the capitate of *Notharctus* seems to be functionally intermediate between the capitates of *Adapis* and *Smilodectes* in terms of midcarpal pronation/supination mobility (Jenkins, 1981; Rose, 1984). Unlike *Smilodectes*, there is some lateral expansion of the capitate body in *Notharctus*, but not to the extent seen in *Adapis*. Also, the lateral facet for the centrale extends onto the dorsal surface of the capitate head in *Notharctus* as it does in *Adapis* but not *Smilodectes*. Taken together these features would appear to indicate a somewhat greater range of midcarpal pronation/supination in *Notharctus* than in *Smilodectes*, but still probably less than that of *Adapis*. The "spiral" facet of the hamate is oriented more mediolaterally in *Notharctus* than it is in either *Smilodectes* or *Adapis*, a trait that has been correlated with limited midcarpal ulnar deviation, but increased midcarpal stability, among some extant primates (e.g., O'Connor, 1975; Beard *et al.*, 1986). In this respect *Notharctus* much more closely approaches the condition observed in some extant palmigrade ceboids and cercopithecoids (O'Connor, 1975) than do either of the other two adapiform genera. Finally, the distal articular surfaces of both the capitate and hamate of *Notharctus* (for the bases of metacarpals III–V) are much more deeply excavated than is the case in either *Smilodectes* or *Adapis*. To our knowledge there is no information in the literature regarding the distribution or possible function of these differences among living primates, although the morphology found in *Notharctus* is biomechanically well-suited to resist shearing stresses at the carpometacarpal joints of digits III–V (such as might occur during the propulsive phase of quadrupedal palmigrade locomotion). The available carpals of *Notharctus* therefore provide tentative evidence that *Notharctus* may have been somewhat intermediate in its locomotor adaptations, perhaps utilizing more palmigrade quadrupedalism than *Smilodectes*, but retaining the leaping adaptations characteristic of notharctines that were evidently lost in *Adapis* (Dagosto, 1983).

Phylogenetic inferences

Although many students of primate phylogeny have generally ignored the postcranium, we fail to see why complex aspects of postcranial anatomy, such as the articular relationships of the separate osseous elements of the wrist, should be less useful for phylogeny reconstruction than are details of craniodental anatomy. Indeed, the multiplicity of conflicting phylogenetic hypotheses previously proposed for adapiforms on the basis of craniodental morphology adequately underscores the need for independent testing based on postcranial anatomy. There are at least four such hypotheses: (1) that adapiforms are ancestral to both extant anthropoids and lemuriforms (Gingerich, 1980, and references therein); (2) that adapiforms cannot be demonstrated to possess a special phylogenetic relationship with either extant anthropoids (Cartmill & Kay, 1978; Rosenberger *et al.*, 1985) or lemuriforms (Cartmill & Kay, 1978; Rasmussen, 1986); (3) that adapiforms are the sister group of extant lemuriforms (Rosenberger *et al.*, 1985; Rosenberger & Strasser, 1985); and (4) that Adapiformes is not a natural, monophyletic taxon, but rather consists of nested clades *within* the radiation of extant lemuriforms (Schwartz & Tattersall, 1979, 1982, 1983, 1985; Schwartz, 1986).

Clearly, the two major differences in wrist anatomy between extant lemuriforms and haplorhines described above comprise alternative character states which, if polarity can be established, should shed light on adapiform affinities by providing an independent test of the preceding hypotheses based on a separate anatomical region. Two of the most common means of establishing morphocline polarity are the analysis of the fossil record and outgroup analysis (e.g., Novacek, 1980), and both of these lines of evidence suggest that the lemuriform midcarpal and ulnocarpal joints are derived for primates. None of the non-primate eutherians (of which tupaiids most closely resemble primates) examined in this study possess midcarpal or ulnocarpal joints similar to those of extant lemuriforms. Likewise, as noted above, *Smilodectes*, *Notharctus*, and *Adapis* all possess midcarpal joints similar to those found in extant haplorhine primates, strongly suggesting that the lemuriform condition in this joint complex is the derived one for primates. If such is indeed the case, lemuriform midcarpal anatomy is almost certainly a synapomorphy uniting lemuriforms with respect to all other living and fossil primates for which the relevant anatomy is known, including the adapiforms *Smilodectes*, *Notharctus*, and *Adapis*. Moreover, it is not insignificant in this context that *Adapis* and the two notharctines *Notharctus* and *Smilodectes* are phylogenetically quite divergent within Adapiformes (e.g., Szalay & Delson, 1979; Beard, in press), the three forms representing approximate end-members of the generally independent European and North American adapiform radiations, respectively. The primitive midcarpal anatomy found in *Adapis*, *Notharctus*, and *Smilodectes* is therefore probably typical of a much larger diversity of Eocene adapiforms (for which carpal anatomy is not yet known), and may possibly be characteristic of the group as a whole. As such, lemuriform midcarpal anatomy, together with the lemuriform toothcomb, argues strongly against recent assertions based on molar morphology (e.g., Schwartz & Tattersall, 1979, 1982, 1983, 1985; Schwartz, 1986) that Eocene adapiforms represent nested clades within the extant radiation of Malagasy primates, as do all of the attendant biogeographic difficulties (i.e., multiple, independent rafting invasions of Madagascar by separate strepsirhine clades) that beset that hypothesis.

The ulnocarpal joints of *Adapis* and *Smilodectes* differ from each other in the same way that extant lemuriforms and haplorhines do, with *Adapis* exhibiting the condition otherwise found only in extant lemuriforms and *Smilodectes* possessing a haplorhine type of ulnocarpal articulation. This distribution of character states is most easily explained by assuming the eutherian/haplorhine/*Smilodectes* condition to be the primitive state for primates, and thus the unusual ulnocarpal articulation of *Adapis* and extant lemuriforms is a likely synapomorphy linking these taxa with respect to haplorhines and *Smilodectes*. Such an hypothesis is corroborated by the facts that the dentition of *Adapis* is much more similar to that of extant lemuriforms than is that of *Smilodectes* or any other known notharctine (Gingerich, 1975; Schwartz & Tattersall, 1985), and that all living and fossil lemuriforms exhibit an Old World distribution (e.g., Tattersall, 1982). Thus while we agree with other workers (Dagosto, 1983; Rosenberger *et al.*, 1985; Rosenberger & Strasser, 1985) that *Adapis* itself possesses uniquely derived features of incisal and postcranial anatomy that exclude it from being directly ancestral to extant lemuriforms, evidence from the ulnocarpal joint may now be added to that previously known from molar morphology suggesting that late Eocene *Adapis* shares more recent common ancestry with extant lemuriforms than either of these taxa does with *Smilodectes*. If so, the last common ancestor of extant lemuriforms may have diverged from an Old World adapine stock relatively late in Paleogene time, perhaps as recently as late Eocene, well after the zenith of the adapiform radiations currently documented in the fossil record.

Summary

The wrist of the middle Eocene notharctine *Smilodectes gracilis* exhibits a combination of traits that individually are characteristic of either extant haplorhines or "prosimians" and is plausibly a close approximation of the euprimate morphotype. Like the rest of the postcranium, there are clear indications in the wrist of *Smilodectes* of reduced ranges of joint mobility compared to the late Eocene adapine *Adapis parisiensis*, possibly indicative of a greater emphasis on climbing and palmigrade quadrupedalism and a reduced emphasis on leaping in the latter form. The carpal anatomy of the middle Eocene notharctine *Notharctus* is currently much more poorly known than that of either *Smilodectes* or *Adapis*, but the available carpals of the former genus provide tentative support for an intermediate use of above-branch palmigrade quadrupedalism compared to the other two Eocene adapiforms. Each of these three Eocene adapiforms lacks the distinctive, clearly derived midcarpal anatomy characteristic of all lemuriforms, which strongly suggests lemuriform monophyly with respect to these adapiform genera. However, the ulnocarpal joint of *Adapis* exhibits the apparently derived morphology otherwise restricted to lemuriforms, suggesting that *Adapis* and lemuriforms share more recent common ancestry with each other than either of these taxa does with *Smilodectes*. This hypothesis is strengthened by the consideration of paleobiogeography and molar morphology. Available evidence tends to support a relatively late origin of lemuriforms from an as yet unknown Old World adapine stock, perhaps as recently as the late Eocene.

Acknowledgements

For access to fossil and Recent specimens, we thank Dr R. W. Thorington, Jr., Dr R. J. Emry, and R. W. Purdy (USNM), and Drs L. Krishtalka and R. K. Stucky (CM). We also thank Drs M. Dagosto, D. L. Gebo, K. D. Rose, and A. Walker for helpful discussion, and E. Strasser and three anonymous reviewers for reading and improving an earlier version of this paper. We are most grateful to Dr H. H. Covert, who kindly provided a copy of his unpublished Ph.D. thesis. Elaine Kasmer skillfully prepared Figure 11 and helped prepare Figure 1. KCB wishes to thank E. Strasser and M. Dagosto for the invitation to participate in this symposium. Research was supported in part by an NSF predoctoral fellowship, the Lucille P. Markey Charitable Trust, and by CNRS ATP "Evolution: aspects biologiques et paléontologiques".

References

Beard, K. C. (in press). New notharctine primate fossils from the early Eocene of New Mexico and southern Wyoming and the phylogeny of Notharctinae. *Am. J. phys. Anthrop.*

Beard, K. C., Teaford, M. F. & Walker, A. (1986). New wrist bones of *Proconsul africanus* and *P. nyanzae* from Rusinga Island, Kenya. *Folia primatol.* **47,** 97–118.

Cartmill, M. & Kay, R. F. (1978). Cranio-dental morphology, tarsier affinities, and primate sub-orders. In (Chivers, D. J. & Joysey, K. A., Eds) *Recent Advances in Primatology. Vol. 3. Evolution*, pp. 205–214. London: Academic Press.

Cartmill, M. & Milton, K. (1977). The lorisiform wrist joint and the evolution of "brachiating" adaptations in the Hominoidea. *Am. J. phys. Anthrop.* **47,** 249–272.

Covert, H. H. (1985). Adaptations and evolutionary relationships of the Eocene primate family Notharctidae. Ph.D. Dissertation, Duke University.

Dagosto, M. (1983). Postcranium of *Adapis parisiensis* and *Leptadapis magnus* (Adapiformes, Primates). *Folia primatol.* **41,** 49–101.

Etter, H.-U. F. (1978). Lorisiform hands and their phylogenetic implications: a preliminary report. In (D. J. Chivers & K. A. Joysey, Eds) *Recent Advances in Primatology. Vol. 3. Evolution,* pp. 161–170. London: Academic Press.

Gazin, C. L. (1958). A review of the middle and upper Eocene primates of North America. *Smithson. Misc. Collns.* **136,** 1–112.

Gingerich, P. D. (1975). Dentition of *Adapis parisiensis* and the evolution of lemuriform primates. In (I. Tattersall & R. W. Sussman, Eds) *Lemur Biology,* pp. 65–80. New York: Plenum.

Gingerich, P. D. (1980). Eocene Adapidae, paleobiogeography, and the origin of South American Platyrrhini. In (R. L. Ciochon & A. B. Chiarelli, Eds) *Evolutionary Biology of the New World Monkeys and Continental Drift,* pp. 123–138. New York: Plenum.

Godinot, M. & Jouffroy, F. K. (1984). La main d'*Adapis* (Primate, Adapidé). In (E. Buffetaut, J. M. Mazin & E. Salmon, Eds) *Actes du Symposium Paléontologique Georges Cuvier,* pp. 221–242. Montbéliard.

Gregory, W. K. (1920). On the structure and relations of *Notharctus,* an American Eocene primate. *Mem. Am. Mus. nat. Hist.* **3,** 49–243.

Jenkins, F. A., Jr. (1981). Wrist rotation in primates: a critical adaptation for brachiators. *Symp. zool. Soc. Lond.* **48,** 429–451.

Jouffroy, F. K. (1962). La musculature des membres chez les lémuriens de Madagascar: étude descriptive et comparative. *Mammalia* **26,** 1–326.

Jouffroy, F. K. (1975). Osteology and myology of the lemuriform postcranial skeleton. In (I. Tattersall & R. W. Sussman, Eds) *Lemur Biology,* pp. 149–192. New York: Plenum.

Lewis, O. J. (1972). Osteological features characterizing the wrists of monkeys and apes, with a reconsideration of this region in *Dryopithecus (Proconsul) africanus. Am. J. phys. Anthrop.* **36,** 45–58.

Lewis, O. J. (1977). Joint remodelling and the evolution of the human hand. *J. Anat.* **123,** 157–201.

Lewis, O. J. (1985). Derived morphology of the wrist articulations and theories of hominoid evolution. Part I. The lorisine joints. *J. Anat.* **140,** 447–460.

Nayak, U. V. (1933). The articulations of the carpus in *Chiromys madagascarensis* with reference to certain other lemurs. *J. Anat.* **68,** 109–115.

Novacek, M. J. (1980). Cranioskeletal features in tupaiids and selected Eutheria as phylogenetic evidence. In (P. W. Luckett, Ed.), *Comparative Biology and Evolutionary Relationships of Tree Shrews,* pp. 35–93. New York: Plenum.

O'Connor, B. L. (1975). The functional morphology of the cercopithecoid wrist and inferior radioulnar joints, and their bearing on some problems in the evolution of the Hominoidea. *Am. J. phys. Anthrop.* **43,** 113–122.

Rasmussen, D. T. (1986). Anthropoid origins: a possible solution to the Adapidae–Omomyidae paradox. *J. hum. Evol.* **15,** 1–12.

Robertson, M. L. (1984). The carpus of *Proconsul africanus:* functional analysis and comparison with selected nonhuman primates. Ph.D. Dissertation, University of Michigan.

Rose, K. D. & Walker, A. (1985). The skeleton of early Eocene *Cantius,* oldest lemuriform primate. *Am. J. phys. Anthrop.* **66,** 73–89.

Rose, M. D. (1984). Hominoid postcranial specimens from the middle Miocene Chinji Formation, Pakistan. *J. hum. Evol.* **13,** 503–516.

Rosenberger, A. L. & Strasser, E. (1985). Toothcomb origins: support for the grooming hypothesis. *Primates* **26,** 73–84.

Rosenberger, A. L., Strasser, E. & Delson, E. (1985). Anterior dentition of *Notharctus* and the adapid–anthropoid fallacy. *Folia primatol.* **44,** 15–39.

Schwartz, J. H. (1986). Primate systematics and a classification of the order. In (Swindler, D. R. & Erwin, J., Eds) *Comparative Primate Biology. Vol. 1. Systematics, Evolution, and Anatomy,* pp. 1–41. New York: Alan R. Liss.

Schwartz, J. H. & Tattersall, I. (1979). The phylogenetic relationships of Adapidae (Primates, Lemuriformes). *Anthrop. Pap. Am. Mus. nat. Hist.* **55,** 271–283.

Schwartz, J. H. & Tattersall, I. (1982). Relationships of *Microadapis sciureus* (Stehlin, 1916) and two new primate genera from the Eocene of Switzerland. *Folia primatol.* **39,** 178–186.

Schwartz, J. H. & Tattersall, I. (1983). A review of the European primate genus *Anchomomys* and some allied forms. *Anthrop. Pap. Am. Mus. nat. Hist.* **57,** 344–352.

Schwartz, J. H. & Tattersall, I. (1985). Evolutionary relationships of living lemurs and lorises (Mammalia, Primates) and their potential affinities with European Eocene Adapidae. *Anthrop. Pap. Am. Mus. nat. Hist.* **60,** 1–100.

Szalay, F. S. & Delson, E. (1979). *Evolutionary History of the Primates.* New York: Academic Press.

Tattersall, I. (1982). *The Primates of Madagascar.* New York: Columbia University Press.

Laurie R. Godfrey

Department of Anthropology,
University of Massachusetts,
Amherst, MA 01003, U.S.A.

Received 16 September 1987
Revision received 19 Janary 1988
and accepted 22 February 1988

Publication date June 1988

Keywords: Locomotor
adaptations, lemurs, postcranial
anatomy, lemur phylogeny

Adaptive diversification of Malagasy strepsirrhines

This overview of the positional behavior and evolutionary history of Malagasy strepsirrhines is based on the humeral and femoral morphology of extinct and extant forms. It argues against the hypothesis that vertical clinging and leaping was ancestral for Malagasy lemurs in general, and for indroids in particular. Special attention is given to the evolutionary history of the Indroidea, and a new interpretation of their relationships is offered. It appears that specialized forms of leaping, as well as specialized slow, quadrupedal climbing, evolved repeatedly within the Malagasy strepsirrhines (i.e., in both Indroidea and Lemuroidea) from a generalized quadrupedal ancestor that would have included a certain amount of leaping and bounding in its positional repertoire, but also slow climbing and occasional hindlimb suspension.

Journal of Human Evolution (1988) **17**, 93–134

Introduction

In 1967, Napier & Walker put forth the notion that the common ancestor of Malagasy strepsirrhines, and, indeed, of post-Paleocene primates in general, was a "vertical clinger and leaper" (VCL)—something akin to extant indrids, *Lepilemur, Hapalemur, Tarsius,* or many galagonids.[1] Their distribution and frequency within primitive Haplorhini and diverse families of Strepsirrhini was taken as evidence that vertical clinging and leaping was preadaptive to all forms of locomotion exhibited by primates today.

There are several difficulties with this hypothesis. First, while "vertical clinging and leaping" has been identified in diverse families, it is manifested in such different forms that common elements appear trivial when compared with distinctions(Cartmill, 1972, 1974*b*; Oxnard, 1973, 1984; Oxnard *et al.*, 1981, in press; Stern & Oxnard, 1973). Animals identified as "vertical clingers and leapers" often exhibit greater behavioral and morphological commonality with relatives classified as "arboreal quadrupeds" than they do with each other. Indeed, it has been recognized that, in many respects, "arboreal quadrupeds" are intermediate in aspects of their morphology and behavior between diverse groups of so-called vertical clingers and leapers (Godfrey, 1977; Jouffroy & Lessertisseur, 1979; Lessertisseur, 1970; Lessertisseur & Jouffroy, 1973). It is true that leaping is quite common among extant Malagasy (and non-Malagasy) strepsirrhines. These animals are also generally quite adept at climbing vertical supports, but vertical supports are not universally preferred. In Madagascar, *Lemur* and *Varecia* exhibit a distinct preference for horizontal over vertical supports, and *Daubentonia* and the cheirogaleids use horizontal and vertical supports at least equally, or show a slight preference for horizontal supports. No living lemur species entirely avoids horizontal supports (although some live in habitats that do not provide many). It is not enough to characterize extant Malagasy

[1] Jenkins (1987) points out consistent spelling errors in the literature on Strepsirrhini. First, one "r" is usually incorrectly omitted from the name of the suborder itself, which was originally spelled Strepsirrhini by E. Geoffroy (1812). Secondly, according to Article 29bII of the International Code of Zoological Nomenclature, family names based on roots that are neither Greek nor Latin must follow the spelling applied by their first users. Thus the family names based on *Indri, Loris,* and *Galago* are not Indriidae, Lorisidae, and Galagidae, but Indridae (Burnett, 1828), Loridae (Gray, 1821) and Galagonidae (Gray, 1825).

strepsirrhines as frequently engaging in leaping or as often favoring vertical supports; living lemurs differ considerably in the frequency with which they utilize horizontal versus vertical supports, and in their manner of negotiating those supports (Martin, 1972; Oxnard, 1973, 1984; Oxnard et al., in press; Petter, 1962; Petter et al., 1977; Pollock, 1977; Richard, 1978).

Secondly, a broad overview of strepsirrhines from the Paleogene (cf., Dagosto, 1983; Szalay, 1972; Szalay & Dagosto, 1980), as well as the Neogene, shows vertical clinging and leaping to be far less common than once presumed. Certainly, the Indroidea, Lemuroidea, and Loroidea all include specialized non-leapers as well as specialized leapers; many of the extinct Malagasy strepsirrhines fall into the former category. The prevalence of specialized saltators among strepsirrhines today may reflect differential survival rather than characteristics of the common ancestor.

Napier & Walker's (1967) ancestral VCL hypothesis made claims not only for an ancestral preference for vertical supports, but for specific preferred positional behaviors on them and on the ground. Clinging to vertical supports was presumed to be the primary resting support posture; hindlimb-propelled leaping between vertical supports was presumed to be the primary mode of arboreal locomotion; and bipedal hopping was presumed to be a preferred mode of ground locomotion, at least for rapid progression. Among the postulated corollaries of these behaviors were: exceptionally elongated hindlimbs and short forelimbs; stout, straight, femoral shafts; narrow, well bordered patellar grooves projecting well anterior to the shaft of the femur; and, greatly anteroposteriorly elongated femoral condyles. These behaviors and characters well describe modern indrids—animals believed, therefore, to exhibit many primitive retentions. This paper takes issue with that notion, and argues that the common ancestor of the Indroidea (and of Malagasy strepsirrhines) was not a specialized "vertical clinger and leaper," sensu Napier & Walker (1967).

Extinct Malagasy strepsirrhines have generally been considered highly derived in their anatomy and positional behavior. Some have been called "modified" VCLs. Walker (1967) reconstructed Megaladapis as koala-like, and, based on films of frog-like hopping in koalas, proposed short, clumsy hopping as part of the positional repertory of this animal. Mesopropithecus was also considered a "modified" VCL, combining slow climbing and frog-like hopping. Other subfossil lemurs were considered more highly deviant from the primitive pattern (Walker, 1967, 1974)—Archaeolemur as baboon-like, Hadropithecus as patas- or gelada-like, and Palaeopropithecus as orang-like. Forelimb elongation was considered a hallmark of deviation from an ancestral VCL pattern.

Since 1967, a number of studies of the postcranial anatomy of the extinct Malagasy strepsirrhines have questioned these interpretations. While Jungers (1976, 1977) accepted a koala model for Megaladapis, he stressed vertical support and slow climbing—not saltation of any sort. Godfrey (1977) saw Archaeolemur as more robust and shorter-legged than the baboon, i.e., as less of a terrestrial cursor. She also noted the presence of derived indroid characteristics in its axial skeleton that signalled some antipronograde postures. She saw Mesopropithecus as sharing forelimb suspensory adaptations with Palaeopropithecus and extant indrids. Jungers (1980) revived the notion, expounded in the early literature on subfossil lemurs (Carleton, 1936, 1937; Lamberton, 1947) that Palaeopropithecus was tree sloth-like in its locomotion (see also MacPhee et al., 1984, MacPhee, 1985), and he followed Lamberton (1948) in suggesting ground sloth-like adaptations for Archaeoindris. Vuillaume-Randriamanantena (1982) concurred with those conclusions and, in addition, reconstructed *Mesopropithecus* as potto-like, a conclusion later supported by Godfrey (1986), who also compared *Mesopropithecus* with *Choloepus hoffmanni*, a two-toed sloth.

Indrids and palaeopropithecids fall at opposite extremes of the spectrum of intermembral indices in lemurs. It is tempting to assume, as Walker (1967) did, that indrids preserve the primitive character states, and palaeopropithecids are highly derived. But indrids and palaeopropithecids are closely related, and share derived traits that are partly responsible for the manner in which "vertical clinging and leaping" is manifested within the former group today. It is essential that the probable positional adaptations of the extinct forms be considered in any attempt to reconstruct the behavior of their common ancestor. This paper will address this problem, based on the morphology of their humeri and femora.

Materials and methods

Relationships between shaft lengths, midshaft diameters, midshaft circumferences, caput surface areas, arcs, and chord lengths were examined for humeri and femora of 63 mammalian genera comprising 110 species (Table 1). Included in the sample were all of the extinct Malagasy lemur species and a comparative sample of extant primates, carnivora, rodents, marsupials, and edentates (including presumed analogues for Malagasy subfossil forms). Functional interpretations of extinct lemurs were drawn from a consideration of general humeral and femoral proportions and morphology, as well as residual analyses. Phylogenetic interpretations were drawn from a standard cladistic analysis of humeral and femoral traits.

Residual analysis provides a useful statistical tool for interpreting morphological variation (Figure 1). For this paper, residuals of nine maximum likelihood estimation regression analyses (Table 2) were calculated for 62 mammalian genera (all except *Archaeoindris* for whom no complete humerus-femur pair exists), and "standardized residual profile" plots were drawn.

Meaningful analysis of multivariate data has always been problematic. With only one or two variables measured on each individual, the way is clear: summarize, describe, and display each, and their relationship (if any). As the number of variables grows, the combinatorial possibilities grow faster yet. Much of modern multivariate analysis is aimed at regaining control of this multivariate combinatorial explosion by replacing the original variables with a few "good" linear combinations of them. This thinking is the essence of such techniques as factor analysis, principal components analysis, correspondence analysis, and most ordination techniques: Project the data onto a lower-dimensional subspace that has "nice" properties (e.g., maximizes variance explained, minimizes χ^2 etc.). Sometimes, however, it is helpful to stay in raw hyperdimensional space and simply deal with the multiplicity of analytical choices available. This maintains the conceptual clarity of the original variables and allows one to select hopefully biologically meaningful relationships to examine, rather than hide them in linear combinations of the original variables that may or may not be biologically meaningful. While both approaches have merit, the more direct approach was selected for this analysis.

Individuals (or centroids for groups of individuals) can be plotted in a space whose axes are the standardized residual scores from a set of regressions. These scores can also be represented by a "standardized residual profile" (SRP) (Figure 2) depicting, for any selected taxon, the positions of its mean residuals (measured in standard deviation units) for each regression. Each residual profile is a simple pictorial representation of how the animal profiled differs from the "norm" for a particular universe of genera.

Table 1   **Genera included in analysis**

Genera	N	
MARSUPIALIA		
MARSUPICARNIVORA		
Didelphidae		
Didelphis	3	Common opossum
Dasyuridae		
Dasyurus	2	Eastern Australian native cat
Thylacinidae		
Thylacinus	1	Tasmanian pouched "wolf"
DIPROTODONTA		
Phalangeridae		
Trichosurus	4	Brush-tailed possum
Phalanger	10	Cuscus
Petauridae		
Dactylopsila	1	Striped possum
Phascolarctidae		
Phascolarctos	5	Koala
Macropodidae		
Dendrolagus	4	Tree kangaroo
EUTHERIA		
EDENTATA		
Myrmecophagidae		
Myrmecophaga	6	Giant anteater
Tamandua	7	Lesser anteater
Bradypodidae		
Bradypus	12	Three-toed sloth
Choloepidae		
Choloepus	14	Two-toed sloth
RODENTIA		
Sciuridae		
Marmota	4	Marmot
Erethizontidae		
Erethizon	5	North American porcupine
Coendou	3	Prehensile-tailed porcupine
Hystricidae		
Hystrix	6	Old World porcupine
CARNIVORA		
Canidae		
Vulpes	4	Fox
Urocyon	3	Gray fox
Ursidae		
Helarctos	3	Sun bear
Ailuropodidae		
Ailuropoda	6	Giant panda
Procyonidae		
Procyon	4	Raccoon
Nasua	8	Coati mundi
Potos	3	Kinkajou
Ailuridae		
Ailurus	6	Lesser panda
Mustelidae		
Gulo	10	Wolverine
Meles	3	European badger
Mellivora	2	Honey badger
Viverridae		
Arctictis	4	Binturong

Table 1—*continued*

Genera	N	
Cryptoproctidae		
Cryptoprocta	1	Fossa
Felidae		
Lynx	5	Bobcat
Felis	5	Cougar
PRIMATES		
Lemuridae		
Hapalemur	14	Bamboo lemur
Lemur	30	Lemur
Varecia	4	Ruffed lemur
Pachylemur	8	
Lepilemuridae		
Lepilemur	16	Sportive lemur
Megaladapidae		
Megaladapis	21	
Indridae		
Avahi	7	Avahi
Propithecus	19	Sifaka
Indri	10	Indri
Palaeopropithecidae		
Mesopropithecus	9	
Palaeopropithecus	17	
Archaeoindris	0	
Archaeolemuridae		
Archaeolemur	8	
Hadropithecus	3	
Daubentoniidae		
Daubentonia	8	Aye-aye
Loridae		
Perodicticus	5	Potto
Atelidae		
Ateles	7	Spider monkey
Brachyteles	1	Woolly spider monkey
Lagothrix	5	Woolly monkey
Alouatta	4	Howler monkey
Pithecia	4	Saki monkey
Chiropotes	1	Bearded saki monkey
Cacajao	2	Uakari monkey
Cercopithecidae		
Theropithecus	2	Gelada baboon
Papio	20	Baboon
Macaca	14	Macaque
Cercocebus	1	Mangabey
Colobus	8	African leaf monkey
Nasalis	9	Proboscis monkey
Presbytis	13	Asiatic leaf monkey
Hylobatidae		
Hylobates	7	Gibbon
Pongidae		
Pongo	7	Orangutan

Values of N listed above are minimums representing individuals with complete data sets. The total N of 464 individuals includes some incomplete pairs that were used in the calculation of some but not all indices and standardized residual profiles. The total *minimum* N is 415 individuals. The Cheirogaleidae (not listed above) were included in the cladistic analysis but not measured.

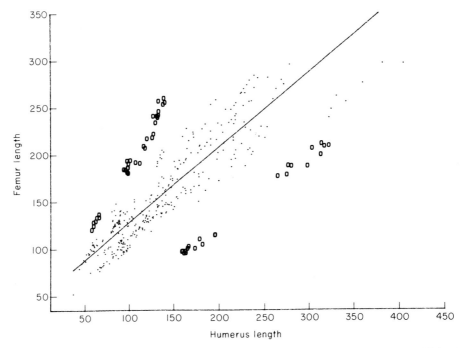

Figure 1. Bivariate plot of femur vs. humerus length for whole sample (Table 1). Positions highlighted (open circles) are: upper left, extant indrids (*Indri, Avahi* and *Propithecus*); lower left, *Bradypus*; and, lower right, *Paleopropithecus*. The solid line is the maximum likelihood regression of femur length on humerus length. Indrids have high positive residuals; *Bradypus* and *Paleopropithecus* have strong negative residuals for this relationship. Note the differences in absolute value of indrid residuals associated with scale. Because indrids vary in a consistent manner over many relationships, their residuals also vary in a consistent manner. The coefficient of correlation between two sets of residuals depends on this consistency, and not on the amplitude of individual values [see Reilman *et al.* (1986) for discussion of maximum likelihood regression analysis].

Table 2	**Maximum likelihood estimation regressions**	
Reg. #	Y	X
1	femoral length	humeral length
2	femoral length	midshift transverse diameter of femur
3	humeral length	midshaft transverse diameter of humerus
4	femoral caput area	humeral caput area
5	femoral length	curvilinear arc length femoral head
6	humeral length	curvilinear arc length (long axis) humeral head
7	humeral caput long. arc	humeral caput longitudinal chord
8	combined humeral plus femoral length	cross-sectional area femoral midshaft
9	femoral midshaft anteroposterior diameter	femoral midshaft transverse diameter

Maximum likelihood regression assumes error in both variables (see Reilman *et al.*, 1986).

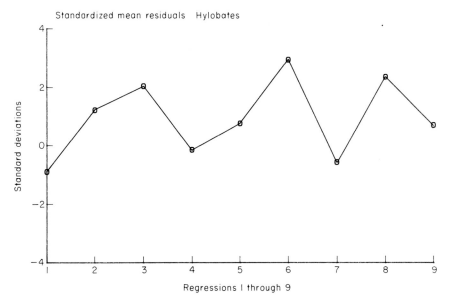

Figure 2. Standardized residual profile of *Hylobates lar*. Open circles are the mean standardized residuals (in standard deviation units) for regressions 1 through 9. The positions of the means are connected to form a standardized residual profile plot.

For the current analysis, eleven raw variables were used to generate nine standardized residual scores for each individual (or centroid for groups of individuals). This means that the whole sample of individuals was "plotted" in 11-dimensional space and the relative positions of group centroids (whole species, whole genera) explored by examining nine of the two-dimensional faces of that space. These two-dimensional subspaces were selected because of their intrinsic biological interest; they emphasize the structural strength of the bones, the generation of force or speed, and the relative sizes and shapes of the joints. The following procedure was used to calculate standardized residual profiles: (1) regression residuals were calculated for all individuals with complete data; (2) residuals were standardized by dividing each by the regression's standard deviation for all residuals; (3) for each regression, mean standardized residual scores were calculated for each genus and species; (4) the procedure was repeated for each of the nine selected bivariate maximum likelihood estimation regressions; (5) correlation matrices for genus' and species' sets of nine standardized residual scores were calculated; (6) standardized residual profile plots (plots of the nine raw standardized residual scores) were drawn from selected species, including all of the Malagasy lemurs and all non-lemurs exhibiting high correlations with any one of the extinct forms.

When genera or species deviate from the norm in a consistent manner, their standardized residual profiles resemble one another. One measure of that resemblance is Pearson's coefficient of correlation. Profiles of functionally similar animals (atelids and hylobatids, choloepids and bradypodids, cercopithecines and colobines, didelpids and phalangerids) are concordant in pattern and thus highly correlated (Figure 3a–b). Functionally similar animals of grossly different body sizes, such as the indrids (*Indri*, *Propithecus*, and *Avahi*), or tamanduas and giant anteaters, have similar profiles despite their differences in body size (Figure 3c). Conversely, animals of grossly different

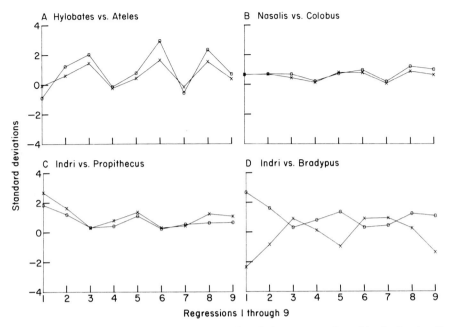

Figure 3. Standardized residual profile comparisons. Correlations are strongly positive for functionally ·similar animals such as A. *Hylobates* and *Ateles* (·97), B. *Nasalis* and *Colobus* (·89) and C. *Indri* and *Propithecus* (·96); correlations are strongly negative for functionally *dissimilar* animals such as D. *Indri* and *Bradypus* (−·90). The residual profiles reveal aspects of the morphology of the genera involved. For example, in A., *Hylobates* (○) and *Ateles* (×) exhibit peaks at regressions 3,6, and 8, and troughs at regressions 1,4, and 7. These profiles can be "read" as follows: Agile arm swingers have long, slender femora and especially humeri; humeral heads are large with respect to femoral heads, but small with respect to humeral shaft lengths.

proportions and morphology have highly negatively correlated profiles (Figure 3d). Correlation matrices (in this case 62 × 62 for genera, 109 × 109 for species) provide a framework for viewing the overall similarities and differences between any genus or species pair; they can be used as diagnostic tools to select profiles to compare. A close reading of the profiles—i.e., of *which* regression yield high or low values for standardized residuals—reveals the anatomical basis for strong or weak correlations.

While it is true that profiles are not individually "corrected" for scale, the correlations between them do offer a linear size-independent tool to assess their similarities (the correlation between Y and X is identical to the correlation between aY and bX). In other words, correlation analysis captures similarities or differences in the *patterns* of deviation, whether those deviations are large or small. The problem of non-linear scaling is examined elsewhere (Godfrey & Sutherland, in prep.). For current purposes, I emphasize that, as long as (1) the effects of scale are essentially linear (which appears, commonly, to be the case) and (2) the main analytical tool is correlation analysis of residual profiles, then it is not necessary to correct for size. Moreover, it is an empirical fact that the effects of size on the values of the indices explored here are negligible in comparison to the effects of function (e.g., the values for bone robusticity given in Table 4 show little size effect; Godfrey & Sutherland, in prep., will discuss this problem and those of linear vs. non-linear scaling of longbone robusticity and other osteometric variables).

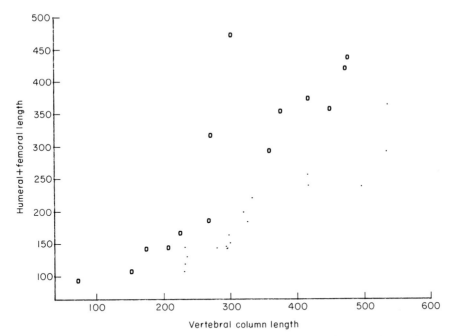

Figure 4. Comparison of general limb and trunk proportions in extant primates and non-primates (marsupials, edentates, carnivores and rodents). Primates (○) have consistently longer limbs relative to trunk length.

Phylogenetic relationships and patterns of evolution within the Indroidea were explored by cladistically determining humeral and femoral characters present at branching points in indroid phylogeny. This procedure was used to test the hypothesis that modern indrid characters which are linked functionally with vertical clinging and leaping were ancestral for the Indroidea. Polarity of homologous character states was evaluated using outgroup comparisons; members of the Lemuroidea and *Daubentonia* (whose superfamilial status is still uncertain) were taken as nearest outgroups to the Indroidea. Shared homologous characters were considered primitive when they are present in nearest outgroups, and derived when they are not. Resemblances were judged to be homoplasious when they are manifested in different forms in the groups concerned, or when the hypothesis of synapomorphy is incongruent with the dominant nested hierarchical pattern of distribution of other apparently derived traits. The primitive morphotype for any character was reconstructed based on outgroup morphology when members of a clade demonstrate divergent character states and the morphologically intermediate condition is common in outgroups.

Functional considerations

General humeral and femoral proportions

Primates can be distinguished from major groups of comparably-sized non-primates by a series of postcranial traits: limbs that are long relative to trunk lengths (Figure 4): humeri and femora that are long relative to their diaphyseal diameters; femoral heads that are large relative to humeral heads. Values for some humeral and femoral indices are given in

Table 3   General proportions

Extinct Malagasy Strepsirrhini	N	1	2	3	4	5
Mesopropithecus pithecoides	1	91·1	24·3	28·8	28·5	27·1
Mesopropithecus globiceps	6	90·3	25·2	30·1	31·7	28·1
Palaeopropithecus ingens	2	152·7	21·6	36·3	23·8	34·0
Palaeopropithecus maximus	8	152·3	21·0	37·3	24·0	34·0
Archaeolemur edwardsi	2	81·0	26·1	23·4	34·2	29·0
Archaeolemur majori	5	82·6	25·1	22·9	30·2	27·3
Hadropithecus stenognathus	1	116·1	21·6	22·0	25·4	27·1
Daubentonia robusta	1	78·5	30·6	22·2	37·7	27·5
Pachylemur insignis	5	83·2	23·9	26·0	29·1	28·6
Pachylemur jullyi	3	79·8	26·2	26·5	32·4	28·6
Megaladapis madagascariensis	2	111·3	27·5	40·0	35·1	36·8
Megaladapis grandidieri	2	110·7	28·4	41·7	36·1	38·5
Megaladapis edwardsi	6	108·9	24·5	36·3	32·5	36·9
Extant Malagasy Strepsirrhini						
Indri indri	10	54·1	21·4	15·2	22·3	15·5
Propithecus verreauxi	14	53·3	22·7	15·4	23·1	16·3
Avahi laniger	7	47·7	22·8	14·6	26·1	15·7
Daubentonia madagascariensis	7	70·0	25·9	18·4	26·6	18·6
Varecia variegata	4	69·8	23·6	20·6	25·1	19·0
Lemur fulvus	17	67·2	23·1	18·7	23·7	19·5
Hapalemur griseus	10	60·1	24·6	17·3	26·9	18·9
Lepilemur mustelinus	4	56·1	26·7	15·5	28·2	18·0
Comparative data: other mammals						
Dactylopsila palpater	1	70·9	31·0	20·6	33·4	25·2
Phascolarctos cinereus	5	79·2	27·7	22·0	29·2	22·8
Phalanger maculatus	6	84·7	26·9	24·9	28·0	27·9
Tamandua tetradactyla	7	98·6	31·4	29·0	37·6	34·4
Choloepus hoffmanni	3	104·0	24·2	26·6	26·4	28·9
Bradypus tridactylus	10	169·3	18·3	32·3	18·6	32·3
Coendou prehensilis	3	88·9	29·9	30·5	31·8	35·9
Ailuropoda melanoleuca	1	96·3	27·9	25·2	32·7	28·4
Theropithecus gelada	2	94·2	20·0	20·1	23·7	21·9
Macaca fascicularis	11	90·7	17·5	18·4	21·8	21·1
Pongo pygmaeus	7	131·5	18·9	26·7	19·8	24·3
Hylobates lar	7	115·4	12·1	17·1	13·8	16·3

1	Humerofemoral length index	(Humeral length × 100 divided by femoral length)
2	Humeral caput dominance index	(Humeral caput longitudinal curvilinear arc × 100 divided by humeral length
3	Femoral caput dominance index	(Femoral caput transverse curvilinear arc × 100 divided by femoral length)
4	Humeral robusticity index	(Humeral midshaft circumference × 100 divided by humerus length)
5	Femoral robusticity index	(Femoral midshaft circumference × 100 divided by femoral length)

For all extant animals, indices shown are means for associated adult humerus-femur pairs. In the case of the subfossil Malagasy lemurs, associations were not always certain. When not known, "associations" were selected based on site, size, and external coloration of the fossils. Ratios calculated for uncertain pairs were then checked against values calculated from larger, whole sample species means.

Table 3 for extinct lemurs (except *Archaeoindris*), and for selected other mammals. Mammals, in general, have midshaft humeral and femoral circumferences which are highly correlated ($r = 0.97$ for the entire sample). Both are strongly constrained by body weight (Aiello, 1981*a,b;* Anderson *et al.*, 1985; Steudel, 1985). Caput surface areas are also strongly correlated with midshaft circumferences ($r = 0.92$ for humeri, 0.93 for femora), suggesting that they too, are largely constrained by body weight. Interspecific variation in relative humeral and femoral caput surface areas might be expected to reflect differential weight bearing or force transmission through fore and hindlimbs, as well as differences in fore and hindlimb movement potential. Indeed, differential force transmission may explain the relatively low humerofemoral caput indices (the ratio of humeral to femoral head surface areas) of primates: primates tend to reduce compressive stresses on forelimbs and rely more strongly than other mammals on hindlimbs for supporting and propelling their bodies (Hildebrand, 1967; Kimura *et al.*, 1979; Reynolds, 1985*a,b*). The fact that humeral and femoral caput surface areas are themselves strongly correlated ($r = 0.92$), but not to the degree exhibited by midshaft circumferences, suggests that they are more strongly influenced by other factors.

Humeral and femoral lengths are relatively weakly correlated ($r = 0.78$, whole sample). Weaker yet is the relationship between bone lengths and caput surface areas (e.g., $r = 0.63$, for the femur). Thus, while midshaft circumferences and, to a lesser extent, caput surface areas, are strongly constrained by body weight, bone lengths are freer to respond to positional demands. This lack of body weight constraint on length, coupled with a strong constraint on midshaft circumference, affects longbone robusticity (the relationship between midshaft circumference and maximum length).

Humeral robusticity tends to covary with femoral robusticity, since both are constrained by the burden of head and trunk weight on the limbs. Heavy-bodied, short-limbed animals tend to exhibit high humeral and femoral robusticities while light-bodied, long-limbed animals tend to exhibit low robusticities. Apart from this and any mutual effect of scaling (or absolute body weight), humeral and femoral robusticities vary independently in response to specializations of fore and hindlimbs. Muscle power-to-load arm ratios vary with the positions of muscle origins and insertions *vis-à-vis* the joints they turn, and with the lengths of the bony links themselves. Short, robust limbs (generally with high muscle power-to-load arm ratios) are adapted to generate force; conversely, long limbs (generally with low muscle power-to-load arm ratios) are adapted to generate speed (Table 4, Hildebrand, 1974; Smith & Savage, 1955). Differential fore and hindlimb use results in deviations from any pattern of covariation in humeral and femoral robusticities, and in deviations in their relative lengths.

Caput surface areas might be expected to vary in accord with the relative lengths of the humerus and femur. They do in extant primates, which display a positive correlation ($r = 0.59$, $N = 192$) for humerofemoral length and caput surface area indices. Indrids have the lowest values for both. Lemurids and lepilemurids fall closest to indrids, followed by colobines, cercopithecines, alouattines, pitheciines, atelines, hylobatines, and finally pongines. In other words, "forelimb dominant" primates (animals relying more strongly on forelimbs for support and propulsion) have both relatively large humeral heads and relatively long humeri, whereas "hindlimb dominant" primates (animals relying more heavily on their hindlimbs for support and propulsion) have both relatively large femoral heads and relatively long femora. Only a few of extant primates sampled here, such as *Daubentonia* and *Perodicticus*, do not appear to "fit" within this spectrum.

L. R. GODFREY

Table 4   **Robusticity**

	Sample humeral robusticity spectrum	
Myrmecophaga	40·8	Diggers, power generators
Tamandua	37·6	
Dactylopsila	33·4	
Marmota	32·6	
Ailuropoda	32·7	Very heavy-bodied climbers
Helarctos	32·1	
Coendou	31·8	
Trichosurus	30·4	
Phalanger	29·6	Slow climbers
Perodicticus	29·5	
Procyon	29·4	
Phascolarctos	29·2	
Didelphis	28·6	Ambulatory quadrupeds
Varecia	25·1	
Papio	23·8	
Theropithecus	23·7	
Lynx	23·6	Cursorial quadrupeds
Nasalis	19·7	
Ateles	16·8	
Brachyteles	14·0	
Hylobates	13·8	Arm swingers
	Sample femoral robusticity spectrum	
Coendou	35·2	
Erethizon	35·2	Slow climbers
Tamandua	34·4	
Bradypus	32·3	
Choloepus	28·9	
Phalanger	27·9	
Didelphis	26·4	Ambulatory quadrupeds
Trichosurus	25·4	
Felis	24·3	
Macaca	21·4	
Papio	20·9	
Lynx	20·9	Cursorial quadrupeds
Vulpes	19·4	
Colobus	19·3	
Lemur	19·2	
Hapalemur	18·8	
Avahi	15·7	Leapers
Indri	15·5	

The effects of scaling on limb robusticity are addressed elsewhere (Godfrey and Sutherland, in prep.); it can be seen from the above two arrays that body size has a relatively minor effect on the value of the robusticity index (midshaft circumference × 100/length). Robusticity is largely a measure of adaptations for power or for speed. Humeral robusticity separates fossorial animals and short-armed, heavy-bodied climbers from cursors and arm swingers. Arboreal sloths, while not agile, exhibit low to intermediate humeral robusticities due to their long arm spans. Femoral robusticity separates slow climber/hangers from cursors and leapers. Some animals with low humerofemoral indices and therefore "short" femora nevertheless exhibit low femoral robusticities. *Ateles* and *Hylobates* have hindlimbs that are short with respect to their forelimbs, but not with respect to their trunk lengths or body weights. In effect, they have long, slender hindlimbs and exceptionally long, slender forelimbs (see Jungers 1984 on the extreme case of *Hylobates*).

The observed correlation between humerofemoral length and caput indices drops to 0·2 when non-primates and extinct Malagasy strepsirrhines are included in the sample. In effect, the positive correlation is weakened by relationships that exist outside the order Primates and, also, among some extinct Malagasy strepsirrhines. In these animals, shaft length proportions and caput surface area proportions may be inversely correlated, for a variety of reasons.

Diggers and probers tend to exhibit relatively short, wide, robust humeri with relatively large heads. Their forelimbs are designed to exert tremendous out-forces—that is, to resist compressive loads and to maximize thrust by adjusting the power-to-load arm ratios of forelimb retractors and forearm flexors, pronators, and supinators Smith & Savage, 1955; (Coombs, 1983; Hildebrand, 1985; Taylor, 1985). Such a combination of traits characterizes anteaters (*Tamandua* and *Myrmecophaga*), sciurid diggers (*Marmota*), mustelid diggers (*Meles*), the *Daubentonia*-like striped possum (*Dactylopsila*), and *Daubentonia* itself. When such animals also leap (e.g., *Daubentonia*, to some extent), or, for some other reason, have long femora, they may exhibit *low* humerofemoral length indices and *high* humerofemoral caput area indices.

Arboreal sloths and other slow climbers exhibit an opposite suite of characters. Because they use their hindlimbs, sometimes in full suspension, as anchors on precarious supports, their femora are short and their hip, knee, and ankle joints highly mobile (Dagosto, 1983; McArdle, 1981; Mendel, 1985*a,b*; Tardieu, 1983). Large femoral heads (set in wide but shallow accetabula) are part of a mobile hip joint complex. Hip flexors (such as iliopsoas) insert distally on a short femur to aid in recovering the torso from full bipedal suspension. The need for a long, flexible trunk, as well as for power in flexing that long torso on the hindlimbs against the pull of gravity, constrains femur/trunk ratios and influences femoral robusticity.

The high humerofemoral length indices of slow climber/suspenders reflect biomechanical constraints of vertical climbing support postures (Cartmill, 1974*b*, 1985; Jungers, 1976, 1977), as well as of torso recovery from full bipedal suspension. The greatest humeral elongation occurs in the slowest, most specialized of slow suspenders—not to increase agility or speed in locomotion, but to increase reach or arm span. The combination of relative *humeral* elongation plus relative dominance of the *femoral* head occurs uniquely in highly specialized slow climbers. It is exemplified by *Bradypus*, the three-toed sloth, and by some extinct Malagasy lemurs, especially *Palaeopropithecus*. In fact, this combination of features separates *Bradypus*, *Palaeopropithecus*, and *Mesopropithecus*, and, to a lesser extent, *Megaladapis*, *Choloepus* (the two-toed sloth) and *Coendou* (the loris-like South American prehensile-tailed porcupine), from *all* other animals sampled, including *Pongo* (which exhibits an enlarged humeral head), and the agile arm-swingers *Ateles* and *Hylobates* (which also exhibit relatively larger humeral heads).

Heavy-bodied, short-limbed, slow climbers tend to exhibit high robusticities of both fore and hindlimb longbones. Indeed, their humeral robusticities may rival or exceed those of fossorial animals. Animals that both climb and dig (e.g., *Tamandua*) have exceptionally high humeral robusticities given their body sizes. Specialized diggers can be distinguished from specialized climbers and browsers not by differences in humeral robusticity but by differences in morphological details (Coombs, 1983); for example, diggers tend to have strong elbow extensors as well as flexors. An animal with a robust humerus with widely flaring brachialis flange and strong deltoids, but a weak olecranon process, is not a digger.

In sum, the combination of relative lengths of humeri and femora, relative caput surface

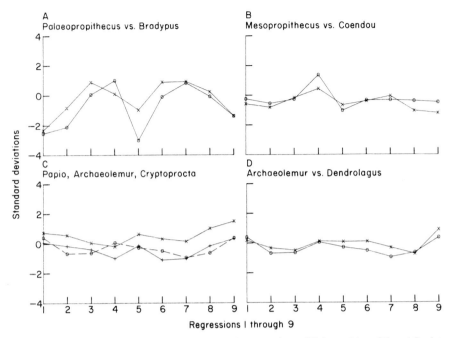

Figure 5. Standardized residual profile comparisons. A, comparison of *Paleopropithecus* (○) and *Bradypus* (×). Highly specialized slow climber/hangers have SRPs that peak at regressions 3,4,6 and 7, and fall at regressions 1,5 and 9. Less specialized slow climbers maintain milder peaks at 4 and 7, and a milder trough at 5. The SRPs of *Paleopropithecus* and *Bradypus* reveal exceptionally short femora; long humeri; wide, anteroposteriorly compressed femora with heads that are large relative to femoral shaft length and humeral caput surface area; and, humeral heads that are strongly curved. B, *Mesopropoithecus* (○) and *Coendou* (×) exhibit flattened versions of the slow climber pattern. C, *Papio* (×) *and Cryptoprocta* (+) are more alike than is either like *Archaeolemur* (○, dashed profile). D, *Archaeolemur* was probably more similar in body proportions to arboroterrestrial marsupials such as *Dendrolagus* than to specialized terrestrial quadrupeds of diverse mammalian orders.

areas, and relative humerofemoral robusticities offers important clues to positional behavior. Bone geometry can also aid in the interpretation of positional behavior in extinct animals (Burr, 1980; Burr *et al.*, 1981, 1982; Currey, 1984; Lanyon & Rubin, 1985; Ruff, 1987; Schaffler *et al.*, 1985). Femoral shafts of leapers are shaped differently from those of quadrupedal runners: leapers have stout, straight femoral shafts designed to resist buckling under the compressive axial loads incurred during leaping, and runners have anteriorly bowed shafts that are better able to resist bending moments induced by eccentric compressive loading (Burr *et al.*, 1982). Cursorial quadrupeds exhibit stronger anterior bowing of the femoral shaft than ambulatory quadrupeds, and suspenders have femora that may actually bow in the opposite direction (i.e., convex posteriorly). Slow vertical climbers and climber/suspenders—animals that often assume vertical trunk postures with strongly flexed and abducted hindlimbs—have wide, flattened femoral shafts, best designed to resist bending loads in the mediolateral plane (Jungers & Minns, 1979).

General morphology, proportions, and standardized residual profiles of extinct Malagasy lemurs

Palaeopropithecus (see Figures 10, 11, 12). *Palaeopropithecus* was the most specialized of the Malagasy lemur slow climber/suspenders. Despite its much larger (approximately chimp-) size, it probably moved like arboreal sloths, climbing and hanging but never

Table 5   **Sample correlations**

	PP	CH	BP	MP	CN
PP					
CH	·90				
BP	·82	·77			
MP	·63	·45	·21		
CN	·60	·62	·49	·71	

PP = *Palaeopropithecus*
CH = *Choloepus*
BP = *Bradypus*
MP = *Mesopropithecus*
CN = *Coendou*

This represents a small portion of the 62 × 62 correlation matrix calculated for standardized residual profiles of genera. These correlations are used as a diagnostic tool to aid in the recognition of profile similarities.

bounding, hopping or leaping. Quadrupedal suspension would have been used in locomotion as well as in feeding. Ground locomotion was probably also similar to that of arboreal sloths, who exhibit an awkward form of quadrupedalism with highly adducted limbs (Mendel, 1981, 1985a).

The SRP profile of *Palaeopropithecus* strongly resembles those of *Choloepus* and *Bradypus* (Figure 5a) and moderately resembles those of shorter-limbed, relatively heavy-bodied slow climbers (such as arboreal porcupines) (Table 5). It highlights a short, wide, and anteroposteriorly compressed femur with disproportionately enlarged head. The humerus is long, and bears a strongly curved head that is, nevertheless, small relative to the length of the humerus and to the size of the femoral head. Humerofemoral length and caput area indices of *Bradypus* and *Palaeopropithecus* are virtually identical, despite the striking differences in body size of these two genera. *Palaeopropithecus* most markedly differs from *Pongo* in that the latter exhibits a relatively longer femur, much larger humeral head, and stouter (less anteroposteriorly compressed) femoral shaft. The SRPs of "arm swingers" or brachiators such as *Ateles*, *Brachyteles*, *Hylobates*, and *Pongo* show little resemblance to that of *Palaeopropithecus* (although *Pongo* resembles *Palaeopropithecus* more than do spider monkeys or gibbons). Only in isolated characteristics (e.g., high humerofemoral index; moderately robust femur with moderately large head) is *Palaeopropithecus* orang-like. *Pongo* has much higher positive SRP correlations with *Hylobates* and *Ateles* than with *Palaeopropithecus*, *Bradypus* and *Choloepus*.

Tree-sloth likenesses are clearly manifested in the morphology of the humerus and femur of *Palaeopropithecus*. Both exhibit loose shoulder, hip and knee joints. The relationships between the femoral head, lesser trochanter and greater trochanter are similar; the femoral head protrudes well above the trochanters, and is almost vertically disposed. There is a small, proximally situated obturator fossa, a strong adductor rugosity, shallow patellar groove with weak crests, narrowly separated and anteroposteriorly compressed condyles widely flanked by strong epicondyles, and a generally reduced but distally elongated third trochanter. Mendel (1985b) points out that the gluteals (maximus and medius) of tree sloths may function as hip flexors, since they have fibers that pass in front of the axis of the joint. This may have been true for *Palaeopropithecus* as well, and may explain the odd disposition of the third trochanter, which, in rugose individuals, protrudes anteriorly. The humeri of sloths match those of *Palaeopropithecus*, not only in proportions, but in the disposition of muscle rugosities, dominance of the capitulum, height and position of the tubercles, and shape and protrusion of the entepicondyle.

Archaeoindris. No standardized residual profile could be calculated for *Archaeoindris*, since no complete adult humerus is available. Morphologically, the humerus and femur of *Archaeoindris*, like those of *Palaeopropithecus*, suggest a huge, *Choloepus*-like animal. The two palaeopropithecines appear to be gigantic "versions" of tree sloths. But while the *Archaeoindris* femur distinctly recalls those of slow climbers, some authors believe this animal, at roughly the size of a gorilla, exceeded the weight limit for climbing and, especially, for arboreal suspension (Jungers, 1980; Vuillaume-Randriamanantena, 1988). *Archaeoindris* was probably not, however, heavier than some bear species, such as the American black bear or the sloth bear, that are adept tree climbers. Its morphology does offer some support for essential terrestriality: the femur is much stockier than that of *Palaeopropithecus*, thus increasing its structural strength in the sagittal plane and its resistance to buckling under compression. But *Archaeoindris* finds no easy analogue within ground sloths. The enormous size of its femoral head, reduced greater trochanter, medially projecting lesser trochanter, and high collodiaphyseal angle, suggest great hip mobility and an unusual mode of terrestrial locomotion for this bizarre strepsirrhine.

Mesopropithecus (see Figures 11, 12, 13). *Mesopropithecus* was a moderately specialized slow climber/suspender. It was similar in size to *Choloepus hoffmanni*. Its range of suspensory activities probably matched that of *Palaeopropithecus*, but *Mesopropithecus* appears to have been more quadrupedal, and less committed to suspensory activities. Extant animals with SRPs most similar to *Mesopropithecus* include *Coendou*, and, to a lesser degree *Erethizon* (the North American porcupine), *Phalanger* (the cuscus or "marsupial sloth"; Dawson & Dagabriele, 1973; Troughton, 1965), and *Choloepus*, the shorter-limbed of the arboreal sloths (but not *Bradypus*) (Figure 5b). None of these animals engages in leaping, hopping, or cursorial quadrupedalism.

The SRP of *Mesopropithecus* describes an animal with a fairly high humerofemoral length index and robust limbs. The femoral shaft is wide given its length; the femoral head is disproportionately large. Most of these characters, with the exception of a robust humerus, are manifested in *Palaeopropithecus* to a greater extreme. *Mesopropithecus* approaches *Palaeopropithecus* morphologically as well, as will become clear below.

The humerofemoral length index of *Mesopropithecus* (90·1) is higher than the norm for lemurs. It closely matches that of *Coendou prehensilis* (88·9), but is not as high as those of *Palaeopropithecus*, *Archaeoindris* (see Vuillaume-Randriamanantena, in press), or the two-toed sloth *Choloepus* (104·2 for both *hoffmanni* and *didactylus*). It is somewhat higher than those of *Phalanger maculatus* (84·7) and *Perodicticus potto* (84·0). While the femur of *Mesopropithecus* exceeds the length of the humerus, residuals for the maximum likelihood estimation regression of femoral length on humeral length are negative for *Mesopropithecus* (just as for *Perodicticus*, *Coendou*, *Phalanger*, *Choloepus*, and, of course, *Palaeopropithecus* and *Bradypus*).

Archaeolemur (see Figures 11, 13). *Archaeolemur* was a versatile, power- rather than speed-driven quadrupedal lemur that may have been largely terrestrial. It was roughly the size of a female baboon. Its SRP is not very distinct from that of *Pachylemur*, the largest and only known extinct lemurid. *Archaeolemur* diverges morphologically from *Pachylemur* in the direction of its sister, *Hadropithecus*, in features related to shoulder, elbow, and hip stability.

The SRP of *Archaeolemur* is only moderately baboon-like; it resembles those of terrestrial quadrupedal marsupials (*Thylacinus*, the Tasmanian wolf) and arboroterrestrial carnivores (*Cryptoprocta*, the fossa) at least as much. But its highest positive correlations are neither with thylacinids, cryptoprocts, nor Old World monkeys; rather, they are with *Pachylemur* and with agile and versatile arboroterrestrial marsupials, such as *Dendrolagus*, the only tree climbing kangaroo (Figures 5c,d). This SRP highlights relatively short limbs, robust humeri and femora, and flattish humeral heads.

Baboons and other Old World monkeys have humeri and femora that are much longer given their head sizes and midshaft diameters, and more greatly bowed. The femoral head of *Archaeolemur* is more dominant on a short, robust shaft. More cursorial animals tend to have longer, less robust, femora, with smaller heads set in relatively deeper acetabula (Jenkins & Camazine, 1977). Nevertheless, the femoral head of *Archaeolemur* is small and the acetabulum deep in comparison to *Pachylemur*. The femoral shaft, while not nearly as anteriorly bowed as those of most cercopithecines, is nevertheless bowed in comparison with other lemurs.

While the humeral and femoral characters of *Archaeolemur* suggest pronograde quadrupedalism (Godfrey, 1977; Jouffroy, 1963; Walker, 1967, 1974), they do not imply cursoriality. There are many ways to be terrestrial, and many ways to be quadrupedal (Jenkins & Camazine, 1977; Rodman, 1979; Rose, 1973), and *Archaeolemur* was certainly not as baboon-like as has been suggested. *Archaeolemur* was a short-limbed quadruped with a wide, heavy trunk (Godfrey 1977; Vuillaume-Randriamanantena 1982); its high limb bone robusticity reflects trunk and limb proportions strikingly unlike those of more cursorial Old World monkeys. Baboon convergences of archaeolemurids are better argued based on craniodental traits, especially for the more specialized *Hadropithecus* (Jolly, 1970; Tattersall, 1973, 1975, 1982).

Not only are femora of *Archaeolemur* more robust than those of Old World monkeys, but major hindlimb propulsive muscles insert relatively further from the fulcrum of the hip joint, signalling adaptations for power but not speed. The greater trochanter is high and wide, the obturator fossa is deep and proximodistally elongated, and the third and lesser trochanters are distally situated on the shaft (twice as distal as in a comparably-sized macaque). The lesser trochanter is knob-like rather than plate-like and oriented posteromedially rather than medially. Indrid-like leaping specializations are absent: the crest for the origin of quadriceps vastus lateralis is neither proximal nor does it overhang the anterior aspect of the shaft; the femoral condyles are not anteroposteriorly elongated; and the patellar groove is not well-bordered and deep.

In contrast with most lemur humeri that exhibit distally prolonged teres major and latissimus dorsi muscle scars (a characteristic of heavy-bodied climbers), *Archaeolemur* humeri have proximally situated latissimus dorsi and teres major scars. Neither overlaps the distal extents of deltoideus and pectoralis. In Old World monkeys, however, the humeral retractors insert even more proximally and are better adapted to effect speedy retraction.

Other humeral characters that signal semiterrestrial quadrupedalism include: a high, anteriorly disposed, robust greater tubercle; lesser tubercle approximately even with the height of the head; somewhat reduced and posteromedially oriented entepicondyle. The entepicondyle is stronger in *Archaeolemur* than in the more terrestrial Old World monkeys, but it exhibits some variation within the genus, being largest in the larger (and probably

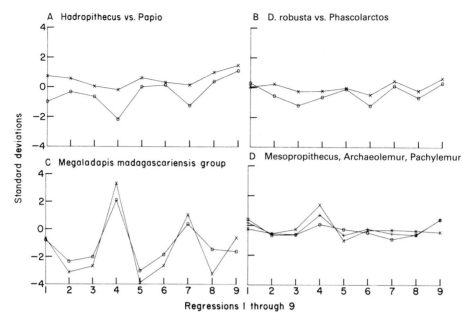

Figure 6. Standardized residual profile comparisons. A, *Hadropithecus* (○) is more *Papio*-like (×) than is *Archaeolemur*, but its humerus and femur were more robust than those of cercopithecines, its humerus longer relative to femur length, and its humeral head flatter and larger relative to femoral head size. B, *Daubentonia robusta* (○) resembles *Phascolarctos* (×) but exhibits a more robust humerus and femur, and a relatively larger humeral head. C, *Megaladapis madagascariensis* (○) is quite similar to its considerably larger sister *M. grandidieri* (×), with distinct SRP peaks at 4 and 7 troughs at 2, 3, 5 and 6. These profiles bear some resemblance to those of extant slow climbers but reveal the much greater longbone robusticities of a heavier-bodied, relatively shorter-limbed animal. *M. edwardsi* differs from its congeners most markedly in the shape of its humeral head. Note the effect of scale on the residual profiles: the values of the residuals generally deviate less from the norm in the smaller *M. madagascariensis* than in the larger *M. grandidieri*, but the deviations are consistently in the same direction. D. *Mesopropithecus* (×) and *Archaeolemur* (○) deviate in opposite directions from *Pachylemur* (+). The SRP of *Mesopropithecus* dips down at regressions 1, 5 and 9, and up at regressions 4 and 7; these deviations are in the direction of *Palaeopropithecus*. *Archaeolemur* deviates, most notably, down at 4 and 7 and up at 5 (in the direction of *Hadropithecus*).

less terrestrial) species from the less arid regions of Madagascar. The trochlea is large and cone-shaped, with its distal edge angled downward from its lateral to medial borders. While this, too, recalls the condition in Old World monkeys, its morphology is actually quite different. There is no medial flange guiding the ulna in flexion and extension, nor is there a lateral flange bordering the olecranon fossa posteriorly. Thus the ulna is not as confined in its fore–aft motion as in cursorial monkeys. A large, wide, and deep olecranon fossa characterizes *Archaeolemur*, bearing testimony (along with the strong, posteriorly deflected ulnar olecranon process) to the relatively greater development of the forearm extensors—again a sign of terrestrial pronogrady.

Hadropithecus (see Figure 13). *Hadropithecus* was slightly larger and more slender-boned than *Archaeolemur*, but still not a cursor *per se*. It is also the only lemur whose SRP significantly resembles those of catarrhines (Old World monkeys and, less so, apes). *Hadropithecus* deviates from Old World monkeys in a consistent manner: its humerus and femur are more robust and its humerus, simply, "too big". It exhibits *Archaeolemur*-like power adaptations and a straight humeral shaft unlike those of terrestrial cursors. *Hadropithecus* is, in effect, an *Archaeolemur*-like animal with a long, oversized humerus.

The SRP of *Hadropithecus* is most strongly correlated with those of quadrupedal monkeys that have moderately long forelimbs (e.g., *Papio*, *Nasalis*; Figure 6a). In fact, *Hadropithecus* exhibits a remarkably high humerofemoral length ratio (116), exceeding those of extant cercopithecines, and falling rather in the range of hylobatines (with whom it exhibits only moderately positive SRP correlations) and gorillas. *Hadropithecus* was a quadruped with unusual limb disproportionality. Furthermore, its humerofemoral caput surface area index exceeds those of forelimb suspenders such as *Pongo*, *Ateles*, and *Hylobates*. This does not mean that *Hadropithecus* engaged in forelimb suspensory activities, however. Given morphological adaptations for shoulder, elbow, hip, and knee stability (e.g., high, anteriorly disposed humeral greater tubercle; reduced, posteromedially deflected humeral entepicondyle; flat humeral head; femur very like that of *Archaeolemur* except for a slightly greater diaphyseal bowing and smaller head), suspension was not likely to have been an important part of its positional repertory. Despite the length of the humerus, this bone is designed for predominant compressive rather than tensile loading. While the axial skeleton of *Hadropithecus* is unknown, one might surmise that *Hadropithecus*, like *Archaeolemur*, had a heavier trunk and shorter legs than terrestrial Old World monkeys. Its unusual forelimb elongation suggests reduced reliance on bounding and leaping. Consistent with the hypothesis that *Hadropithecus* was more ground committed than *Archaeolemur* (Walker, 1967, 1974; Godfrey, 1977; Tattersall, 1973) is the slightly greater anterior bowing of the femoral shaft and smaller femoral caput surface area.

The reasons for unusual forelimb elongation in an animal with terrestrial adaptations are not obvious; one possibility is that *Hadropithecus* habitually foraged in the manner of *Theropithecus gelada*, in a sitting or crouching position (Jolly, 1970; Tattersall, 1982). Such feeding postures may help to explain apparent adaptations for vertical sitting in the axial skeleton of *Archaeolemur* as well (Godfrey, 1977; Tattersall, 1982). However, it should be cautioned that no positive humeral and femoral associations exist for *Hadropithecus*, and femora attributed to *Hadropithecus* are far more *Archaeolemur*-like than are humeri. Other possible explanations for the disproportionality must tentatively include incorrect generic attribution or incorrect association of humeri and femora belonging to conspecific individuals of grossly different body sizes. Hadropithecine postcranial material is rare, and hindlimb material exceptionally so. Only the discovery of new, associated specimens will resolve this dilemma.

Daubentonia robusta. *Daubentonia robusta*—an animal that exceeded the size of *Mesopropithecus*—was a quadrupedal climber with specializations for generating unusual forelimb power. Its SRP highlights short limbs and a very robust humerus with enlarged head. The positional repertory of *D. robusta* may have resembled that of *D. madagascariensis*, but *D. robusta* was clearly less agile than its morphologically similar but less robust congener.

The extant *Daubentonia madagascariensis* extracts insects from beneath bark (Petter & Peyriéras, 1969; Petter *et al.*, 1977) much in the manner of the considerably smaller striped possum from New Guinea, *Dactylopsila palpater* (Ziegler, 1977). As Cartmill (1974a) noted, these behavioral convergences are registered in craniodental convergences. Perhaps, then, it is unsurprising to find convergences in the forelimb anatomy of *Daubentonia* and *Dactylopsila*.

Humeral robusticity is higher than the norm for lemurs in both daubentoniids, but exceptionally so in *D. robusta*. Force-generating features of the humerus are better developed in *D. robusta* than in *D. madagascariensis*, although they are well developed in both. The extinct form has the relatively shorter, distally wider humerus with larger head; the extant form has the relatively longer femur. Greater relative elongation of the humerus of the larger *Daubentonia* is a prediction of the vertical climbing support model (Jungers, 1980).

Force-generating features of the daubentoniid humerus include: a prominent, distally extensive ridge for the insertion of teres major; a strong, medially projecting entepicondyle which expands proximally for increased mechanical advantage of pronator teres and the digital flexors; a strong, wing-like brachialis flange, providing for very lateral and proximal origins of supinator and brachioradialis. The daubentoniid femur is slender in comparison to the humerus, and the femoral head is small relative to the size of the humeral head (especially in *D. robusta*).

Neither daubentoniid is exceptionally *Dactylopsila*-like in its overall SRP (although *D. madagascariensis* has a moderately high *Dactylopsila* correlation). This is because the peculiar feeding specilizations that so strongly affect the morphology of the humeri of *Daubentonia* and *Dactylopsila* are not necessarily associated with any particular mode of locomotion. It is the humerus more than the femur that renders these lemurs similar to *Dactylopsila*. Like *Daubentonia*, *Dactylopsila* possesses a short wide humerus with a large head. Also like *Dactylopsila*, *Daubentonia* exhibits the unique *combination* of relatively high humerofemoral caput indices and low humerofemoral length indices.

In other characteristics (e.g., the relative stoutness of the femoral shaft), *Daubentonia* deviates strongly from *Dactylopsila*, which more closely resembles its relatives such as *Phalanger*. The femur of *Dactylopsila* is flatter and wider than that of *Daubentonia*. The femur of *Daubentonia* is more slender and lemur-like, suggesting greater propensity for leaping. Indeed, while *D. madagascariensis* frequently climbs or engages in slow, quadrupedal progression on horizontal arboreal supports, it also occasionally leaps, landing on all fours. The femoral morphology of extinct as well as extant *Daubentonia* (high greater trochanter, moderately well developed femoral condyles, moderately narrow and deep patellar groove with strong borders) does not preclude deliberate leaping or short, awkward jumping, although slow quadrupedism and vertical climbing were undoubtedly the dominant elements of the extinct species' positional repertory. The SRP of *D. robusta* has its highest correlations with robust, heavy-bodied, short-limbed quadrupeds and vertical climbers such as *Ailuropoda* (the giant panda), which is more robust, and *Phascolarctos* (the koala), which is less so (Figure 6b).

Megaladapis (see Figure 7). *Megaladapis*, roughly the size of the smallest of bear species (e.g., *Helarctos*, the sun bear; 50–65 kg) but with shorter, more robust hindlimbs, was also a slow, quadrupedal climber. In many features, this giant lemur converged with *Mesopropithecus*, with whom it exhibits its highest SRP correlation. Extant animals with residual profiles most similar to *Megaladapis* are all short-limbed, heavy-bodied slow climbers with fairly long forelimbs, such as arboreal porcupines (*Coendou* and *Erethizon*) and cuscuses (*Phalanger*). While *Megaladapis* lacks the particular suite of humeral specializations associated with forelimb suspension in *Mesopropithecus* and *Palaeopropithecus*, it does appear to have used its forelimb in tension, probably in browsing and vertical climbing (Jungers, 1976, 1977). A relatively high humerofemoral length index (cf., those of *Mesopropithecus*, *Coendou*, *Erethizon*, and *Phalanger*, Table 3) is a prediction of the vertical climbing support model for animals of very large body size.

The SRPs of the three species of *Megaladapis* highlight exceptionally large femoral heads

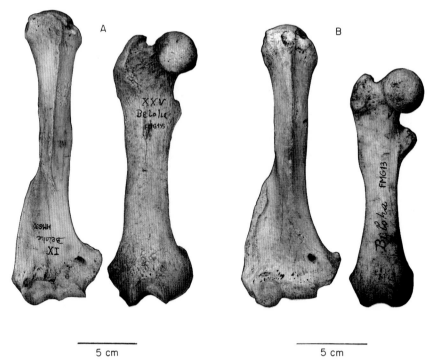

Figure 7. *Megaladapis edwardsi* (A) compared with *M. madagascariensis* (B). Right humerus (left), right femur (right). Medial right, distal bottom. Anterior view. Specimens in the collections of the Académie Malgache.

(both relative to femoral length and to the size of the humeral head); femora and humeri with large midshaft diameters; unusually anteroposteriorly flattened femoral shafts; and, femora shorter than "expected" given humeral length. They exhibit the combination of traits that distinguishes *Palaeoprophithecus* and *Bradypus* from all other animals sampled here: long humerus, short femur, small humeral head, large femoral head. But in *Megaladapis*, as in *Mesopropithecus*, it is the large femoral head that most contributes to setting these animals apart, whereas, in *Bradypus* and *Palaeopropithecus*, it is the relative elongation of the humerus. The femoral head of *Mesopropithecus* is very large relative to the size of the humeral head, as in *Coendou* and *Choloepus*, and more so than in the longer-armed *Bradypus* and *Palaeopropithecus*.

There are some striking differences in the cranial and postcranial morphology of the three species of *Megaladapis* (Vuillaume-Randriamanantena, 1982; Vuillaume-Randriamanantena & Godfrey, in prep.). *Megaladapis grandidieri* and *M. madagascariensis* appear to be scaled versions of one another (Figure 6c)—distinct from *M. edwardsi* in numerous morphological details (Figure 7).

The smaller *M. madagascariensis* and its larger sister *M. grandidieri* (an animal which overlapped in body size with *Megaladapis edwardsi*) exhibit adaptations for greater power generation in the humerus and for looser shoulder, hip and knee joints. In comparison with *Megaladapis edwardsi*, the humeri of *M. madagascariensis* and *M. grandidieri* have lower tubercles, more widely flaring brachialis flanges, and larger, more strongly curved heads. The femora also have larger heads; their greater trochanters and obturator fossae are

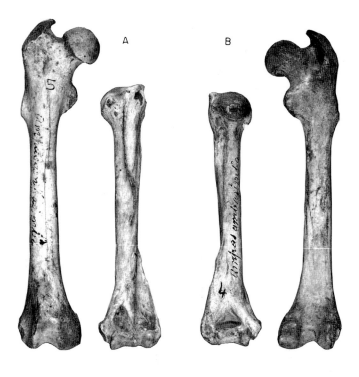

5 cm

Figure 8. Left humerus and right femur of a quadrupedal lemuroid, *Pachylemur jullyi*. A, anterior view.
Distal bottom. B, posterior view. Specimens in the collection of the Laboratoire d'Anatomie Comparée,
Muséum National d'Histoire Naturelle, Paris.

smaller, and their lesser trochanters stronger, more plate-like, and more medially
projecting. The medial and lateral femoral condyles are more anteroposteriorly
compressed, the patellar grooves shallower and more poorly bordered. They appear to
have been more versatile climbers, subjecting fore and hindlimbs to a greater variety of
tensile stresses, perhaps on a greater variety of arboreal supports. Their SRP correlations
with diggers (such as *Myrmecophaga* and *Tamandua*) and specialized hangers (such as
Choloepus and *Palaeopropithecus*) are considerably higher than are those of *Megaladapis
edwardsi*.

Pachylemur (see Figure 8). *Pachylemur*, the largest and only extinct member of the family
Lemuridae (about equal in size to *Daubentonia robusta*), was a robust, somewhat *Varecia*-like
arboroterrestrial quadruped. Its highest SRP correlations are with quadrupedal
marsupials such as *Phalanger, Trichosurus* (the brushtailed possum), and climber/leapers
such as *Dendrolagus*. Its highest lemurid SRP correlation is with *Varecia*; its lowest with
Hapalemur. As is typical of lemurids, the femur of *Pachylemur* exhibits a moderate
collodiaphyseal angle, high but not wide greater trochanter, moderately strong quadriceps
vastus lateralis crest, moderately distal lesser trochanter, narrow, moderately deep,
well-bordered patellar groove, moderately anteroposteriorly elongated femoral
condyles—in other words, characters indicative of general versatility in positional behavior

(quadrupedal walking and running, bounding, leaping, climbing). As in other lemurids, also, the humerus of *Pachylemur* has a moderately flaring brachialis flange, small capitulum, and broad trochlea. The entepicondyle is moderately well developed but not proximally extensive; deltoideus inserts on a distinctive deltoid "V" with a wide, anteriorly flattened area for clavicular fibers (Jouffroy, 1962); and, teres major and latissimus dorsi insert distally on the shaft.

Pachylemur differs from other lemurids in its much greater humeral and femoral robusticities, larger femoral head (relative to humeral head size and to femoral shaft length), and higher power-to-load arm ratios for propulsive muscles. It also displays the highest lemurid humerofemoral length index, followed by *Varecia*, then *Lemur*, and finally *Hapalemur*. *Varecia* resembles *Pachylemur* in having a relatively large femoral head. This may reflect a propensity for hindlimb suspension; *Varecia* does seem to engage in bipedal suspension more frequently than other extant lemurids (Oxnard *et al.*, in press). Leaping and bounding (in the manner of *Varecia*) may have been part of the positional repertory of *Pachylemur*, as was, certainly, slow quadrupedal climbing.

The SRPs of *Pachylemur* and *Varecia* occupy a central position *vis-à-vis* other Malagasy lemurs, with *Varecia* more similar to gracile quadrupeds and leapers, and *Pachylemur* to slower, more robust, quadrupedal climbers (Figure 6d). Deviating from *Pachylemur* in one direction (with more slender femora sporting smaller heads) are *Archaeolemur* and then *Hadropithecus* (with longer, more slender humeri), whose residual profiles signal increased reliance on terrestrial quadrupedalism. Deviating in another direction (with longer humeri but shorter, wider femora with larger heads) are *Mesopropithecus* and *Palaeopropithecus*, as well as *Megaladapis*, whose positional repertories appear to have included a greater component of slow quadrupedal climbing, and to have excluded leaping. *Lemur*, *Hapalemur*, and *Lepilemur* exhibit longer femora and shorter humeri; their positional repertories include a greater leaping component. The indrids have much longer, almost cylindrical femora, as well as shorter humeri. *Daubentonia* exhibits its most extreme modifications in the forelimb, deviating in still another direction.

Heritage considerations

Historical background: Malagasy strepsirrhine phylogeny
Humeri and femora have played a minor role in interpretations of sister relationships of Malagasy lemurs, which have generally relied, instead, on craniodental data (Cartmill, 1975; Mahé, 1976; Tattersall & Schwartz, 1974; Seligsohn, 1977; Szalay & Delson, 1979; Tattersall, 1973, 1975, 1982; Schwartz & Tattersall, 1985). Postcranial studies of extinct forms have tended to emphasize function and not sister taxon relationships.

Craniodental studies have consistently recognized at least two clades—an indroid clade consisting of archaeolemurids, palaeopropithecids and indrids, and a lemuroid clade comprising all the rest. The Cheirogaledae are increasingly accepted as the sister to lorises (Cartmill, 1975; Szalay & Delson, 1979; Szalay & Katz, 1973; Tattersall, 1982), and therefore members of the superfamily Loroidea. Beyond these interpretations, little consensus exists, based on craniodental data, concerning Malagasy primate sister relationships.

Within the Indroidea, only *Archaeolemur* and *Hadropithecus*, and *Palaeopropithecus* and *Archaeoindris*, are unequivocally regarded as sisters (Lamberton, 1929, 1934; Tattersall, 1973, 1982). *Mesopropithecus* is considered, incorrectly I believe, the sister to *Propithecus*

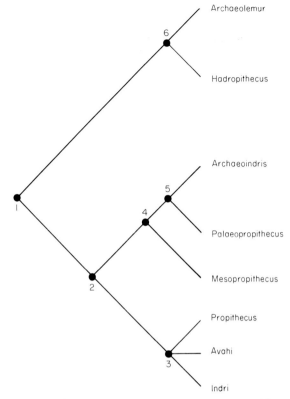

Figure 9. Phylogeny of the Indroidea. Nodes defined in Table 6 and discussed in text.

(Tattersall 1971, 1982; Tattersall & Schwartz, 1974; Mahé, 1976), and therefore an extinct indrid. There has been no consistency in interpretations the phylogeny of extant indrids, or of the three indroid families. The Palaeopropithecidae and Archaeolemuridae have both been interpreted as sisters of the indrid clade (Schwartz & Tattersall, 1985; Tattersall, 1973, 1982). *Daubentonia* has also been considered an indroid or a member of some broad "*Indri*-group" (Schwartz & Tattersall, 1985: Szalay & Delson, 1979; Tattersall & Schwartz, 1974, but see Mahé, 1976; Oxnard, 1981, 1984), although the presence of numerous autapomorphies forces this conclusion to depend on a rather small number of presumed synapomorphies.

Within the superfamily Lemuroidea, again based on presumed craniodental synapomorphies, there are few unequivocal linkages. *Lepilemur* and *Megaladapis* are often considered sister taxa (whether placed in a single family or two closely related families), and *Hapalemur* and other, non-Malagasy strepsirrhines are sometimes regarded as closely aligned with them (Schwartz & Tattersall, 1985; Tattersall, 1982; Tattersall & Schwartz, 1974). Alternatively, *Hapalemur* has been considered the sister of *Lemur*, *Varecia* and *Pachylemur* (Mahé, 1976; Petter *et al.*, 1977). The latter interpretation has gained increasing acceptance in recent years, even among earlier proponents of a *Hapalemur-Lepilemur* clade (Schwartz & Tattersall, 1985). *Lemur*, *Varecia*, and *Pachylemur* are universally regarded as comprising a single clade (Mahé, 1976; Tattersall, 1982); *Pachylemur* is believed to be the

Table 6   **Morphotypes for nodes in Figure 9**

Ancestral indroid morphotype: characters at Node 1

1. Femur longer than humerus; humerofemoral length index
   ca. 70–75.   x
2. Femoral head moderately large; femoral caput dominance
   index ca. 23–25(?).   x
3. Femoral collodiaphyseal angle moderate, ca. 125–130°.   x
4. Greater trochanter approximates or slightly exceeds level of
   head.   x
5. Crest for origin of quadriceps vastus lateralis overhangs
   anterior aspect of shaft.   x
6. Lesser trochanter strong, projects medially.   x
7. Third trochanter isolated from greater trochanter, moderately
   developed, top fifth of shaft.   x
8. Obturator fossa deep, well developed, extends below head.   x
9. Femoral shaft straight, robusticity index ca. 20–25.   x
10. Femoral epicondyles do not protrude medially and laterally
    beyond condyles.   x
11. Femoral condyles moderately anteroposteriorly expanded,
    widely separated, medial condyle broader than lateral in
    posterior view.   x
12. Patellar groove narrow and moderately deep, well bordered,
    lateral crest higher than medial.   x
13. Humeral tubercles approximate level of head.   x
14. Humeral head curvature moderate, caput dominance index
    ca. 23–25.   x
15. Latissimus dorsi and teres major insert relatively distally on
    humeral shaft; deltopectoral insertion area strong.   x
16. Brachialis flange "born" proximally on shaft; relatively narrow.   D
17. Medial epicondyle strong; projects medially.   x
18. Capitulum large.   D
19. Trochlea does not exceed width of capitulum.   D
20. Olecranon fossa moderate in width and depth.   x

*Ancestral morphotype for the common ancestor of the palaeopropithecids and the
indrids: characters at Node 2*

1. Femur longer than humerus; humerofemoral length index
   ca. 70–75.   R
2. Femoral head moderately large; femoral caput dominance
   index ca. 23–25(?).   R
3. Femoral collodiaphyseal angle moderate, ca. 125–130°.   R
4. Greater trochanter approximates or slightly exceeds level of
   head, somewhat reduced in proximodistal development.   D
5. Crest for origin of quadriceps vastus lateralis overhangs
   anterior aspect of shaft.   R
6. Lesser trochanter strong, projects medially.   R
7. Third trochanter isolated from greater trochanter, moderately
   developed, top fifth of shaft.   R
8. Obturator fossa somewhat reduced in proximodistal
   development but deep.   D
9. Femoral shaft straight, robusticity index ca. 20–25.   R
10. Femoral epicondyles do not protrude medially and laterally
    beyond condyles.   R
11. Femoral condyles moderately anteroposteriorly expanded,
    widely separated, medial condyle broader than lateral in
    posterior view.   R
12. Patellar groove narrow and moderately deep, well bordered,
    lateral crest higher than medial.   R
13. Humeral tubercles below level of head; lesser tubercle medially
    extensive.   D

Table 6—*continued*

14.	Humeral head strongly curved along sagittal axis, caput dominance index ca. 23–25.	D
15.	Latissimus dorsi and teres major insert relatively distally on humeral shaft.	R
16.	Brachialis flange "born" proximally on shaft; relatively narrow.	R(1)
17.	Medial epicondyle very strong, medially projecting and proximally extensive.	D
18.	Capitulum large.	R(1)
19.	Trochlea small; distal edge proximal to that of capitulum.	D
20.	Olecranon fossa narrow, shallow.	D

Ancestral indrid morphotype: characters at Node 3

1.	Femur much longer than humerus, humerofemoral length index ca. 55.	D
2.	Femoral head large relative to humeral head but not femoral shaft; femoral caput dominance index low (ca. 15).	D
3.	Femoral collodiaphyseal angle very low, ca. 110° or lower.	D
4.	Greater trochanter high but proximally compressed.	D
5.	Anterior crest for quadriceps vastus lateralis obliquely oriented, greatly overhangs anterior aspect of shaft.	D
6.	Lesser trochanter proximally situated, medially directed.	D
7.	Third trochanter isolated, proximally situated, top 15% of shaft.	D
8.	Obturator fossa deep though very proximal.	D
9.	Femoral diaphysis stout and straight; femoral robusticity index low (ca. 16).	D
10.	Femoral epicondyles do not protrude medially and laterally beyond condyles.	R
11.	Femoral condyles very anteroposteriorly expanded, knee massive; medial condyle broader than lateral in posterior view.	D
12.	Patellar groove very narrow and deep, borders high; lateral lip much higher than medial.	D
13.	Humeral tubercles below level of head; lesser tubercle medially extensive.	R(2)
14.	Humeral head strongly curved sagittally but relatively small; caput dominance 23 or lower.	D
15.	Latissimus dorsi and teres major insert relatively distally on humeral shaft.	R
16.	Brachialis flange "born" proximally on shaft; relatively narrow.	R(1)
17.	Medial epicondyle very strong, medially projecting and proximally extensive.	R(2)
18.	Capitulum large.	R(1)
19.	Trochlea very small, almost square; distal edge proximal to that of capitulum.	D
20.	Olecranon fossa narrow and shallow.	R(2)

Ancestral palaeopropithecid morphotype: characters at Node 4

1.	Femur relatively short, humerofemoral length index ca. 90.	D
2.	Femoral head very large, femoral caput dominance high (ca. 30).	D
3.	Femoral collodiaphyseal angle high, ca. 140°.	D
4.	Greater trochanter reduced and proximally compressed; tip below top of head.	D
5.	Crest for quadriceps vastus lateralis barely overhangs anterior aspect of shaft; adductor crest strong and distally prolonged.	D
6.	Lesser trochanter large, plate-like; protrudes strongly medially.	D
7.	Third trochanter reduced but proximodistally prolonged; exceeds top fifth of shaft.	D
8.	Proximally limited obturator fossa narrow and constricted.	D

Table 6—*continued*

9. Femoral diaphysis anteroposteriorly flattened, robust (robusticity index ca. 27); shaft waisted at level of third trochanter, widens distally to epicondyles.	D
10. Femoral epicondyles slightly wider than condyles.	D
11. Femoral condyles narrowly separated and anteroposteriorly compressed; medial condyle slightly wider than lateral in posterior view.	D
12. Patellar groove narrow, shallow, and low; medial and lateral reduced.	D
13. Humeral tubercles below level of head; lesser tubercle medially extensive.	R(2)
14. Humeral head wide as well as strongly curved sagittally, caput dominance unchanged.	D
15. Latissimus dorsi and teres major insert relatively distally on humeral shaft; deltopectoral insertion area distally extensive.	D
16. Brachialis flange "born" proximally on shaft; relatively narrow.	R(1)
17. Medial epicondyle very strong, medially projecting and proximally extensive.	R(2)
18. Capitulum large.	R(1)
19. Trochlea relatively small; distal edge proximal to that of capitulum.	R(2)
20. Olecranon fossa narrow and shallow.	R(2)

Ancestral palaeopropithecine morphotype: characters at Node 5

1. Femur very short, humerofemoral length index >100.	D
2. Femoral head very large, femoral caput dominance index well above 30.	D
3. Femoral collodiaphyseal angle very high, >150°.	D
4. Greater trochanter greatly reduced; tip much lower than top of head.	D
5. Crest for quadriceps vastus lateralis negligible; no rugosity for medial vastus.	D
6. Lesser trochanter very large, plate-like. distal, and directed medially to attain or exceed internal margin of head.	D
7. Third trochanter reduced to elongated rugosity or groove; distally emplaced.	D
8. Obturator fossa proximally limited; barely discernible.	D
9. Femoral diaphysis very anteroposteriorly flattened robusticity index well above 30; shaft waisted at level of third trochanter, then widens distally to epicondyles.	D
10. Femoral epicondyles very strongly developed.	D
11. Femoral condyles very narrowly separated and anteroposteriorly compressed; medial and lateral condyles more equal in width.	D
12. Patellar groove exceptionally narrow, shallow, and low, with poorly defined borders; medial and lateral borders approximately coequal in height.	D
13. Humeral tubercles well below level of head.	D
14. Humeral head enlarged in proximal view, strongly curved along sagittal axis, although caput dominance is low (21–22) (due to elongation of shaft).	D
15. Latissimus dorsi and teres major insert relatively distally on humeral shaft; deltopectoral insertion area very distally prolonged.	D
16. Brachialis flange "born" proximally on shaft; relatively narrow.	R(1)
17. Medial epicondyle very strong, medially projecting and proximally extensive.	R(2)
18. Capitulum large.	R(1)

Table 6—*continued*

19. Trochlea relatively small; distal edge proximal to that of
    capitulum.   R(2)
20. Olecranon fossa narrow, fairly shallow, although deep enough
    to accommodate weak olecranon process in full extension.   D

Ancestral archaeolemurid morphotype: characters at Node 6

1. Femur longer than humerus, humerofemoral length index
   ca. 80.   D
2. Femoral caput dominance index ca. 23 (slightly reduced from
   primitive condition?)   D?
3. Femoral collodiaphyseal angle under 120°; neck long.   D
4. Greater trochanter wide and proximodistally elongated; tip
   well surpasses head.   D
5. Crest for quadriceps vastus lateralis proximodistally elongated;
   medial vastus rugosity distal.   D
6. Lesser trochanter knob-like, posteromedially directed, but
   distally emplaced.   D
7. Third trochanter very distal on shaft (at one-fourth total length)   D
8. Obturator fossa proximodistally elongated, well excavated.   D
9. Femoral diaphysis very slightly anteriorly bowed; robusticity
   ca. 27.   D
10 Femoral epicondyles not medially or laterally expanded
    beyond condyles.   R
11. Femoral condyles set wide apart; intercondylar groove wide;
    condyles moderately anteroposteriorly expanded.   D
12. Patellar groove very wide, fairly shallow, and low; crests
    moderate, lateral higher than medial.   D
13. Greater tubercle large, well exceeds head, protrudes sharply
    anteriorly; lesser tubercle at level of head, reduced.   D
14. Humeral head wide but flat; caput dominance ca. 25.   D
15. Latissimus dorsi and teres major insert relatively proximally on
    humeral shaft relative to strong deltopectoral insertion area.   D
16. Brachialis flange "born" proximally on shaft; very narrow.   D
17. Medial epicondyle strong; projects posteromedially.   D
18. Capitulum large.   R(1)
19. Trochlea cone-shaped, enlarged; distal edge angled downward
    from lateral to medial border.   D
20. Olecranon fossa very large, wide, and deep.   D

x Reconstructed from outgroup morphology
D Derived (at least in extreme form) at given node
R Retained from earlier, unspecified node
R(1) Retained from node numbered in parentheses

sister of *Varecia*, and is sometimes considered its congener (Seligsohn, 1977; Seligsohn & Szalay, 1974; Walker, 1974).

Humeral and femoral synapomorphies can help clarify sister relationships of Malagasy strepsirrhines, especially among the Indroidea (Figure 9, Table 6).[2] Derived characters distinguish the palaeopropithecids and indrids, making it likely that palaeopropithecids, and not archaeolemurids, are the sister of the indrid clade. Within the palaeopropithecid-indrid clade, *Mesopropithecus* forms the sister of a *Palaeopropithecus-Archaeondris* clade (Godfrey, 1986), and is best considered a primitive palaeopropithecid (rather than an indrid), distinguishable from *Archaeoindris* and

[2] I am preparing a review of craniodental characters (Godfrey, in prep) which supports the cladogram presented in Figure 9, and further resolves the branching sequence at Node 3. The latter is not possible without craniodental data, and is omitted from this paper.

Palaeopropithecus at the subfamilial level. The Indridae (*Propithecus*, *Indri*, and *Avahi*) are distinguished by characters heretofore treated as primitive but probably derived. *Daubentonia* is difficult to place; its phylogenetic linkages will not be considered in this paper.

Humeral and femoral data are less helpful in sorting out lemuroid relationships. There are few evident megaladapid-lepilemurid synapomorphies. However, humeral and femoral synapomorphies do define sister relationships *within* the Megaladapidae, and corroborate a *Hapalemur-Lemur* link.

Lepilemur has been considered a good model for the common ancestor of the Indroidea and the Lemuroidea—indeed, close to the morphotype of the common ancestor of all Malagasy strepsirrhines (Godfrey, 1977). The not-insignificant postcranial similarities of *Lepilemur* to modern indrids seem to support the notion that some form of vertical clinging and leaping was symplesiomorphic for Malagasy strepsirrhines. However, if the common ancestors of both the palaeopropithecid-indrid and indroid clades were quadrupeds, then shared lepilemurid-indrid adaptations for leaping are not primitive retentions, but are convergences.

Node 1. Ancestral indroid: generalized quadrupedal climber

Perhaps the reason why indroid postcrania have been virtually ignored in reconstructing indroid phylogeny is that they are extraordinarily diverse; their nested hierarchical pattern of synapomorphies is partially hidden within a pattern of extreme functional divergence (Figures 10–13). Furthermore, there are few postcranial synapomorphies linking the Archaeolemuridae, Palaeopropithecidae, and Indridae; instead, this linkage depends on numerous craniodental traits. The ancestral morphotype of the humerus and femur of Indroidea must be largely reconstructed from the morphology of neighboring outgroups (Figures 7–8; Table 6, Node 1).

For example, consider the mean values of the humerofemoral length indices of indroids: *Avahi* (48), *Propithecus* (53), *Indri* (54), *Archaeolemur* (82), *Hadropithecus* (116), *Mesopropithecus* (90), *Archaeoindris* (over 100), and *Palaeopropithecus* (152). These can be contrasted with figures ranging between 60 and 85 in the Lemuridae, Cheirogaleidae, and Daubentoniidae. The only neighboring outgroup that exhibits nearly as great a range is the lepilemurid-megaladapid clade, with values ranging from the high 50s (*Lepilemur*) to about 110 (*Megaladapis*). The most plausible inference that can be drawn from these data is that extreme values, whether low or high, are derived, and values of around 70–75 are primitive. This would mean that, of indroids, *Archaeolemur* exhibits the humerofemoral ratio closest to that of the common ancestor at Node 1. Given the dominant pattern of synapomorphies at later nodes, deriving indroids from a modern indrid-like ancestor would require repeated independent reversals of a highly specialized condition.

Similar arguments can be made for other characters. Indrids have stout, long, and narrow femoral shafts; palaeopropithecids have short, wide, and anteroposteriorly compressed femoral shafts. Indrids have massive knees with anteroposteriorly elongated femoral condyles and deep patellar grooves bordered by prominent crests; palaeopropithecids have anteroposteriorly compressed femoral condyles and shallow, poorly bordered patellar grooves. Indrids have low femoral collodiaphyseal angles (little above 100°); *Palaeopropithecus* and *Archaeoindris* have exceptionally high femoral collodiaphyseal angles (above 150°). For each of these traits, indrid and palaeopropithecid morphotypes represent the extremes for Malagasy strepsirrhines; nearest outgroups, including the Archaeolemuridae, exhibit an intermediate, and probably primitive, condition.

But archaeolemurids are not consistently primitive in their morphology. In many characters, they simply deviate from palaeopropithecids, indrids, and neighboring outgroups in a unique direction. Thus, for example, their femora display high, wide, and proximodistally elongated greater trochanters that contrast with the proximally compressed greater trochanters of indrids, and the compressed and reduced greater trochanters of palaeopropithecids. These represent deviations in opposite directions from an intermediate (and probably primitive) condition exhibited by lemurids, lepilemurids, and cheirogaleids.

While some indrid likenesses (stout femoral shaft, anteroposterior development of the femoral condyles, low femoral collodiaphyseal angle) are manifested in other specialized leapers such as *Lepilemur* and, to some extent, *Hapalemur*, details of the femoral morphology of leapers are distinct. For most traits, *Lepilemur* more closely resembles lemurids (for example, in the configuration of the greater trochanter and relative placement of the vasti). Specialized leaping seems to have evolved independently in Lepilemuridae, Indridae, and some Lemuridae, and quadrupedalism was only truly abandoned in Indridae.

The few probable indroid synapomorphies that do exist do not contravene the hypothesis that the common ancestor at Node 1 was a quadrupedal climber. The humeral capitulum is stout in all indroids, indicating increased weight support or force transmission through the radius. The brachialis flange is relatively narrow, flaring less than in lemurids, cheirogaleids, megaladapids, daubentonilds, and lepilemurids. This is probably due to greater habitual elbow extension in indroids. Humeri that have widely projecting epicondyles and flaring brachialis flanges are well adapted for habitual elbow flexion. However, the brachialis flange is generally "born" proximally on the shaft of indroids. This proximal (but not lateral) development of the brachialis flange is not a sign of reduction of brachioradialis and brachialis, which are massive and originate quite proximally on the humeri of sloths and indrids (Jouffroy, 1962; Mendel, 1985*b*), and were undoubtedly similarly massive in *Palaeopropithecus* (which exhibits an exceptionally narrow brachialis flange). Whatever the habitual range of motion at the elbow, a proximal origin of brachialis and brachioradialis indicates that these animals are well adapted to resist elbow extension against the pull of gravity (Dagosto, 1983; Miller, 1943).

Node 2. Indrid-palaeopropithecid ancestor: generalized quadrupedal climber/hanger

Indrid-palaeopropithecid synapomorphies are primarily humeral rather than femoral. (There are other postcranial indrid-palaeopropithecid synapomorphies not considered here; for example, elongated, hook-like hands and feet.) One possible *femoral* synapomorphy is an enlarged head. Both indrids and palaeopropithecids have mobile hip joints—large femoral heads that fit into shallow acetabula. But this represents a minor deviation from the condition present in lemurids and lepilemurids. It is likely that an enlarged femoral head was symplesiomorphic for Malagasy strepsirrhines, and that frequent femoral abduction and hindlimb suspension were primitive corollaries of that condition. Such capacities may have been enhanced in the common ancestor of indrids and palaeopropithecids, although not to the degree exhibited by *Mesopropithecus*, *Palaeopropithecus*, and *Archaeondris*.

Humeral synapomorphies are specializations for climbing and hanging (Table 6, Node 2). Despite their small size, indrid humeri and, indeed, entire forelimbs, are well adapted

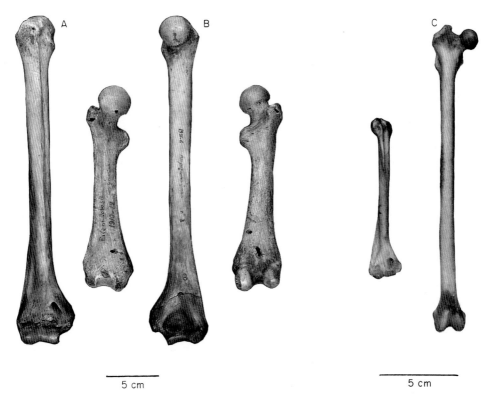

5 cm   5 cm

Figure 10. Right humeri and femora of *Paleopropithecus maximus* (A, B) compared with *Propithecus verreauxi* (C). A and C, anterior view, medial right, distal bottom. B, posterior view. Fossil specimens in the collection of the Laboratoire d'Anatomie Comparée, Muséum National d'Histoire Naturelle, Paris.

for tensile stress (Oxnard, 1973, 1984; Roberts, 1974; Roberts & Davidson, 1975). The humeral head, unknown in *Archaeoindris*, is large, strongly curved sagitally, and dominant over greater and lesser tubercles in all other indrids and palaeopropithecids. The trochlea is reduced and boxlike. The entepicondyle is both medially and proximally enlarged, thereby increasing the mechanical advantage of pronator teres and the digital flexors. All indrid and palaeopropithecid humeral synapomorphies are tree sloth convergences.

Node 3. Ancestral indrid: non-quadrupedal hanger/leaper
In comparison with *Mesopropithecus* and *Palaeopropithecus*, extant indrids exhibit humeral suspensory adaptations in miniature. Despite adaptations for tensile stress, indrid humeri are dwarfed by exceptionally long femora.

Femoral elongation and concomitant modifications for thigh-powered leaping appear to have evolved in an indrid ancestor already engaging in hindlimb suspension—an animal with a strong iliopsoas for flexing the trunk on the thigh, a mobile hip joint well adapted for abduction, flexion and suspension, and pedal adaptations for grasping and climbing (Gebo & Dagosto, 1988). Rather than depending on pedal thrust (as do tarsiers, galagonids, and non-indrid Malagasy VCLs), the descendant indrids achieved vertical saltation via forceful extension at the hip and knee, accompanied by strong medial rotation at the hip (Godfrey, 1977). It is this femoral elongation that renders normal quadrupedal ground progression virtually impossible. The ancestral indrid could climb, hang, leap, or hop, but not walk or run on all fours. Modifications for thigh-powered leaping partially masked adaptations for hindlimb suspension, greatly reducing, for example, the apparent dominance of the femoral head on the shaft, as reflected in the femoral caput dominance index (Table 3). Nevertheless, indrids retain some hindlimb (as well as forelimb) adaptations for hanging: e.g., large femoral head articulating with shallow acetabulum; strong, medially directed lesser trochanter; elongated hook-like feet (though not as elongated or hook-like as those of *Palaeopropithecus*). They remain capable of sloth-like (quadrupedal), bipedal, and bimanual suspension, and readily adopt suspensory postures in feeding, and occasionally, in locomotion. Extant indrids truly are *hanger*/leapers.

Vertical saltation in indrids depends on myological specializations that include hypertrophy of the knee extensors (especially quadriceps vastus lateralis), iliopsoas, and medial rotators of the leg (the medial hamstrings) (Godfrey, 1977; Jouffroy, 1962). Rectus femoris originates quite proximally on the innominate, and inserts, along with the vasti, well anterior to the fulcrum of the knee. The peripheral vasti themselves have proximal origins. Power-to-load arm relationships of the hamstrings, gluteals, and iliopsoas shift to effect rapid take-off and hindlimb recovery. These myological specializations are manifested in a suite of derived osteological modifications (Table 6, Node 3).

Node 4. Ancestral palaeopropithecid: moderately specialized, slow climber/hanger

While *Mesopropithecus* was not as specialized as either *Palaeopropithecus* or *Archaeoindris*, its morphology reflects what might be presumed to represent an early stage of deviation in that direction. The belief that *Mesopropithecus* is the sister taxon to *Propithecus* is contradicted by numerous femoral palaeopropithecid synapomorphies. No humeral or femoral features specifically link *Mesopropithecus* with *Propithecus*, nor even with indrids as a group. All features shared with indrids are also shared, sometimes in more extreme form, by *Palaeopropithecus*, and were undoubtedly present at Node 2 (see Table 6, Node 4).

Like *Archaeoindris* and *Palaeopropithecus*, *Mesopropithecus* exhibits a femur that is shorter and more robust than those of archaeolemurids, and much more so than those of indrids. The size of the femoral head (relative to shaft length) also deviates from the lemur norm in the direction of *Palaeopropithecus* and *Archaeondris*.

The femoral collodiaphyseal angle of *Mesopropithecus* (approximately 140°), while not as high as in *Palaeopropithecus* or *Archaeondris* (>150°), is nevertheless higher than those of all other indroids. Associated with this shift in collodiaphyseal angle is a greater medial protrusion of the lesser trochanter. The greater trochanter of *Mesopropithecus* is low and reduced, although not to the degree exhibited by *Archaeoindris* and, especially, *Palaeopropithecus*. The third trochanter is reduced to an elongated rugosity or groove, generally only slightly more prominent than that of *Palaeopropithecus*, and less protrusive than those of archaeolemurids and extant indrids. The obturator fossa, while not nearly as reduced as in the palaeopropithecines, is nevertheless constricted, again deviating from the lemur norm in the direction of *Palaeopropithecus* and *Archaeoindris*.

The femoral shaft of *Mesopropithecus* is anteroposteriorly compressed, as in *Palaeopropithecus*. The fossae for tibial as well as fibular collateral ligaments are both enlarged, as in *Palaeopropithecus*, and the femoral condyles are narrowly separated and greatly anteroposteriorly compressed. As in *Palaeopropithecus*, the femoral shaft is narrow at the level of the third trochanter, and then widens to the medial and lateral epicondyles. The patellar groove itself is narrow, shallow and low, although its medial and lateral rims are

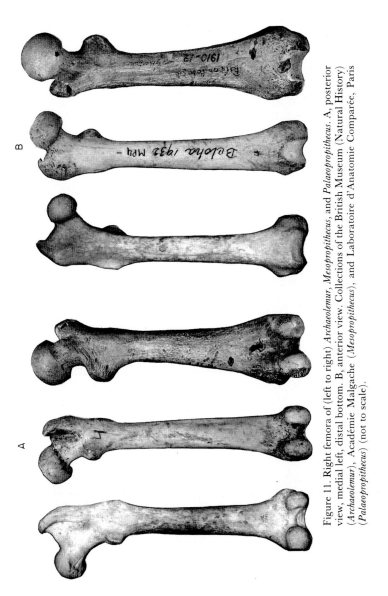

Figure 11. Right femora of (left to right) *Archaeolemur*, *Mesopropithecus*, and *Palaeopropithecus*. A, posterior view, medial left, distal bottom. B, anterior view. Collections of the British Museum (Natural History) (*Archaeolemur*), Académie Malgache (*Mesopropithecus*), and Laboratoire d'Anatomie Comparée, Paris (*Palaeopropithecus*) (not to scale).

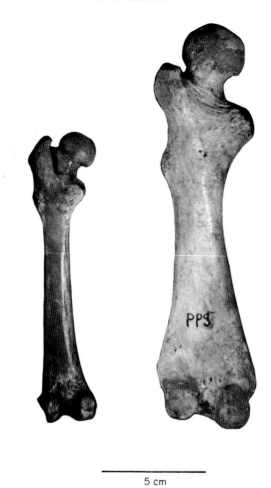

5 cm

Figure 12. Left femora of *Mesopropithecus* (left) and robust specimen of *Palaeopropithecus* (right). Posterior view, medial right, distal bottom (collection of the Académie Malgache).

stronger than in *Palaeopropithecus*. In sum, the morphology of the femur of *Mesopropithecus* suggests great hip and knee mobility, though not to the extreme exhibited by *Palaeopropithecus*.

The humerus reveals fewer palaeopropithecid synapomorphies, because synapo-morphies exist at the broader level of the palaeopropithecid-indrid clade, and because *Mesopropithecus* exhibits several humeral autapomorphies (very strong entepicondyle, strong triceps-vastus lateralis ridge). These appear to be related to greater reliance on slow climbing and lesser reliance on below-the-branch activities. Nevertheless, even on the humerus, palaeopropithecid synapomorphies exist. Like that of *Palaeopropithecus*, the humeral head is wide and dominant in proximal view. The deltopectoral crest is distally elongated in *Mesopropithecus*, although not to the extent of *Palaeopropithecus* or *Archaeoindris*. Most striking is the increase in the relative length of the humerus. *Mesopropithecus* deviates from the lemur norm in the direction of *Archaeondris* and *Palaeopropithecus*.

Node 5. Ancestral palaeopropithecine: highly specialized, sloth-like, climber/hanger
Femoral synapomorphies of *Archaeoindris* and *Palaeopropithecus* are striking (Table 6, Node 5). The shaft is very short, wide, and anteroposteriorly compressed (especially in *Palaeopropithecus*); the head is huge. The collodiaphyseal angle is exceptionally high—greater than 150°—shifting laterally the line of weight transmission (the line connecting the center of the head, orthogonally, with the bicondylar plane). The greater trochanter is greatly reduced and closely approximates the head in superior view. Its tip for piriformis is well below the top of the head. The anterior border of the greater trochanter, which supports the origin of quadriceps vastus lateralls, is less protrusive than in leapers; it barely protrudes anteriorly beyond the anterior aspect of the shaft. The third trochanter is greatly reduced. The obturator fossa is also very reduced, forming a mere depression. The lesser trochanter for insertion of iliopsoas, on the other hand, is very large, plate-like, distally positioned on the shaft, and strongly medially directed. Its medial border extends as far medially as the internal margin of the femoral head. The patellar groove is exceptionally narrow, shallow, and low, with poorly defined borders. The femoral condyles are anteroposteriorly compressed, narrowly separated, and flanked by wide epicondyles (increasing the mechanical advantage of rotators of the lower leg).

While no complete adult humerus of *Archaeoindris* is known, the available fragment reveals humeral synapomorphies as well (Vuillaume-Randriamanantena, in press). The most obvious is the lengthening of the humeral shaft to exceed the femur, a feature which reaches its most extreme condition in *Palaeopropithecus*. The deltopectoral crest is not well delineated, but it is distally extensive in both. The distal humerus is wider in the more massive *Archaeoindris* than in *Palaeopropithecus*, but otherwise similar. The proximal portion of the humerus of *Archaeoindris* is unknown. The proximal humerus of *Palaeopropithecus* exhibits specializations for increased use in tension. It differs from *Mesopropithecus* and indrids in having a larger, more rounded head, even lower tubercles, and a deeper, more cranially oriented infraspinatus fossa.

Node 6. Ancestral archaeolemurid: terrestrial (?) quadruped
The archaeolemurids share neither the forelimb suspensory adaptations of palaeopropithecids and indrids, nor the ricochetal leaping specializations of indrids. Instead, they exhibit a suite of unique, derived postcranial traits that signal an evolutionary trend in a direction distinct from those of other indroids—involving increased pronograde quadrupedalism and, probably, increased terrestriality (Table 6, Node 6).

Archaeolemurid synapomorphies underscore their quadrupedalism. The neck of the femur is longer than is usual for lemurs, and it is anteromedially directed. The collodiaphyseal angle is relatively low. The femoral head is relatively small; the acetabulum into which it fits, relatively deep. The tip of the greater trochanter well surpasses the head. In anterior view, the proximal femur is wide; the greater trochanter is widely separated from the head, and the surface area for attachment of gluteus minimus, scansorius, and gluteus medius is proximodistally elongated. The associated crest for quadriceps vastus lateralis does not overhang the anterior face of the shaft; nor is it proximally compressed into nearly horizontal orientation as it is in extant indrids. Instead, it, too, is elongated proximodistally, bordering the insertion area for gluteus medius, and turning medialward for a short distance proximally. The crest marking the proximal-most attachment of quadriceps vastus medialis is also quite distal in archaeolemurids and rectus femoris crosses the hip closer to its axis than in indrids.

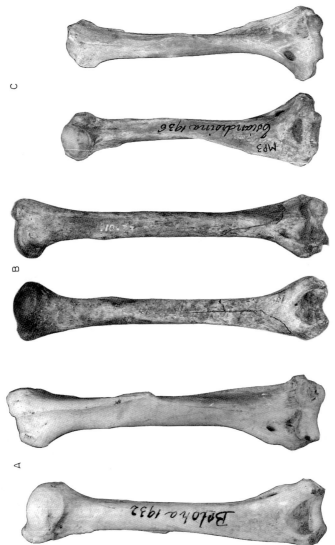

Figure 13. Left humeri of *Archaeolemur* (A), *Hadropithecus* (B) and *Mesopropithecus* (C). Left posterior view, medial right, distal bottom. Right anterior view. Collections of the Académie Malgache (*Archaeolemur* and *Mesopropithecus*) and Paris Muséum Nationale d'Histoire Naturelle (*Hadropithecus*) (not to scale). Note greater robusticity of *Archaeolemur* cf. *Hadropithecus*.

The archaeolemurid obturator fossa is unusually large and deep; it extends well distal to the inferior margin of the head. The lesser trochanter (for iliopsoas and, posteriorly, quadratus femoris) is knob-like (rather than plate-like), posteromedially rather than medially directed (enhancing the extensor function of quadratus femoris), and distally emplaced. The third trochanter is also distal on the shaft, and not as reduced as that of palaeopropithecids. The patellar groove is wide, shallow, and low. Its crests are weakly developed though not as ill-formed as those of the palaeopropithecids, including *Mesopropithecus*. The decreased development of the patellar crests, increased width and decreased depth and height of the groove itself (cf., lemurids and lepilemurids as well as indrids), are derived archaeolemurid traits. The femoral condyles are set wide apart in archaeolemurids; the intercondylar groove is wide. Anterior bowing of the femoral shaft, while slight, is, nevertheless, derived.

The archaeolemurid humerus is distinguished by the great height and large size of the greater tubercle, and relative reduction of the lesser tubercle (especially in *Hadropithecus*). The greater tubercle protrudes sharply anteriorly in superior view, well surpassing the head of the humerus in anterior view. Latissimus dorsi inserts more proximally (especially in *Hadropithecus*) than is typical for lemurs. Teres major also inserts proximally and does not overlap the distal extent of deltoideus and pectoralis as is common in other lemurs. Another humeral synapomorphy is the posteromedial deflection of the entepicondyle. In contrast with palaeopropithecids and indrids, the archaeolemurid trochlea is large and cone-shaped; the olecranon fossa is wide and deep.

Synopsis

The evidence of the humerus and the femur fails to support the notion that indrid leaping specializations are primitive retentions. The nearest sister to the Indridae is the Palaeopropithecidae, and the most primitive palaeopropithecid (*Mesopropithecus*) was clearly a quadrupedal climber/hanger. Derived palaeopropithecid-indrid traits are suspensory adaptations that must have constrained the later evolution of saltation in indrids. Quadrupedal adaptations of Archaeolemuridae also suggest that the common ancestor of the Indroidea was a generalized quadrupedal climber. The virtual absence of quadrupedalism in indrids appears to be related to a *unique* and extraordinary femoral elongation that evolved in association with thigh- rather than foot-powered leaping.

It appears that indroid phylogeny entailed a diversification of three specialized families from a more generalized common ancestor. At Node 3 (and Node 3 only), a specialized form of saltation evolved—one that depended on the prior existence of specializations for climbing and hanging. Ancestors at Nodes 4, 5, and 6, as well as Nodes 1 and 2, were not vertical clingers and leapers.

Conclusions

1. Climbing and slow quadrupedalism were important elements of the positional repertories of many of the *extinct* lemurs of Madagascar. None of the extinct Malagasy lemurs were specialized leapers; *Megaladapis, Mesopropithecus, Palaeopropithecus,* and *Archaeoindris* undoubtedly did not leap at all.

2. The common ancestor of the Indroidea was not a vertical clinger and leaper, but a versatile and probably arboroterrestrial quadruped with limb proportions and a positional repertory not very different from those of *Varecia* or *Pachylemur*. An initial split divided the

Indroidea into two clades, one of which (the Archaeolemuridae) specialized in terrestrial quadrupedalism while the other (the palaeopropithecid-indrid clade) specialized in slow quadrupedal climbing and hanging. The latter subsequently split into two clades, one of which (the Palaeopropithecidae) sacrificed rapid locomotion of any sort to perfect deliberate hanging skills, the other of which (the Indridae) sacrificed quadrupedalism to develop a new form of "vertical clinging and leaping" while retaining sloth-like hanging skills.

3. *Mesopropithecus* is a primitive palaeopropithecid and not an indrid.

Acknowledgements

I would like to thank E. Strasser and M. Dagosto for inviting me to prepare this paper and F. S. Szalay and two anonymous reviewers for their constructive comments on the submitted version. This work would not have been possible without the assistance and support of numerous people, to whom I am greatly indebted. I thank the curators, heads, and staffs of the institutions that generously made materials available to me, especially: C. Rabenoro, Président, Académie Malgache; B. Rakotosamimanana, Ministère de l'Enseignement Supérieur, Madagascar; R. Ralaiarison-Raharizelina, Service de Paléontologie, Université de Madagascar; M. Rutzmoser, Mammals Department, Museum of Comparative Zoology, Harvard University; R. Thorington, Mammal Division, Smithsonian Institution; G. Musser, Department of Mammalogy, American Museum of Natural History: C. Smeenk, Mammals Department, Rijksmuseum van Natuurlijke Historie; T. Molleson, I. Bishop, P. Jenkins, and P. Andrews, Mammals and Paleontology Departments, British Museum (Natural History); J. Anthony, R. Saban, J. Roche, M. Tranier, F. and G. Petter, the late A. Gordon (and many others, too numerous to mention), Laboratoire d'Anatomie Comparée, Oiseaux et Mammifères, and Institut de Paléontologie, Muséum National d'Histoire Naturelle, Paris; E. Simons and P. Chatrath, Duke University Primate Center. Photographs were taken by P. Godfrey, A. Gordon, and P. Randriamanantsoa.

Warm thanks go to my colleagues M. Vuillaume-Randriamanantena, G. Albrecht, M. Sutherland, E. Simons, R. Dewar, and I. Tattersall, for stimulating lemur discussions over the past few years. Further development of arguments made here will be published in joint articles by M. Vuillaume-Randriamanantena and myself, and I gratefully acknowledge her role in developing some of these ideas. M. Sutherland was responsible for developing the statistical programs used in this research. H. Ball generously donated some vertebral column data used in Figure 4.

This research was carried out under the protocol of accord for collaborative research (Duke University Primate Center and the Université de Madagascar). I thank the Ministre de l'Enseignement Supérieur de Madagascar, I. Rakoto, and the Recteur de l'Université de Madagascar, who permitted the realization of this protocol. This research was funded, in part, by a Wenner-Gren Foundation grant-in-aid of research (no. 4429), a Mary Ingraham Bunting Science Scholar Fellowship, and a Faculty Research Grant from the University of Massachusetts, Amherst.

References

Aiello, L. C. (1981*a*). Locomotion in the Miocene Hominoidea. In (C. B. Stringer, Ed.) *Aspects of Human Evolution* (Symposia for the Study of Human Biology Vol. 21), pp. 63–97. London: Taylor and Francis.

Aiello, L. C. (1981*b*). The allometry of primate body proportions. *Symp. Zool. Soc. Lond.* **48,** 331–358.

Anderson, J. F., Hall-Martin, A., & Russell, D. A. (1985). Longbone circumference and weight in mammals, birds and dinosaurs. *J. Zool., Lond.* **207,** 53–61.

Burr, D. B. (1980). The relationships among physical, geometrical and mechanical properties of bone, with a note on the properties of non-human primate bone. *Yrbk. Phys. Anthrop.* **23,** 109–146.

Burr, D. B., Piotrowski, G. & Miller, G. J. (1981). Structural strength of the macaque femur. *Am. J. phys. Anthrop.* **54,** 305–319.

Burr, D. B., Piotrowski, G., Martin, R. Bruce & Cook, P. N. (1982). Femoral mechanics in the lesser bushbaby (*Galago senegalensis*): Structural adaptations to leaping in primates. *Anat. Rec.* **202,** 419–429.

Carleton, A. (1936). The limb bones and vertebrae of the extinct lemurs of Madagascar. *Proc. Zool. Soc. Lond.* **106,** 281–307.

Carleton, A. (1937). Note on the osteology of *Palaeopropithecus*. *Proc. Zool. Soc. Lond.* **107,** 295–297.

Cartmill, M. (1972). Arboreal adaptations and the origin of the order Primates. In (R. Tuttle, Ed.) *The Functional and Evolutionary Biology of Primates*, pp. 97–122. Chicago: Aldine-Atherton.

Cartmill, M. (1974*a*). *Daubentonia, Dactylopsila*, woodpeckers and klinorhynchy. In (R. D. Martin, G. A. Doyle & A. C. Walker, Eds) *Prosimian Biology*, pp. 655–670. London: Duckworth.

Cartmill, M. (1974*b*). Pads and claws in arboreal locomotion. In (F. A. Jenkins, Jr., Ed.) *Primate Locomotion*, pp. 45–84. New York: Academic Press.

Cartmill, M. (1975). Strepsirhine basicranial structures and the affinities of the Cheirogaleidae. In (W. P. Luckett & F. S. Szalay, Eds) *Phylogeny of the Primates: A Multidisciplinary Approach*, pp. 313–354. New York: Plenum Press.

Cartmill, M. (1985). Climbing. In (M. Hildebrand, D. M. Bramble, K. F. Liem & D. B. Wake, Eds) *Functional Vertebrate Morphology*, pp. 73–88, 386–389. Cambridge, MA: Belknap.

Charles-Dominique, P. & Martin, R. D. (1970). Evolution of lorises and lemurs. *Nature* **227,** 257–260.

Coombs, M. C. (1983). Large mammalian clawed herbivores: a comparative study. *Trans. Am. phil. Soc.,* Vol. 73 (Part 7).

Currey, J. D. (1984). *The Mechanical Adaptations of Bones*. Princeton: Princeton University Press.

Dagosto, M. (1983). Postcranium of *Adapis parisiensis* and *Leptadapis magnus* (Adapiformes, Primates): Adaptational and phylogenetic significance. *Folia primatol.* **41,** 49–101.

Dawson, T. J. & Dagabriele, R. (1973). The cuscus (*Phalanger maculatus*)—a marsupial sloth? *J. Comp. Phys.* **83,** 41–50.

Gebo, D. & Dagosto, M. (1988). Foot anatomy, climbing, and the origin of the Indriidae. *J. hum. Evol.* **17,** 135–154.

Godfrey, L. R. (1977). Structure and function in *Archaeolemur* and *Hadropithecus* (subfossil Malagasy lemurs): the postcranial evidence. Ph.D. Dissertation, Harvard University.

Godfrey, L. R. (1986). What were the subfossil indriids of Madagascar up to? *Am. J. phys. Anthrop.* **69,** 204 (abstract).

Godfrey, L. R. (in prep.) Evolution and systematics of the Indroidea.

Godfrey, L. R. & Sutherland, M. R. (in prep.). Descriptive allometry: models and muddles.

Hildebrand, M. (1967). Symmetrical gaits of primates. *Am. J. phys. Anthrop.* **26,** 119–130.

Hildebrand, M. (1974). *Analysis of Vertebrate Structure*. New York: John Wiley and Sons.

Hildebrand, M. (1985). Digging of quadrupeds. In (M. Hildebrand, D. M. Bramble, K. K. Liem & D. B. Wake, Eds) *Functional Vertebrate Morphology*, pp. 89–109, 389–391. Cambridge, MA: Belknap.

Jenkins, F. A. Jr & Camazine, S. M. (1977). Hip structure and locomotion in ambulatory and cursorial carnivores. *J. Zool. Lond.* **181,** 351–370.

Jenkins, P. D. (1987). *Catalogue of Primates in the British Museum (Natural History) and Elsewhere in the British Isles. Part IV. Suborder Strepsirrhini, including the Subfossil Madagascan Lemurs and Family Tarsiidae*. London: British Museum (Natural History).

Jolly, C. J. (1970). *Hadropithecus*: a lemuroid small-object feeder. *Man* **5,** 619–626.

Jouffroy, F. K. (1962). La musculature des membres chez les lémuriens de Madagascar: étude descriptive et comparative. *Mammalia* **26** (Suppl. 2), 1–326.

Jouffroy, F. K. (1963). Contribution à la connaissance du genre *Archaeolemur*, Filhol 1895. *Annals Paléont.* **49,** 129–155.

Jouffroy, F. K. & Lessertisseur, J. (1979). Relationships between limb morphology and locomotor adaptations among prosimians: An osteometric study. In (M. E. Morbeck, H. Preuschoft & D. N. Gomberg, Eds) *Environment, Behavior, and Morphology: Dynamic Interactions in Primates*, pp. 143–182. New York: Gustav Fischer.

Jungers, W. L., Jr. (1976). Osteological form and function: the appendicular skeleton of *Megaladapis*, a subfossil prosimian from Madagascar (Primates, Lemuroidea). Ph.D. Dissertation, University of Michigan.

Jungers, W. L., Jr. (1977). Hindlimb and pelvic adaptations to vertical climbing and clinging in *Megaladapis*, a giant subfossil prosimian from Madagascar. *Yrbk. Phys. Anthrop.* **20** (for 1976), 508–524.

Jungers, W. L., Jr. (1980). Adaptive diversity in subfossil Malagasy prosimians. *Z. morph. Anthrop.* **71,** 177–186.

Jungers, W. L., Jr. (1984). Scaling of the hominoid locomotor skeleton with special reference to the lesser apes. In (H. Preuschoft, D. Chivers, W. Brockelman & N. Creel, Eds) *The Lesser Apes: Evolutionary and Behavioral Biology*, pp. 146–169. Edinburgh: Edinburgh University Press.

Jungers, W. L., Jr. & Minns, R. J. (1979). Computed tomography and biomechanical analysis of fossil long bones. *Am. J. phys. Anthrop.* **50**, 285–290.

Kimura, T., Okada, M. & Ishida, H. (1979). Kinesiological characteristics of primate walking: Its significance in human walking. In (M. E. Morbeck, H. Preuschoft & D. N. Gomberg, Eds) *Environment, Behavior, and Morphology: Dynamic Interactions in Primates*, pp. 297–311. New York: Gustav Fischer.

Lamberton, C. (1929). Sur les *Archaeoindris* de Madagascar. *C. r. hebd. Séanc. Acad. Sci., Paris* **188** (24), 1572–1574.

Lamberton, C. (1934). Contribution à la connaissance de la faune subfossile de Madagascar: lémuriens et ratites. *L'Archaeoindris fontoynonti* Stand. *Mém. Acad. malgache* **17**, 9–39.

Lamberton, C. (1947) (for the years 1944–45). Contribution à la connaissance de la faune subfossile de Madagascar. Note XVI: *Bradytherium* ou Palaeopropithèque? *Bull. Acad. malgache* (nouv. sér.) **26**, 89–140.

Lamberton, C. (1948) (for the year 1946). Contribution à la connaissance de la faune subfossile de Madagascar. Note XX: Membre postérieur des Néopropithèques et des Mésopropithèques. *Bull. Acad. malgache* (nouv. sér.) **27**, 24–28.

Lanyon, L. E. & Rubin, C. T. (1985). Functional adaptation in skeletal structures. In (M. Hildebrand, D. M. Bramble, K. F. Liem & D. B. Wake, Eds) *Functional Vertebrate Morphology*, pp. 1–25, 381–383. Cambridge, MA: Belknap.

Lessertisseur, J. (1970). Les proportions du membre postérieur de l'homme comparées à celles des autres primates. *Bull. Mém. Soc. Anthrop., Paris* **6**, 227–241.

Lessertisseur, J. & Jouffroy, F. K. (1973). Tendances locomotrices des primates traduites par les proportions du pied. *Folia primatol.* **20**, 125–160.

MacPhee, R. (1985). Primates. *1986 Yearbook of Science and Technology*, pp. 362–364. New York: McGraw Hill.

MacPhee, R. (1986). Environment, extinction, and Holocene vertebrate localities in southern Madagascar. *National Geographic Research* **2**, 441–455.

MacPhee, R., Burney, D. & Wells, N. (1985). Early Holocene chronology and environment of Ampasambazimba, a Malagasy subfossil lemur site. *Internat. J. Primatol.* **6** (5), 463–489.

MacPhee, R., Simons, E. Wells, N. & Vuillaume-Randriamanantena, M. (1984). Team finds giant lemur skeleton. *Geotimes* **29** (1): 10–11.

Mahé, J. (1976). Craniométrie des lémuriens: Analyses multivariables—phylogénie. *Mém. Mus. natn. Hist. nat., Paris*, Sér. C **32**, 1–342.

Martin, R. D. (1972). Adaptive radiation and behavior of the Malagasy lemurs. *Phil. Trans. Roy. Soc. Lond.*, Series B. **264**, 295–332.

McArdle, J. E. (1981). Functional morphololgy of the hip and thigh of the Lorisiformes. *Contrib. Primatol.* **17**, 1–132.

Mendel, F. C. (1981). Use of hands and feet of two-toed sloths (*Choloepus hoffmanni*) during climbing and terrestrial locomotion. *J. Mamm.* **62**(2), 413–421.

Mendel, F. C. (1985a). Use of hands and feet of three-toed sloths (*Bradypus variegatus*) during climbing and terrestrial locomotion. *J. Mamm.* **66**(2), 359–366.

Mendel, F. C. (1985b). Adaptations for suspensory behavior in the limbs of two-toed sloths. In (G. Gene Montgomery, Ed.) *The Evolution and Ecology of Armadillos, Sloths, and Vermilinguas*, pp. 151–162. Washington: Smithsonian Institution Press.

Miller, R. A. (1943). Adaptations in the forelimb of slow lemurs. *Am. J. Anat.* **73**, 153–184.

Napier, J. R. & Walker, A. C. (1967). Vertical clinging and leaping—a newly recognized category of primate locomotion. *Folia primatol.* **6**, 204–219.

Oxnard, C. E. (1973). Some locomotor adaptations among lower Primates: implications for primate evolution. *Symp. Zool. Soc. Lond.* **33**, 255–299.

Oxnard, C. E. (1981). The uniqueness of *Daubentonia*. *Am. J. phys. Anthrop.* **54**, 1–22.

Oxnard, C. E. (1984). *The Order of Man: A Biomathematical Anatomy of the Primates*. New Haven: Yale University Press.

Oxnard, C. E., Crompton, R. H. & Lieberman, S. S. (in press). *Animal Lifestyles and Anatomies: The Case of Prosimian Primates*. Seattle: University of Washington Press.

Oxnard, C. E., German, R. & McArdle, J. E. (1981). The functional morphometrics of the hip and thigh in leaping prosimians. *Am. J. phys. Anthrop.* **54**, 481–498.

Petter, J. J. (1962). Recherches sur l'écologie et l'éthologie des lémuriens malgaches. *Mém. Mus. natn. Hist. nat., Paris* **27**, 1–146.

Petter, J. J., Albignac, R. & Rumpler, Y. (1977). *Mammifères Lémuriens (Primates Prosimiens) Faune de Madagascar 44*, pp. 1–513. Paris: ORSTOM/CNRS.

Petter, J. J. & Peyriéras, A. (1969). Nouvelle contribution à l'étude d'un Lémurien malgache: Le Aye-Aye. *Mammalia* **34**, 167–193.

Pollock, J. I. (1977). Field observations on *Indri indri*, a preliminary report. In (I. Tattersall & R. W. Sussman, Eds) *Lemur Biology*, pp. 287–312. New York: Plenum Press.

Reilman, M. A., Gunst, R. F. & Lakshminarayanan, M. Y. (1986). Stochastic regression with errors in both variables. *J. of Quality Technology* **18**(3), 162–169.

Reynolds, T. R. (1985*a*). Mechanics of increased support of weight by the hindlimbs in Primates. *Am. J. phys. Anthrop.* **67**, 335–349.

Reynolds, T. R. (1985*b*). Stresses on the limbs of quadrupedal primates. *Am. J. phys. Anthrop.* **67**, 351–362.

Richard, A. (1978). *Behavioral Variation: Case Study of a Malagasy Lemur*. Lewisburg: Bucknell University.

Roberts, D. (1974). Structure and function of the primate scapula. In (F. A. Jenkins, Jr., Ed.) *Primate Locomotion*, pp. 171–200. New York: Academic Press.

Roberts, D. & Davidson, I. (1975). The lemur scapula. In (I. Tattersall & R. W. Sussman, Eds) *Lemur Biology*, pp. 125–147. New York: Plenum Press.

Rodman, P. S. (1979). Skeletal differentiation of *Macaca fascicularis* and *Macaca nemestrina* in relation to arboreal and terrestrial quadrupedalism. *Am. J. phys. Anthrop.* **51**, 51–62.

Rose, M. D. (1973). Quadrupedalism in primates. *Primates* **14**, 337–357.

Ruff, C. (1987). Structural allometry of the femur and tibia in Hominoidea and *Macaca*. *Folia primatol.* **48**, 9–49.

Schaffler, M. B., Burr, D. B., Jungers, W. L. & Ruff, C. B. (1985). Structural and mechanical indicators of limb specialization in Primates. *Folia Primatol.* **48**, 9–49.

Schwartz, J. H. & Tattersall, I. (1985). Evolutionary relationships of living lemurs and lorises (Mammalia, Primates) and their potential affinities with European Eocene Adapidae. *Anthrop. Pap. Am. Mus. nat. Hist.* **60**, 1–100.

Seligsohn, D. (1977). Analysis of species-specific molar adaptations in strepsirhine Primates. *Contrib. Primatol.* **11**, 1–116.

Seligsohn, D. & Szalay, F. S. (1974). Dental occlusion and the masticatory apparatus in *Lemur* and *Varecia*. Their bearing on the systematics of living and fossil primates. In (R. D. Martin, G. A. Doyle, and A. C. Walker, Eds) *Prosimian Biology*, pp. 543–561. London: Duckworth.

Smith, J. Maynard & Savage, R. J. G. (1955). Some locomotory adaptations in mammals. *J. Linn. Soc. Zool.* **42**, 603–622.

Standing, H. F. (1910, for the year 1909). Note sur les ossements subfossiles provenant des fouilles d'Ampasambazimba. *Bull. Acad. malgache* **7**, 61–64.

Stern, J. & Oxnard, C. E. (1973). Primate locomotion: Some links with evolution and morphology. *Primatologia* **4**, 1–93.

Steudel, K. (1985). Allometric perspectives on fossil catarrhine morphology. In (W. L. Jungers, Jr, Ed.) *Size and Scaling in Primate Biology*, pp. 449–475. New York: Plenum Press.

Szalay, F. S. (1972). Paleobiology of the earliest primates. In (R. Tuttle, Ed.) *The Functional and Evolutionary Biology of the Primates*, pp. 3–35. Chicago: Aldine-Atherton.

Szalay, F. S. & Dagosto, M. (1980). Locomotor adaptations as reflected on the humerus of Paleogene primates. *Folia primatol.* **34**, 1–45.

Szalay, F. S. & Delson, E. (1979). *Evolutionary History of the Primates*. New York: Academic Press.

Szalay, F. S. & Katz, C. (1973). Phylogeny of lemurs, galagos and lorises. *Folia primatol.* **19**, 88–103.

Tardieu, C. (1983). *L'Articulation du Genou: Analyse Morpho-fonctionelle chez les Primates et les Hominidés Fossiles*. Paris: CNRS.

Tattersall, I. (1971). Revision of the subfossil Indriinae. *Folia primatol.* **16**, 257–269.

Tattersall, I. (1973). Cranial anatomy of the Archaeolemurinae (Lemuroidea, Primates). *Anthrop. Pap. Am. Mus. nat. Hist.* **52**, 1–110.

Tattersall, I. (1975). Notes of the cranial anatomy of the subfossil Malagasy lemurs. In (I. Tattersall & R. W. Sussman, Eds) *Lemur Biology*, pp. 111–124. New York: Plenum Press.

Tattersall, I. (1982). *The Primates of Madagascar*. New York: Columbia University Press.

Tattersall, I. & Schwartz, J. H. (1974). Craniodental morphology and the systematics of the Malagasy lemurs (Primates, Prosimii). *Anthrop. Pap. Am. Mus. nat. Hist.* **52**, 139–192.

Taylor, B. K. (1985). Functional anatomy of the forelimb in vermilinguas (anteaters). In (G. Gene Montgomery, Ed.) *The Evolution and Ecology of Armadillos, Sloths, and Vermilinguas*, pp. 163–172. Washington: Smithsonian Institution Press.

Troughton, E. (1965). *Furred Animals of Australia*. Eighth ed. Sydney: Angus & Robertson.

Vuillaume-Randriamanantena, M. (1982). Contribution à l'étude des os longs des lémuriens subfossiles malgaches. Thèse de Doctorat de 3e Cycle. Université de Madagascar.

Vuillaume-Randriamanantena, M. (in press). The taxonomic attributions of giant subfossil lemur bones from Ampasambazimba: *Archaeoindris* and *Lemuridotherium*. *J. hum. Evol.*, **17**.

Vuillaume-Randriamanantena, M. & Godfrey, L. R. (1988) Morphology, taxonomy and paleoecology of *Megaladapis*, giant subfossil lemur from Madagascar.

Walker, A. C. (1967). Locomotor adaptations in recent and fossil Madagascan lemurs. Ph.D. Dissertation, University of London.

Walker, A. C. (1974). Locomotor adaptations in past and present prosimian primates. In (F. A. Jenkins, Jr., Ed.) *Primate Locomotion*, pp. 349–381. New York: Academic Press.

Ziegler, A. C. (1977). Evolution of New Guinea's marsupial fauna in response to a forested environment. In (B. Stonehouse & D. Gilmore, Eds) *The Biology of Marsupials*. Baltimore: University Park Press.

Daniel L. Gebo

Department of Anthropology, Northwestern Illinois University, DeKalb, IL 60115, U.S.A.

Marian Dagosto*

Departments of Anthropology and Cell Biology and Anatomy, Northwestern University, Evanston, IL 60208, USA).

Received 15 June 1987
Revision received 27 December 1987
and accepted 6 January 1988

Publication date June 1988

Keywords: Indriidae, foot, locomotion, climbing.

Foot anatomy, climbing, and the origin of the Indriidae

Living indriids share a suite of derived osteological and myological features of the foot which are related to three of the most important components of their locomotor repertoire: climbing (primarily on vertical supports), vertical clinging, and leaping. However, the overall structure of the foot bones and musculature reflects the requirements for climbing more than any other behavior. All indriid subfamilies have postcranial morphologies and locomotor behaviors which are derived compared to the probable indriid–lemurid common ancestor. It seems most likely that the ancestral indriid was an "arboreal quadruped" (i.e. lemur-like), not a vertical clinger and leaper.

Journal of Human Evolution (1988) **17**, 135–154

Introduction

In 1967, Napier & Walker proclaimed their bold ideas concerning the evolution of primate locomotor systems. In their scheme, indriids were grouped with galagos, tarsiers, *Lepilemur*, and various fossil taxa as examples of the locomotor category "vertical clinging and leaping" (VCL). They further hypothesized that VCL was the ancestral locomotor type of primates. Although the behavioral evidence clearly shows that the living indriids, like galagos and tarsiers, commonly adopt the posture of vertical clinging and move between vertical supports by leaping, indriid foot anatomy (especially relative tarsal length) does not conform to the VCL adaptations identified for galagos and tarsiers. The contrast in foot anatomy between indriids and tarsiers and galagos has been cited in arguments questioning: (1) the association of morphological features with leaping; (2) the applicability of the label "vertical clinging and leaping" to fossil primates; and (3) Napier & Walker's proposition that VCL was the ancestral primate locomotor mode (Cartmill, 1972; Martin, 1972; Szalay, 1972; Stern & Oxnard, 1973).

Oxnard, German & McArdle (1981) proposed three reasons why different morphologies exist in VCL primates: (1) different original genetic bases and parallel evolution of leaping in different groups; (2) different biomechanical modes of leaping; and (3) the influence of other postural and locomotor behaviors. They hypothesized that different biomechanical modes of leaping was the best explanation for the differences. These are not, however, mutually exclusive explanations. In our view, an explanation for the unique construction of the indriid foot includes elements of all three factors. We believe that frequent leaping (i.e., >40%) evolved in parallel in indriids, bushbabies, and tarsiers. Other behaviors which are important in indriid locomotion, especially climbing on vertical supports, necessitate a compromise morphology for indriids that has resulted in a biomechanically different mode of leaping compared to other VCL primates. Another important factor is body size. Indriids are considerably larger than galagos or tarsiers (although *Avahi*, by far the smallest indriid, is similar in weight to *Galago crassicaudatus*, the largest galago). This

* Order of names does not imply seniority of authorship.

difference in body size is also likely responsible for some of the morphological differences between indriids and other VCL primates, especially differences in foot proportions (see below).

We believe indriid foot anatomy has taken its present form because indriid feet are adapted primarily for climbing (especially on vertical supports) and vertical clinging; the few slight modifications for frequent leaping were acquired secondarily after indriids attained a relatively large body size.

Indriids thus possess a foot which is an interesting compromise for climbing, vertical clinging, and leaping. This paper, which is a synthesis of our respective dissertation work, explains how we independently arrived at a functional and adaptive explanation for the unique features of the indriid foot. In this paper we will discuss locomotor behavior, muscle anatomy, bone and joint anatomy, and foot function in indriids as contrasted with (1) lemurids, the sister taxon of the Indriidae (Szalay & Delson, 1979; Tattersall, 1982), and (2) with other VCL prosimians (tarsiers, galagos, and *Lepilemur*).

Materials and methods

The results presented here are based on comparative analyses of the strepsirhine foot skeleton, muscles, and locomotor behavior presented in Gebo (1985, 1986, 1987*a,b*) and Dagosto (1986). Osteological descriptions and measurements are based on 268 individuals representing all extant strepsirhine species with the exception of *Allocebus* (Table 1). The measurements discussed in this paper are described in Dagosto (1985, 1986) and illustrated in Figure 1. Dissection techniques and the muscle-weighing procedure have been discussed by Gebo (1985, 1987*a*). Locomotor observations at the Duke Primate Center were recorded using the focal animal approach of Fleagle (1977) and Fleagle & Mittermeier (1980). A more detailed description of this procedure is provided in Gebo (1987*b*).

Locomotor behavior

Indriid locomotor behavior has been studied in wild and captive settings since 1935 (Rand, 1935). Of the four extant indriid species (*Propithecus verreauxi*, *Propithecus diadema*, *Indri indri*, and *Avahi laniger*) only the locomotor behavior of *Propithecus verreauxi* has been studied intensively (Rand, 1935; Petter, 1962; Jolly, 1966; Walker, 1974, 1979; Pollock, 1975, 1977; Petter *et al.*, 1977; Richard, 1978; Tattersall, 1982; Gebo, 1986, 1987*b*). There is very little locomotor information about the three other indriid species other than that in Petter *et al.* (1977) and Pollock (1975) who note general similarity to the locomotion of *Propithecus verreauxi*. All indriids are considered to be vertical clingers and leapers but wild (Jolly, 1966; Walker, 1974; Richard, 1978) and captive studies (Gebo, 1986, 1987*b*) have shown a wider locomotor repertoire for *Propithecus verreauxi*. In addition to vertical clinging and leaping *Propithecus verreauxi* is known to be (1) a frequent climber on vertical supports; (2) a frequent practitioner of bimanual suspensory movements akin to brachiation; (3) able to hang below a branch utilizing only its hindfeet; (4) a practitioner, although to a lesser extent, of below-branch quadrupedal suspensory movements; and (5) a capable bipedal hopper when on the ground. *Propithecus verreauxi* has never been observed to utilize loris-like bridging movements in the wild nor in captivity. Table 2 displays the locomotor frequencies for *Propithecus verreauxi* that were observed at the Duke Primate Center. The percentage of use of climbing is similar to frequencies reported for lorisines and some lemurs (Gebo, 1987*b*); however, almost all of the recorded observations for *Propithecus*

Table 1   **List of specimens of extant primates used in this study. The first column of numbers lists the sample size of specimens on which osteological measurements were taken. The second column gives the sample size of specimens on which dissections were made and muscle weights were taken**

Lemur fulvus	20	3
Lemur mongoz	3	1
Lemur rubriventer	1	–
Lemur macaco	3	1
Lemur coronatus	3	1
Lemur catta	9	1
Varecia variegata	14	1
Hapalemur griseus	14	2
Hapalemur simus	1	–
Lepilemur mustelinus leucops	13	–
Daubentonia madagascariensis	6	–
Avahi laniger	10	–
Indri indri	11	–
Propithecus verreauxi	7	1
Propithecus diadema	11	–
Microcebus murinus	14	2
Mirza coquereli	2	1
Cheirogaleus medius	5	3
Cheirogaleus major	8	–
Phaner furcifer	1	–
Galago senegalensis	6	2
Galago moholi	9	–
Galago gallarum	2	–
Galago elegantulus	16	–
Galago matschei	1	–
Galagoides demidovii	9	–
Galagoides zanzibaricus	1	–
Galagoides alleni	5	–
Galago crassicaudatus	6	2
Galago garnetti	3	2
Perodicticus potto	16	1
Arctocebus calabarensis	11	–
Nycticebus coucang	12	1
Loris tardigradus	6	1
Tarsius syrichta	3	2
Tarsius bancanus	3	–
Tarsius spectrum	3	–
Total	268	27

represent upward vertical climbing whereas lorisine and lemurid climbing movements are more varied. Although the percent of time spent in various postures was not recorded, it should be noted that all prosimians will cling to vertical supports and that for most prosimians, vertical clinging is but a stopping position in a vertical climbing sequence. Of course, indriids and a few other prosimians utilize vertical clinging for prolonged time periods which is precisely why they were originally placed in a special locomotor/postural category. Vertical clinging is well documented in *Propithecus verreauxi* which has been noted to vertically cling or sit for 75% of the time when feeding (Richard, 1978).

In contrast, lemurids are far more likely to adopt the horizontal body postures associated with quadrupedalism (Table 2). Quadrupedalism, leaping, and climbing represent the most frequent movements of the lemurid locomotor repertoire (Petter, 1962; Jolly, 1966; Sussman, 1972, 1974; Walker, 1974, 1979; Tattersall & Sussman, 1975; Tattersall, 1977,

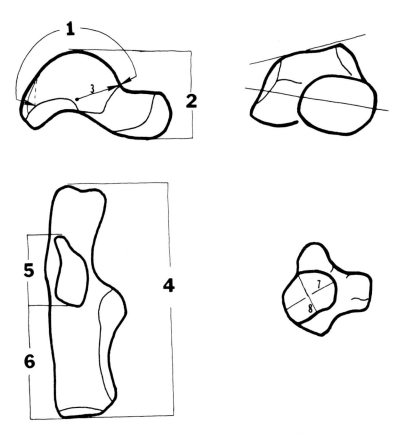

Figure 1. Measurements discussed in this paper. Lateral (top row, left) and anterior (top row, right) views of the talus. Dorsal (bottom row, left) and anterior (bottom row right) views of the calcaneus. (1) arc length of upper ankle joint; (2) talar body height; (3) radius of the upper ankle joint; (4) total calcaneal length; (5) length of the posterior talocalcaneal facet (PAC); (6) length of the anterior part of the calcaneus; (7) width of calcaneocuboid facet; (8) height of calcaneocuboid facet. The anterior view of the talus illustrates the lines used to measure the orientation of the talar head.

1982; Gebo, 1986, 1987*b*). As a group, lemurids tend to utilize more below-branch quadrupedal and bipedal suspensory movements and more frequent bridging movements than does *Propithecus verreauxi* at the Duke Primate Center, but observers of this species in the wild report that these behaviors do occur. Only *Lemur macaco* uses bimanual suspensory movements with any significant frequency, although not to the same degree as indriids (Gebo, 1987*b*). Lemurids, with the exception of *Hapalemur*, leap less often than *Propithecus* (Table 2).

Lemurids prefer above-branch horizontal supports to move about rather than the vertical supports that indriids prefer. *Hapalemur* and *Lepilemur* utilize vertical supports more often than do the other lemurids, but in general lemurids prefer horizontal supports. Lastly, one ecological difference between living indriids and lemurids is the lack of any terrestrially adapted species (i.e., *Lemur catta*) among the living indriids.

A comparison of locomotor frequencies for *Propithecus verreauxi*, a large bodied (about 4000 g) vertical clinger and leaper with *Galago senegalensis*, *Tarsius syrichta*, and *Tarsius bancanus*, small-bodied vertical clingers and leapers (about 250, 125 and 125 g, respectively) shows that all four species climb about the same amount (26–30%), and utilize quadrupedalism in very small amounts, especially tarsiers (0–3%) (see Table 2). However, the small-bodied vertical clingers and leapers do not utilize bimanual suspension or "brachiating" as often as *P. verreauxi*, and tarsiers and galagos leap far more often.

In sum, in contrast to lemurids, indriids use vertical supports more often, leap and vertically climb more frequently, and utilize bimanual suspensory movements more often. In contrast to the small-bodied vertical clingers and leapers, indriids leap less often but use bimanual suspension more often. Both types of vertical clingers and leapers climb in the same proportion and avoid quadrupedalism (especially tarsiers).

Table 2 Locomotor frequencies (in percent) for taxa observed at the Duke Primate Center. Q, quadrupedalism; L, leaping; CL, climbing; S, suspension; BR, bridging; BIM, bimanual suspension; BIP, bipedal; BOUTS, number of locomotor bouts observed

Taxon	Q	L	CL	S	BR	BIM	BIP	BOUTS
Lemur catta	51	22	10	4	1	<1	7	642
L. fulvus	39	34	17	9	1	<1	<1	1273
L. mongoz	29	37	22	8	2	1	1	2164
L. coronatus	33	30	31	2	1	<1	3	842
L. macaco	30	31	28	6	1	4	1	525
Vareica variegata	35	21	31	11	1	1	1	525
Hapalemur griseus	24	56	15	7	1	<1	1	2118
Propithecus verreauxi	6	46	30	5	0	10	3	2149
Galago senegalensis	5	63	26	2	<1	<1	2	2245
Tarsius syrichta	3	61	28	8*	0	0	1	512
Tarsius bancanus	0	65	30	5*	0	0	0	549

* Suspension in tarsiers is not the same as below branch suspensory movements. This value is for leaning out and capturing insects while holding onto the vertical support with only two feet.

Foot anatomy

Hallux. When comparing the gross foot shape of indriids with lemurids, one is struck by the larger size of the hallux and the great cleft between the big toe and the four lateral digits (Figure 2). Mivart (1867) and Jouffroy & Lessertisseur (1979) have documented greater hallucal length compared to foot length in indriids when compared to lemurids. From this evidence alone one might infer that indriids have emphasized a strong grasping foot above and beyond what is observed for lemurids. The wider cleft in indriid feet allows for greater independence of the hallux from the other digits as well as for an increased range of abduction. This allows an indriid foot to encompass a larger diameter support than can a lemurid foot of the same size.

Among other prosimians, lorisines also exhibit a relatively long hallux (Mivart, 1867; Jouffroy & Lessertisseur, 1979), although differences between lorisines and lemurids are less than those between indriids and lemurids. Like indriids, lorisines increase the span of the grasp by increasing the distance between the hallux and the other digits, but the anatomical basis for this differs in the two groups. In lorisines, the hallux is reoriented medially and proximally through realignment of the talonavicular and navicular-entocuneiform joints (Dagosto, 1986; Gebo, 1986). In indriids and lemurids, the lateral digits are set off from the hallux by modifications of the ectocuneiform and by the more lateral slant of the entocuneiform-metatarsal two joint which angles the lateral digits away from the hallux (Gebo, 1985). Indriids are further distinguished by the more laterally

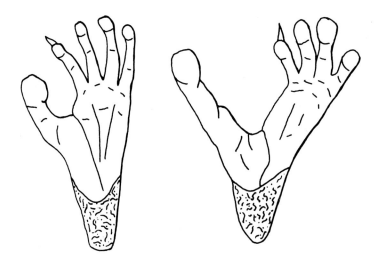

Figure 2. Sole of left foot of *Lemur* (left) and *Indri* (right). *Indri* after Biegert, 1963. Note the long and divergent big toe in *Indri*.

rotated mesocuneiform–ectocuneiform articulation which increases the angle between the hallux and lateral digits.

Retinacula. Indriids and lemurids possess similar crural and tarsal retinacular patterns with one exception. Indriids possess an extra band above the superior extensor bands of the leg which holds the anterior tibialis and the extensor hallucis longus tendons in place (Gebo, 1986, 1987*a*). This extra band may be a functional response to the more frequent use of vertical supports since extra retinacular bands are also reported in tarsiers and a modified band pattern is documented for galagos (see Gebo, 1987*a* for details). Other leapers (lemurids, cheirogaleids) do not exhibit these extra bands.

Musculature. Indriids differ from lemurids and from other vertical clingers and leapers in the patterns of relative weight of extrinsic foot musculature. Although the total weight of flexor musculature compared to total extrinsic foot musculature is about the same in lemurids and indriids (Table 3), indriids have better developed deep flexors (flexor hallucis longus, flexor digitorum longus, and tibialis posterior) while lemurids have a larger superficial flexor musculature (gastrocnemius, soleus, and plantaris). Lemurids are similar to cheirogaleids, galagines and tarsiers in this pattern of superficial flexor development over deep flexor development while indriids are similar to lorisines, animals which utilize slow-climbing movements and have never been observed to leap. When one compares specific relative muscle weights for *Propithecus verreauxi*, the only indriid species dissected, with several lemurid species (Table 4), several distinctions emerge which substantiate the contrasting indriid–lemurid muscular pattern noted above. In *Propithecus verreauxi* flexor hallucis longus is very well developed and even exceeds the relative weight value for the gastrocnemius. The relative weight value for flexor hallucis longus is the highest value recorded for all prosimians dissected including cheirogaleids, lorisids, and tarsiers. Peroneus longus, an adductor of the hallux, and thus a muscle important in grasping, is also well developed in *Propithecus*. Relative weight values of anterior tibialis and flexor digitorum longus are similar in *Propithecus* and lemurids. Thus, in extrinsic foot musculature indriids differ from lemurids, tarsiers, and galagos in their emphasis on muscles that flex the digits versus muscles that plantarflex the foot, especially in the great size of flexor hallucis longus and peroneus longus. This is a pattern which is more mechanically advantageous for enhanced digital flexion and grasping than leaping, and which is also characteristic, in a more extreme form, of lorisines (Gebo, 1986).

The intrinsic muscle development in the foot of indriids and lemurids agrees more closely since both families share a greatly developed adductor hallucis muscle and a large dorsal interosseous muscle on digit two. This contrasts with the more common pattern observed in cheirogaleids, lorisids, and tarsiids, which places emphasis on muscles concentrated on digits one and five (Gebo, 1985). The functional consequence of the lemurid and indriid pattern of intrinsic muscle development is an enhanced hallucal grasp which has been designed the I–II grasp and which is likely an adaptation for vertical clinging in non-clawed animals of relatively large body size (Gebo, 1985).

Table 3 Relative weight values for extrinsic foot flexors. Determined by dividing muscle group weight by total extrinsic muscle weight. Superficial flexors are gastrocnemius, plantaris, and soleus. Deep flexors are tibialis posterior, flexor hallucis longus, and flexor digitorum longus. L, lemurids; I, indriids; C, cheirogaleids; LO, lorisines; G, galagines; T, tarsiers; N, number of species dissected within each group (see Table 1 for species dissected and sample sizes)

N	L	I	C	LO	G	T
	7	1	4	3	3	1
Superficial flexors	0·37	0·26	0·33	0·18	0·33	0·42
Deep flexors	0·28	0·37	0·31	0·47	0·29	0·22
Posterior compartment total	0·65	0·64	0·64	0·65	0·62	0·64

Table 4 Relative weight of extrinsic foot muscles. Computed by dividing individual muscle weight by total extrinsic foot muscle weight (abbreviations as in Table 3)

Muscle		L	I	C	LO	G	T
	N	7	1	3	3	3	1
Gastrocnemius	X	0·25	0·19	0·24	0·08	0·20	0·26
	range	0·22–0·25		0·20–0·27	0·05–0·10	0·18–0·24	
Soleus		0·07	0·02	0·06	0·08	0·06	0·08
		0·05–0·09		0·05–0·07	0·06–0·10	0·03–0·11	
Plantaris		0·05	0·05	0·03	0·02	0·07	0·08
		0·04–0·07		0·02–0·04	0·01–0·04	0·06–0·07	
Flexor hallucis longus		0·18	0·24	0·16	0·16	0·17	0·13
		0·16–0·19		0·14–0·20	0·13–0·20	0·13–0·20	
Flexor digitorum longus		0·08	0·10	0·12	0·27	0·09	0·05
		0·06–0·11		0·09–0·15	0·19–0·40	0·07–0·10	
Tibialis posterior		0·02	0·03	0·03	0·04	0·03	0·04
		0·02–0·04		0·02–0·03	0·04–0·04	0·03–0·04	
Tibialis anterior		0·13	0·11	0·15	0·12	0·15	0·16
		0·10–0·15		0·12–0·17	0·10–0·13	0·12–0·18	
Extensor hallucis longus		0·03	0·03	0·03	0·04	0·04	0·05
		0·03–0·04		0·02–0·04	0·03–0·06	0·03–0·05	
Extensor digitorum longus		0·04	0·04	0·03	0·04	0·02	0·04
		0·03–0·04		0·02–0·03	0·03–0·06	0·02–0·03	
Peroneus longus		0·12	0·14	0·10	0·10	0·14	0·06
		0·10–0·15		0·09–0·13	0·06–0·12	0·13–0·16	
Peroneus brevis		0·04	0·04	0·05	0·05	0·04	0·04
		0·03–0·05		0·03–0·06	0·04–0·06	0·03–0·06	

Osteology. The general shape of the talocrural joint is like that of lemurids. The trochlea of the talus is moderately deep in *Indri* and *Propithecus*, but very shallow in *Avahi*. The most striking difference from lemurids is the greater development (posterior extension) of the posterior trochlear shelf, which also has a marked medio-plantar twist (Figure 3A). This should increase the degree of lateral rotation of the tibia which accompanies plantarflexion. As a result of this twisting, the groove for flexor hallucis longus is plantarmedially angled in indriids rather than transverse as in lemurids and the posterior talocalcaneal facet of the talus is lifted dorsally (Figure 3A). This causes the posterior part of the talus to be more strongly buttressed in the inverted rather than everted position, and is possibly related to vertical clinging, although other vertical clingers (bushbabies, *Lepilemur*, *Tarsius*) do not share this feature.

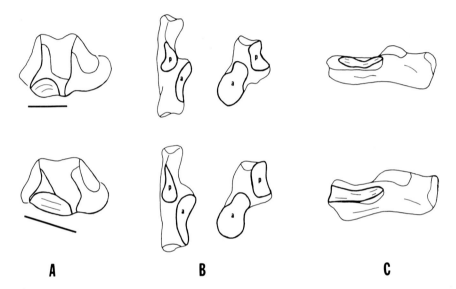

Figure 3. A, posterior view of talus of *Lemur* (top) and *Avahi* (bottom). Line below bone indicates level of groove for flexor hallucis longus. Note the oblique or twisted appearance of the posterior talus in indriids. B, dorsal view of calcaneus (left) and ventral view of talus (right) in *Lemur* (top) and *Indri* (bottom). Facets of talocalcaneal articulation are marked in dark outline; p, posterior facet; a, anterior facet; C, medial view of calcaneus of *Lemur* (top) and *Indri* (bottom). The anterior talocalcaneal articulation is marked in dark outline.

The functional significance of the posterior trochlear shelf is obscure. Some development is probably a primitive character for all cuprimates (Decker & Szalay, 1974). An especially large posterior trochlear shelf is found in all extant indriids and notharctines, although in the latter the shelf is even larger and is not twisted. Reduction or loss of the shelf is characteristic of lorisids, most omomyids, and *Tarsius*. Distribution alone provides no clue to its function. Dagosto (1983) suggested that the shelf provides a buttress for a plantarflexed foot in push off, but the absence of a shelf in galagines and *Tarsius* weakens this hypothesis.

The medial tibiotalar joint of indriids is slightly less medially curved than in lemurids. Indriids also have somewhat less tibial malleolar rotation than lemurids, as do other frequent leapers like *Lepilemur* and galagines (Table 5; Dagosto, 1985, 1986). These two features suggest that upper ankle joint dorsi- and plantarflexion are accompanied by less

Table 5 Measures discussed in text. Definitions of measurements from Dagosto, 1986 (see Figure 1 for illustration of measurements). N, sample size; PAC, posterior talocalcaneal facet; sd, standard deviation; STL, skeletal trunk length; UAJ, upper ankle joint; X, mean

	Rotation of medial malleolus		Radius UAJ/ STL		Arc length of UAJ/STL		Talar body height/STL		PAC length/ length calcaneus	
	X	N	X	N	X	N	X	N	X	N
Avahi laniger	28·0	1	2·41	9	7·17	9	3·10	6	30·93	10
Propithecus verreauxi	31·0	3	2·14	5	6·30	5	2·86	3	31·08	6
Propithecus diadema	22·0	1	2·09	10	6·39	9	3·16	3	31·84	11
Indri indri	32·0	1	2·18	10	6·62	10	3·14	7	33·78	9
	X	S.D.	X	S.D.	X	S.D.	X	S.D.	X	S.D.
		N		N		N		N		N
Indriidae	29·17	5·0	2·21	0·22	6·66	0·71	3·09	0·2	32·05	2·4
		6		33		33		19		36
Lemuridae	34·05	3·4	1·82	0·17	5·55	0·57	2·65	0·12	27·57	4·4
		18		36		36		27		67
Cheirogaleidae	30·30	6·0	1·61	0·11	4·96	0·43	2·59	0·19	22·34	4·9
		13		14		14		6		27
Galaginae	27·64	3·4	2·07	0·16	6·64	0·56	3·17	0·17	14·00	3·1
		14		33		33		10		55
Lorisinae	39·80	7·6	1·57	0·33	4·19	0·9	2·34	0·23	32·77	5·7
		10		26		26		24		44
Tarsius	13·00	2·4	–	–	–	–		–	12·37	0·66
										9

	PAC length/ STL		Orientation of talar head		Relative anterior calcaneal length		Calcaneocuboid facet ht/width	
	X	N	X	N	X	N	X	N
Avahi laniger	3·34	5	15·00	10	40·99	10	86·78	3
Propithecus verreauxi	2·95	5	10·86	7	43·68	6	81·81	4
Propithecus diadema	3·07	10	18·25	8	45·78	11	80·33	2
Indri indri	3·40	9	13·67	9	44·30	11	80·41	1
	X	S.D.	X	S.D.	X	S.D.	X	S.D.
		N		N		N		N
Indriidae	3·16	0·27	14·44	7·5	43·77	3·3	82·86	6·6
		34		36		38		10
Lemuridae	2·49	0·28	9·19	3·9	44·00	3·6	71·76	7·1
		42		56		43		26
Cheirogaleidae	2·40	0·24	(11·48)	8·5	56·12	6·5	85·98	8·5
		16		25		27		20
Galaginae	2·98	0·28	(15·58)	7·6	70·54	4·6	82·56	8·1
		38		56		45		17
Lorisinae	2·06	0·49	(17·83)	8·1	40·21	4·4	77·44	5·3
		30		41		38		14
Tarsius	–		10·78	7·1	75·47	0·61		
				9		9		

conjunct and concomitant rotation than in lemurids. Indriids have a higher radius of curvature and a longer arc length of the tibiotalar joint, and higher talar bodies than lemurids (Table 5). These are also characteristic of other specialized leapers (Dagosto, 1986).

Indriids are perhaps the most distinctive strepsirhine taxon in the construction of the subtalar joint. They are distinguished by relatively longer, and thus apparently narrower posterior talocalcaneal facets (Figure 3B). The facet on the calcaneus is long compared to either calcaneal length or skeletal trunk length (Table 5). In medial view, the facet tends to flatten out posteriorly, making the joint less perfectly circular than in lemurids and limiting the degree to which the foot plantarflexes and supinates during inversion.

The anterior talocalcaneal facet of the calcaneus extends to the distal end of the bone reflecting the serial condition of the tarsus. (In a serial tarsus, the distal extent of the calcaneus and talus is equivalent and the transverse tarsal joint makes a straight line across the foot.) The entire facet is bent plantad so that its surface lies more on the medial side of the bone than on its dorsal side (Figure 3C). This is probably related to the more marked dorsolateral orientation of the talar head. The angle of the facet must accommodate this angulation of the talus. This probably provides greater support for the talar head in the inverted position, and may be related to vertical clinging with the feet strongly inverted, but other vertical clingers (bushbabies, *Tarsius*) do not share this trait).

In the transverse tarsal joint, the significant morphological differences from lemurids are as follows: the talonavicular facet of the talus is larger, forming a greater section of a sphere in indriids than in lemurids (Figure 4). The head of the talus has a slightly more marked dorsolateral orientation than in lemurids, being 15 degrees rather than 10; this difference is statistically significant ($P < 0.001$) (Table 5; Figure 4). The calcaneocuboid facet on the calcaneus is higher and narrower than in lemurids (Table 5; Figure 4). This is a feature found in other frequent leapers (galagines, some cheirogaleids). The eminence of the cuboid pivot is not as well developed as in lemurids, giving a much flatter calcaneocuboid articulation.

The more spherical talonavicular joint would theoretically increase the degree of axial rotation possible, and thus the range of pronation and supination. A somewhat greater degree of dorsal sliding at the calcaneocuboid joint may be allowed due to the reduction of the cuboid pivot. This may also help increase the degree of rotational movement, since in lemurids one of the main stops to medial rotation (pronation) is the conjunct dorsal sliding of the cuboid which is eventually blocked by the impact of the cuboid pivot against the calcaneus. In addition, the fact that in indriids the calcaneus extends only slightly more distally than the talus (forming a serial tarsus) whereas in lemurids it is more noticeably so (with the exception of *Varecia*) may also increase the degree of axial rotation possible. In lemurids, lateral rotation (supination) is blocked by the impact of the navicular and the calcaneus. In indriids, since the transverse tarsal joint is not as stepped (essentially forming a serial tarsus) contact of the navicular and the calcaneus is not as much of a problem. This modification is shared by *Daubentonia*, *Varecia*, and lorisines. It thus seems to be associated with climbers and larger sized strepsirhines (though large size and increased climbing ability are almost certainly related (Cartmill, 1974; Jungers, 1979). In indriids, the consequence of all of this is a foot which is freer in axial rotation at the transverse tarsal joint than is the lemurid foot. It seems that indriids are selectively increasing the range of eversion more than inversion. By reducing the cuboid pivot and increasing the dorsolateral orientation of the talonavicular joint, a more everted position of the tarsus can be achieved.

Joint mobility data indicate that indriids do have a greater range of eversion at the transverse tarsal joint (Gebo, 1986). Unfortunately, we do not know which specific indriid behavior is associated with this greater eversion capability. We postulate that this action may occur when pushing off a support while leaping but only a high speed film sequence can test this hypothesis.

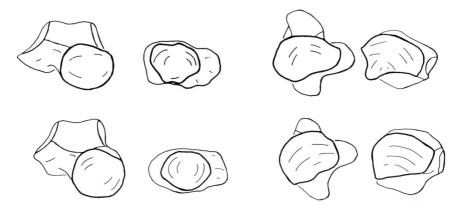

Figure 4. The transverse tarsal joint in *Lemur* (top) and *Indri* (bottom). From left to right, anterior view of talus, posterior view of articulating navicular, anterior view of calcaneus, posterior view of articulating cuboid.

Limb proportions

The hindlimb is long compared to vertebral column length or body weight in all vertical clingers and leapers (Jouffroy & Lessertisseur, 1979; Jungers, 1985), but the proportions of the hindlimb differ. In indriids the femoral portion is longest and the foot is relatively short. In galagids and tarsiers the foot makes up the largest percentage of the hindlimb (Jouffroy & Lessertisseur, 1979). The intrinsic proportions of the foot also distinguish indriids from small bodied vertical clingers and leapers. In indriids the tarsus is relatively short while the metatarsus and especially the phalangeal portion of the foot is relatively long (Jouffroy & Lessertisseur, 1979). These proportions are more like those of lorisines than any other prosimians. Indriids do not exhibit the extensive lengthening of the anterior calcaneus which is characteristic of galagines and *Tarsius*; their values are more similar to those of lemurids (Table 5; Dagosto, 1986; Gebo, 1986).

If lengths of foot segments are compared to an extrinsic variable such as body weight, a similar story emerges. Tarsus length of indriids is similar (if slightly lower) to what one would expect in a lemurid of the same body size, but much shorter than expected in a galagine, and much longer than in a lorisine (Figure 5A; Tables 6, 7) (Table 7 compares actual and predicted values for these variables in strepsirhines of small and large body weights). Relative anterior calcaneal lengthening scales negatively in all strepsirhines as expected (Walker, 1970), but less so in indriids than in other taxa (Figure 5A; Tables 6, 7). Relative length of the anterior calcaneus decreases with increased body size in all families except indriids, but indriids still have shorter anterior calcanei than would be expected if they were simply scaled up versions of galagos (Figure 5B; Table 7). Indriids (and tarsiers) have slightly longer digits than would be expected in any other strepsirhine with comparable body weight, including lorisines (Figure 5C).

Table 6   Slope, intercept, and correlation coefficient (*r*) for relationships illustrated in Figure 5

For log(n) tarsus length and log (n) body weight (Figure 5A):

	Strepsirhini	Lemuridae	Indriidae	Galaginae	Lorisinae
Slope	0·155	0·273	0·248	0·218	0·379
Intercept	2·358	1·451	1·642	2·338	0·440
r	0·565	0·941	0·988	0·914	0·999

For log(n) relative anterior calcaneal length and log(n) body weight (Figure 5B):

	Strepsirhini	Lemuridae	Indriidae	Galaginae	Lorisinae
Slope	−0·125	−0·076	0·050	−0·066	−0·115
Intercept	4·783	4·391	3·359	4·613	4·462
r	0·732	0·666	0·922	0·903	0·908

For log(n) digit length and log(n) body weight (Figure 5C):

	Strepsirhini	Lemuridae	Indriidae	Galaginae	Lorisinae
Slope	0·358	0·187	0·385	0·348	0·292
Intercept	1·117	2·408	1·027	1·173	1·434
r	0·947	0·801	0·994	0·994	0·888

Table 7   Species means for some variables discussed in the text for strepsirhines of similar body weights, but from different families. Body weights (g) from Jungers, 1985; foot measures from Dagosto, 1986; muscle weights from Gebo, 1986. For "large" taxa, hypothetical galagine and lorisine values are predicted based on parameters in Table 6. Tarsus length equals calcaneus + cuboid length; digit length equals length of metatarsal and proximal phalanx of fourth digit; RLAC, relative length of anterior calcaneus; Superficial flexors, weight (g) of superficial flexor mass; Deep flexors, weight (g) of deep flexor mass

Species	Weight	Tarsus length	Digit length	RLAC	Superficial flexors	Deep flexors
Small taxa						
Hapalemur griseus	880	28·06	41·53	46·96	3·72	3·35
Avahi laniger	920	28·70	44·73	40·09	—	—
Galago garnetti	760	45·70	31·60	65·80	4·62	4·51
Nycticebus coucang	920	20·53	29·90	38·87	1·01	4·60
Large taxa						
Varecia variegata	3800	41·95	57·82	41·26	17·66	15·33
Propithecus verreauxi	3780	37·55	62·00	43·68	9·42	13·29
Galagine (predicted)	(3800)	62·48	56·91	58·49	—	—
Lorisine (predicted)	(3800)	35·30	46·57	33·58	—	—

Discussion

Compared to other strepsirhines, the living indriids are distinguished by several derived features of the foot. All indriids have a large, posteriorly expanded, twisted posterior trochlear shelf. The extension of the shelf is probably related to the increased extent of the posterior talocalcaneal facets (Decker & Szalay, 1974). Indriids are characterized by a more spherical and dorsolaterally tilted talonavicular joint, a high, flat calcaneocuboid joint, and a serial tarsus, all of which contribute to freer rotation at the transverse tarsal joint, especially for eversion. Indriids are also distinguished by a plantarly bent anterior

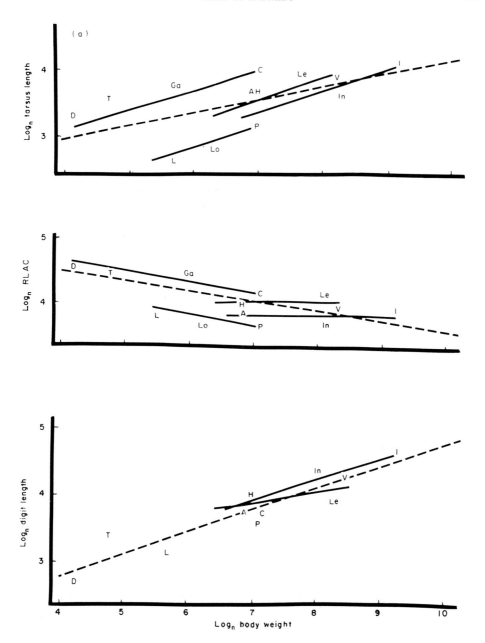

Figure 5. Least squares regression lines calculated for species means of natural log body weight (g) against A, tarsus length; B, relative anterior calcaneal length; C, length of digit 4. Variables are tabulated in Table 5. Values for slope, intercept, and correlation coefficients are given in Table 5. Dotted line, regression line for all strepsirhines; solid lines, regression lines for lemurids (Le), indriids (In), galagines (Ga), lorisines (Lo). Position of the smallest and largest member of each group is marked by small letters: A, *Avahi laniger*; C, *Galago crassicaudatus*; D, *Galago demidovii*; H, *Hapalemur griseus*; I, *Indri indri*; L, *Loris tardigradus*; P, *Perodicticus potto*; T, *Tarsius*; V, *Varecia variegata*.

talocalcaneal facet on the calcaneus. The cuboideonavicular joint is more sellar shaped than in lemurids, and the cuboid-mesocuneiform contact is greatly expanded. Gebo (1985) has described modifications of the tarsometatarsal joints and intrinsic foot musculature which increase grasping power in indriids. Likewise, the pattern of extrinsic foot muscle development in indriids indicates an emphasis on grasping over leaping.

Vertical clinging behavior in indriids is possibly evidenced by the twisted posterior trochlear shelf, the plantarly-bent anterior talocalcaneal facet, the extra retinacular bands of the leg, and the distinctive tarsometatarsal structure and intrinsic foot musculature noted by Gebo (1985), although the mechanical connection between the first three of these traits and vertical clinging has not been strongly established.

The indriid foot shows a few specializations reflecting frequent leaping such as the less curved talocrural joints and the less flaring talofibular joint. The flattening of the subtalar joint facets may act to stabilize this joint. The naviculocuboid joint is also slightly more stabilized than in lemurids, but the difference is not great. The pedal adaptations of indriids which can be mechanically correlated to leaping are only subtly different from lemurids, and tend to be concentrated at the upper ankle joint. Indriids do not show the extreme modifications of tarsus lengthening seen in galagines and *Tarsius* or the modifications of the transverse tarsal joint, naviculocuneiform joints, and naviculocuboid joint which characterize these groups (Dagosto, 1986; Gebo, 1986, 1987a).

Most of the other characteristically indriine features (the serial tarsus, the spherical talonavicular joint, the high, flat calcaneocuboid joint, and the large cuboid-mesocuneiform contact) are adaptations which increase mobility and/or folding of the foot in the midtarsal region. The relatively short tarsus, long digits, and long hallux are surprising features in a specialized prosimian leaper. The relative weights of muscles in the leg shows an emphasis on grasping over leaping musculature. These aspects of the indriid foot are likely to be more mechanically advantageous for climbing and/or grasping than leaping.

Indriids thus have a foot which is an interesting compromise for leaping, climbing, and vertical clinging. The adaptations for leaping are largely confined to the upper ankle joint whereas climbing and vertical clinging are reflected in the transverse tarsal joint, distal tarsal and metatarsal structure, and leg and foot musculature.

Although allometric considerations are important, they are not the total explanation for differences between indriids and small vertical clingers and leapers. Indriids are not scaled up versions of galagos. The fact that indriids do not follow galago–tarsier trends suggests that the indriid ancestor adopted frequent leaping only after it was relatively large (i.e., *Propithecus verreauxi*-sized). In such a heavy animal, lengthening the tarsal bones is not an ideal solution to deal with the demands of leaping. In an indriid with a calcaneus and navicular as proportionately long as those of a galago, the bending forces that the distal end of the bone would be subject to (which are proportional to the animals weight and the length of the bone) are likely to be proportionally larger than in a galago. To resist these forces, the bones of the indriid would have to be proportionately much larger in diameter, or very thick and heavy. If this reasoning is correct, *Avahi* is best interpreted as a dwarfed species.

Godfrey (1977) recognized three postural/locomotor activities which have influenced the form of the indriid skeleton: vertical sitting and clinging, hanging, and leaping. The features which Godfrey associated with hanging (see below) might equally well be related to climbing. Adaptations for vertical posture are largely confined to the vertebral column (i.e., increased number of lumbar and sacral vertebrae) (Godfrey, 1977). Besides the features of the foot we have identified, many other features in the indriid skeleton can be associated with climbing or hanging. The infraspinous fossa of the scapula is better developed in indriids (Ashton & Oxnard, 1964; Roberts, 1974), the greater and lesser tuberosities of the humerus are below the humeral head, the infraspinous pit is directed craniolaterally instead of laterally, and the humeral head is transversely compressed, elongated sagittally and faces backwards and upwards (Godfrey, 1977). The hand and the digits of the hand are long (Bishop, 1964; Jouffroy, 1975; Jouffroy & Lessertisseur, 1979) and the thumb, like the big toe, is well set off from the other digits (Cartmill, 1974).

As noted above, there are very few features of the indriid foot which suggest that leaping is a more frequent component of locomotion than in lemurids. If isolated indriid feet were found as fossils, they would no doubt be attributed to a climbing or suspensory animal. At the same time, it is obvious that leaping is indeed a frequent and important component of indriid locomotion. In contrast to tarsiers and galagines which use the elongate tarsus as an extra lever element to increase the maximum distance attainable in a leap (Hall-Craggs, 1965; Jouffroy *et al.*, 1984; Gebo, 1986, 1987*a*), indriids use primarily only the femur for this purpose. Thus the leaps of the small vertical clingers and leapers are at least partly "foot-powered", while those of indriids are "thigh-powered". Although we have no data on hip and thigh muscle weights, Jouffroy (1962) notes a better developed iliopsoas and gluteus medius in indriids as well as expanded attachment areas for these muscles.

Indriid evolutionary history

The family Indriidae, which includes the extant indriines and the subfossil genera *Archaeolemur, Hadropithecus, Palaeopropithecus, Archaeoindris*, and *Mesopropithecus*, is probably the most diverse family of Malagasy prosimians. A tremendous range of body size is covered, and not unexpectedly, there is a wide range of locomotor types (Lamberton, 1939; Walker, 1967; Jungers, 1980). All of these locomotor types are apomorphic compared to the probable common ancestor of lemurids and indriids, making an assessment of the history of locomotor evolution in the Indriidae difficult at best. Some of the multiple possible interpretations are discussed below.

An hypothesis of phylogenetic relationships in the Indriidae and related taxa is presented in Figure 6 (after Schwartz & Tattersall, 1985). There is some disagreement among authors as to the relative placement of the Archaeolemurinae and the Paleopropithecinae (compare Tattersall, 1982 with Schwartz & Tattersall, 1985; Szalay & Delson, 1979), but for the purposes of this discussion, the difference is irrelevant.

Current evidence suggests that indriids and lemurids form a sister group relative to other lemuriforms (Dene *et al.*, 1976; Sarich & Cronin, 1976; Szalay & Delson, 1979; Tattersall, 1982; but see Rumpler *et al.*, 1983). The common ancestor of this group (Node 1 in Figure 6) was, in terms of postcranial anatomy and locomotor behavior, more similar to a lemur than to any indriid (Gebo, 1985; Dagosto, 1986); that is to say, it was an arboreal quadruped capable of leaping and climbing, but not as specialized for these activities as are indriids. As discussed above, the extant indriids (the Indriinae) exhibit specializations for vertical climbing, vertical clinging, and leaping which are derived relative to this common ancestor. Among the subfossil indriids are arboreal/terrestrial pronograde quadrupeds (archaeolemurines), arboreal sloth-like forms (*Palaeopropithecus*), and ground sloth-like forms (*Archaeoindris*) (Lamberton, 1939; Walker, 1967; Godfrey, 1977; Jungers, 1980). None resemble our concept of the lemurid-indriid common ancestor or indriines.

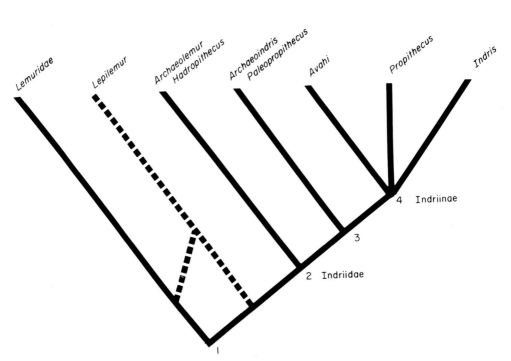

Figure 6. Hypothesis of relationships of indriids and lemurids. Node 1, lemurid-indriid common ancestor. Node 2, indriid common ancestor. Node 3, indriine-paleopropithecine common ancestor. Node 4, indriine common ancestor.

Tarsal bones have only been described for archaeolemurines (Lamberton, 1939). These subfossil indriids share none of the special pedal features which distinguish indriines from lemurids (Dagosto, 1986). The tali of *Archaeolemur* and *Hadropithecus* differ from those of indriines in that: the posterior trochlear shelf is shorter and not twisted, the talofibular facet is not as vertical, the medial talotibial facet is deeper and more medially curved, and the talar head is less spherical and is not as strongly dorsolaterally oriented. The calcanei differ in that the heel is proportionately longer; RLAC is very short; the posterior talocalcaneal facet is shorter and wider and is not aligned close to the long axis of the bone; the anterior talocalcaneal facet faces dorsally rather than medially; and the calcaneocuboid facet is longer, lower, and deeper than in indriines. In most of these features, archaeolemurines more closely resemble lemurids than indriines. We interpret this to mean that the ancestral indriid (Node 2 in Figure 6) was not very different from the lemurid–indriid common ancestor in postcranial morphology or locomotion. Thus, the three subfamilies of indriids independently diverged from this common ancestor in three different ways.

This simple hypothesis is complicated by several factors. One is the observation by Godfrey (1977) that archaeolemurines and palaeopropithecines share features with the indriines which are related to vertical posture (i.e., increased number of lumbar and sacral vertebrae and expansion of iliac blade). This implies that vertical clinging (and probably vertical climbing) was a feature of early indriids modified by archaeolemurine and palaeopropithecine lineages. If indriine-like frequent leaping was also part of the ancestral indriid adaptation, independent reversal of all derived foot (and many other postcranial) features in both subfossil groups is required.

The second complication concerns the phyletic position of *Lepilemur*. *Lepilemur* is usually considered to be highly derived lemurid, but there is evidence that suggests a potential derivation from the indriid line. In 1873 Mivart (p. 490) wrote that "*Lepilemur* seems to be that genus of the Lemurinae which most approximates the Indrisinae". He based his conclusions on the fact that these taxa share fusion of the os centrale with the scaphoid (but they remain separate in *Propithecus*), and some resemblances in the vertebral column. Szalay & Dagosto (1980) describe some similarities in morphology of the distal humerus (e.g., short, high humeral trochlea), and Godfrey (1977) lists other vertebral features shared by *Lepilemur* and indriids (e.g., increased number of lumbar and sacral vertebrae). *Lepilemur* also shares some derived pedal features with indriines (Dagosto, 1986) including reduction of curvature of the components of the upper ankle joint, a high talar body, a long upper ankle joint arc, a large cuboid-mesocuneiform articulation, a more sellar-shaped cuboidnavicular articulation, a strong dorsolateral orientation of the talar head, and a strong plantar bend to the anterior talocalcaneal facet. *Lepilemur* does not, however, share the large, posteriorly bent trochlear shelf, the foot proportions, the serial tarsus, or the spherical talar head of indriines. Since *Lepilemur* exhibits only a part of the indriine complex of postcranial features, it is possible that the similarities are the result of convergent evolution due to similar locomotor modes. However, a close relationship between *Lepilemur* and indriids is supported by craniodental morphology (Schwartz & Tattersall, 1985), albumin and transferrin immunological distance (Dene *et al.*, 1976), but not by whole serum immunological distance (Sarich & Cronin, 1976).

If *Lepilemur* is indeed an early offshoot of the indriid line, the ancestral indriid was likely to have been a vertical clinger and leaper. This scenario, like the previous one, requires the features associated with vertical clinging and leaping to have been lost/modified independently in the archaeolemurine and palaeopropithecine lineages. These types of morphological changes would not be unexpected given the distinctive locomotor adaptations in these forms.

A further complication arises with *Mesopropithecus*, which is probably closely related to *Propithecus* (Tattersall, 1982; but see Godfrey, 1986). Its postcranial bones indicate an arboreal quadrupedal climber, not a specialized leaper (Walker, 1967; Jungers, 1980; Godfrey, 1986; Dagosto, 1986). In either of the two above scenarios, *Mesopropithecus* would thus have to have modified many indriine traits. Including *Daubentonia* as a member or closer relative of the Indriidae (Tattersall & Schwartz, 1974; Schwartz & Tattersall, 1985; but see Groves, 1974) introduces even further complications since it does not share derived foot or postcranial features with indriids (Dagosto, 1986).

Clearly, insufficient evidence is available to resolve this issue. Pending further data, we prefer the first "simple" hypothesis and thus we view *Lepilemur* as a lemurid with postcranial features convergent with indriines. Just when increased body size and more strongly developed morphological adaptations for climbing became typical for indriids is not clear, but the Palaeopropithecinae and Indriinae share features related to climbing and hanging suggesting that their common ancestor (Node 3, Figure 6) may have been more suspensory than the ancestral indriid (Godfrey, 1977).

Conclusion

The differences in foot structure between the extant indriids and other prosimian vertical clingers and leapers is related to differences in overall locomotor pattern. Compared to tarsiers and galagos, indriines leap less often, and use bimanual suspension more often. They are also much larger. The foot anatomy of indriines differs from that of other vertical clingers and leapers (and lemurids) in its long and divergent hallux, more spherical talonavicular joint, flatter calcaneocuboid articulation, serially arranged tarsus, and better developed deep flexor muscles. They differ from tarsiers and galagos, but not lemurids, in the relative length of the tarsus and degree of anterior calcaneal elongation. This set of features, plus other indriine autapomorphies, is more likely to be advantageous for climbing/grasping than any other locomotor behavior. There are a few indriine features which may be related to vertical clinging (e.g., retinacular pattern and slope of sustentaculum) and a few differences from lemurids which reflect more frequent leaping (e.g., less curved and flared upper ankle joint).

A myriad of complicating factors prevents a conclusive resolution of the history of indriid locomotor behavior. In the simplest hypothesis, the similarity between archaeolemurine and lemurid foot bones suggests that the indriid ancestor was more probably lemur-like than indriine-like. The three subfamilies of indriids, which are all distinctive postcranially and in terms of presumed locomotor behavior from the indriid common ancestor and from each other, diverged independently from this ancestor. *Lepilemur* is convergent with indriines in some aspects of the postcranial skeleton.

The alterative hypothesis, which is supported by some similarities in the trunk skeleton among all indriids, and the similarities between *Lepilemur* and indriines, suggests that VCL was the ancestral indriid adaptation. This requires independent reversal of features related to VCL in both subfossil subfamilies. At this time, we prefer the first hypothesis. More intensive scrutiny of other areas of the postcranium and increased knowledge of the morphology, behavior, and relationships of the subfossil indriids will help to resolve these issues.

Acknowledgements

We would like to thank Dr E. L. Simons and the staff of the Duke University Primate Center for their generous help and assistance. Drs G. G. Musser (Department of Mammalogy, American Museum of Natural History) and R. Thorington (Division of Mammals, National Museum of Natural History) provided access to specimens. Anonymous reviewers improved the quality of the manuscript. This work was partly supported by NSF BNS82-03982 to M.D. and Johns Hopkins University Postdoctoral fellowships to M.D. and D.L.G.

References

Ashton, E. H. & Oxnard, C. E. (1964). Locomotor patterns in primates. *Proc. Zool. Soc. London.* **142,** 1–28.

Biegert, J. (1963). The evaluation of characteristics of the skull, hands, and feet for primate taxonomy. In (S. L. Washburn, Ed.) *Classification and Human Evolution*, pp. 116–145. Chicago: Adline-Atherton.

Bishop, A. (1964). Use of the hand in lower primates. In (J. Buettner-Janusch, Ed.) *Evolutionary and Genetic Biology of Primates*, Vol. 2, pp. 133–225. New York: Academic Press.

Cartmill, M. (1972). Arboreal adaptations and the origin of the Order Primates. In (R. Tuttle, Ed.) *The Functional and Evolutionary Biology of Primates*, pp. 97–122. Chicago: Aldine-Atherton.

Cartmill, M. (1974). Pads and claws in arboreal locomotion. In (F. Jenkins, Ed.) *Primate Locomotion*, pp.. 45–84, New York: Academic Press.

Dagosto, M. (1983). Postcranium of *Adapis parisiensis* and *Leptadapis magnus* (Adapiformes, Primates). *Folia primatol.* **41,** 49–101.

Dagosto, M. (1985). The distal tibia of primates with special reference to the Omomyidae. *Int. J. Primatol.* **6,** 45–75.

Dagosto, M. (1986). The joints of the tarsus in the strepsirhine primates: functional, adaptive, and evolutionary implications. Ph.D. Dissertation, City University of New York.

Decker, R. L. & Szalay, F. S. (1974). Origin and function of the pes in the Eocene Adapidae (*Lemuriformes, Primates*). In (F. A. Jenkins, Ed.) *Primate Locomotion*, pp. 261–291, New York: Academic Press.

Dene, H., Goodman, M., Prychodko, W. & Moore, G. W. (1976). Immunodiffusion systematics of the primates II. The Strepsirhini. *Folia primatol.* **25,** 35–61.

Fleagle, J. G. (1977). Locomotor behavior and skeletal anatomy of sympatric Malaysian leaf-monkeys (*Presbytis obscura* and *Presbytis melalophos*). *Yrbk phys. Anthrop.* **20,** 440–453.

Fleagle, J. G. & Mittermeier, R. A. (1980). Locomotor behavior, body size, and comparative ecology of seven Surinam monkeys. *Am. J. phys. Anthrop.* **52,** 301–314.

Gebo, D. L. (1985). The nature of the primate grasping foot. *Am. J. phys. Anthrop.* **67,** 269–278.

Gebo, D. L. (1986). The anatomy of the prosimian foot and its application to the primate fossil record. Ph.D. Dissertation, Duke University.

Gebo, D. L. (1987*a*). Functional anatomy of the tarsier foot. *Am. J. phys. Anthrop.* **73,** 9–31.

Gebo, D. L. (1987*b*). Locomotor diversity in prosimian primates. *Am. J. Primatol.* **13,** 271–281.

Godfrey, L. (1977). Structure and function in *Archaeolemur* and *Hadropithecus*. Ph.D. Dissertation, Harvard University.

Godfrey, L. (1986). What were the subfossil indriids of Madagascar up to? *Am. J. phys. Anthrop.* **69,** 204 (abstract).

Groves, C. P. (1974). Taxonomy and phylogeny of prosimians. In (R. Martin, G. Doyle & A. Walker, Eds) *Prosimian biology*, pp. 449–473, London: Duckworth.

Hall-Craggs, E. C. B. (1965). An osteometric study of the hindlimb of the Galagidae. *J. Anat.* **99,** 119–126.

Jolly, A. (1966). *Lemur Behavior*, Chicago: University of Chicago Press.

Jouffroy, F. K. (1962). La musculature des membres chez les lémuriens de Madagascar. *Mammalia* **26,** 1–326.

Jouffroy, F. K. (1975). Osteology and myology of the lemuriform postcranial skeleton. In (I. Tattersall & R. W. Sussman, Eds) *Lemur Biology*, pp. 149–192. New York: Plenum Press.

Jouffroy, F. K. & Lessertisseur, J. (1979). Relationships between limb morphology and locomotor adaptations among prosimians: an osteometric study. In (M. E. Morbeck, H. Preuschoft & N. Gomberg, Eds) *Environment, Behavior and Morphology: Dynamic Interactions in Primates*, pp. 143–182. New York: Gustav Fischer.

Jouffroy, F. K., Berge, C. & Niemitz, C. (1984). Comparative study of the lower extremity in the genus *Tarsius*. In (C. Niemitz, Ed.) *Biology of Tarsiers*, pp. 167–190, New York: Gustav Fischer.

Jungers, W. L. (1979). Locomotion, limb proportions, and skeletal allometry in lemurs and lorises. *Folia primatol.* **32,** 8–28.

Jungers, W. L. (1980). Adaptive diversity in subfossil Malagasy prosimians. *Z. Morph. Anthrop.* **71,** 177–186.

Jungers, W. L. (1985). Body size and scaling of limb proportions in Primates. In (W. L. Jungers, Ed.) *Size and Scaling in Primate Biology*, pp. 345–381. New York: Plenum.

Lamberton, C. (1939). Des os du pied de quelques Lémuriens subfossiles malgache. *Mem. de l'Acad. Malgache* **27,** 75–139.

Martin, R. D. (1972). Adaptive radiation and behavior of the Malagasy lemurs. *Phil. Trans. Roy. Soc. Lond.* **264,** 295–352.

Mivart, St. G. (1867). On the appendicular skeleton of the primates. *Phil. Trans. Zool. Soc. London* **157,** 299–429.

Mivart, St. G. (1873). On *Lepilemur*, and *Cheirogaleus*, and on the zoological rank of the Lemuroidea. *Proc. Zool. Soc. London,* 484–510.

Napier, J. R. & Walker, A. C. (1967). Vertical clinging and leaping: a newly recognized category of primate locomotion. *Folia primatol.* **6,** 204–219.

Oxnard, C. E., German, R. & McArdle, J. E. (1981). The functional morphometrics of the hip and thigh in leaping prosimians. *Am. J. phys. Anthrop.* **54,** 481–498..

Petter, J. J. (1962). Recherches sur écologie et l'éthologie des lémuriens malgaches. *Mem. Mus. Nat. d'Hist. Nat.* **27,** 1–146.

Petter, J. J., Albinac, R. & Rumpler, Y. (1977). Mammifères lémuriens (Primates, Prosimii). *Faune Madagascar* **44,** 1–513.

Pollock, J. I. (1975). The social behavior and ecology of *Indri indri*. Ph.D. Dissertation, University of London.

Pollock, J. I. (1977). Field observations on *Indri indri*, a preliminary report. In (I. Tattersall & R. W. Sussman, Eds) *Lemur Biology*, pp. 287–312. New York: Plenum.

Rand, A. L. (1935). On the habits of some Madagascar mammals. *J. Mammal.* **16,** 89–104.

Richard, A. (1978). *Behavioral Variation: Case Study of a Malagasy Lemur.* Lewisburg: Bucknell University Press.

Roberts, D. (1974). Structure and function of the primate scapula. In (F. A. Jenkins, Ed.) *Primate Locomotion*, pp. 171–200, Academic Press: New York.

Rumpler, Y., Couturier, J., Warter, S. & Dutrillaux, B. (1983). Chromosomal evolution in Malagasy lemurs. *Cytogenet. Cell Genet.* **36,** 542–546.

Sarich, V. M. & Cronin, J. E. (1976). Molecular Systematics of the Primates. In (M. Goodman, R. E. Tashian and J. E. Tashian, Eds) *Molecular Anthropology*, pp. 141–170. New York: Plenum Press.

Scwartz, J. H. & Tattersall, I. (1985). Evolutionary relationships of living lemurs and lorises (Mammalia, Primates) and their potential affinities with European Eocene Adapidae. *Anth. Papers Am. Mus. nat. Hist.* **60,** 1–100.

Stern, J. & Oxnard, C. E . (1973). Primate locomotion: some links with evolution and morphology. *Primatologia* **4,** 1–93.

Sussman, R. W. (1972). *An ecological study of two Madagascan primates: Lemur fulvus rufus* Audebert and *Lemur catta* Linnaeus. Ph.D. Dissertation, Duke University.

Sussman, R. W. (1974). Ecological distinctions in sympatric species of *Lemur*. In (R. D. Martin, G. A. Doyle & A. C. Walker, Eds) *Prosimian biology*, pp. 75–108. London: Duckworth.

Szalay, F. S. (1972), Paleobiology of the earliest primates. In (R. Tuttle, Ed.) *The Functional and Evolutionary Biology of Primates*, pp. 3–35. Chicago: Aldine-Atherton.

Szalay, F. S. & Dagosto, M. (1980). Locomotor adaptations as reflected on the humerus of Paleogene primates. *Fol. primatol.* **34,** 1–45.

Szalay, F. S. & Delson, E. (1979). *Evolutionary History of the Primates.* New York: Academic Press.

Tattersall, I. (1977). Ecology and behavior of *Lemur fulvus mayottensis* (Primates, Lemuriformes). *Anthrop. Papers Amer. Mus. nat. Hist.* **54,** 421–482.

Tattersall, I. (1982). *The Primates of Madagascar.* New York: Columbia University Press.

Tattersall, I. &·Schwartz, J. H. (1974). Craniodental morphology and the systematics of the Malagasy lemurs. *Anthrop. Papers Am. Mus. nat. Hist.* **52,** 139–192.

Tattersall, I. & Sussman, R. W. (1975). Observations on the ecology and behavior of the mongoose lemur *Lemur mongoz mongoz* Linnaeus (Primates, Lemuriformes) at Ampijoroa, Madagascar. *Anthrop. Papers Am. Mus. nat. Hist.* **52,** 193–216.

Walker, A. C. (1967). Locomotor adaptations in recent and fossil Madagascan lemurs. Ph.D. Dissertation, University of London.

Walker, A. C. (1970). Postcranial remains of the Miocene lorisidae of East Africa. *Am. J. phys. Anthrop.* **33,** 249–261.

Walker, A. C. (1974). Locomotor adaptations in past and present prosmian primates. In (F. A. Jenkins, Ed.) *Primate Locomotion*, pp. 349–381. New York: Academic Press.

Walker, A. C. (1979). Prosimian locomotor behavior. In (G. A. Doyle & R. D. Martin, Eds) *The Study of Prosimian Behavior*, pp. 543–565. New York: Academic Press.

Susan M. Ford

Department of Anthropology,
Southern Illinois University,
Carbondale, IL 62901, U.S.A.

Received 9 July 1987 Revision
received 23 December 1987 and
accepted 20 January 1988

Publication date June 1988

Keywords: Platyrrhini,
postcranial anatomy,
adaptations, euprimates, Fayum.

Postcranial adaptations of the earliest platyrrhine

A model of the earliest platyrrhine primate is generated as the result of a cladistic analysis of platyrrhine relationships. The morphological features of the postcranium hypothesized as present in this model are analyzed in order to suggest the adaptations and behavioral repertoire of the first platyrrhine. A consistent pattern emerges indicating a grasping, arboreal quadruped, emphasizing walking and running on horizontal supports with only limited leaping or suspensory activities, approximately 1000 g in body mass. It is suggested that this pattern, with only minor changes, also characterized the earliest euprimates. The major derivations in the first platyrrhines were increased body size from a smaller euprimate ancestor, and improvements in stability as arboreal quadrupeds. The fossil record indicates that platyrrhines evolved as a distinct group before reaching South America, with possible relatives remaining behind in Africa.

Journal of Human Evolution (1988) **17**, 155–192

Introduction

Studies on aspects of the morphology of New World monkeys have indicated that specializations in the postcranial skeleton are key to defining this monophyletic group of primates (Ford, 1980*a*, 1986*a*). This suggests that associated novel aspects of locomotor behavior of the earliest platyrrhine may have been central factors in the divergence of this group from other early anthropoids. The intent of this paper is to generate a description of what the earliest platyrrhine primate may have looked like in terms of a number of key postcranial features and, based on this, to try to reconstruct how this monkey may have moved through the trees and utilized its environment.

Much interest has been expressed in the origin of the Anthropoidea, ways in which the anthropoids differ from other primates, and what the ancestor of anthropoids may have looked like and how it may have behaved (e.g., Gingerich, 1980; Rosenberger & Szalay, 1980; Rosenberger, Strasser & Delson, 1985; Rasmussen, 1986; Schwartz, 1986). In the postcranium, this debate has largely centered around the question of a vertical clinging and leaping ancestor (Napier & Walker, 1967) versus a hindlimb dominated arboreal quadruped utilizing primarily horizontal supports (Martin, 1972; Szalay & Dagosto, 1980; Rollinson & Martin, 1981). Recently, Gebo (1986*a*,*b*) has supported the idea of a pre-anthropoid ancestor utilizing vertical supports, but with more vertical climbing as opposed to leaping. He hypothesizes a switch to horizontal supports in early anthropoids, perhaps associated with increased body size.

All of these models have been drawn primarily from study and consideration of locomotor and postural habits in modern strepsirhine primates and Eocene fossils. Yet the functionally generalized nature of many platyrrhine skeletons has often been commented on (Fleagle & Simons, 1978; Fleagle, 1983; Rose, 1983). A better understanding of the structure and possible associated habits of the earliest platyrrhines has great bearing on the issue of the adaptive strategies of the earliest anthropoids and euprimates.

Methods, or why and how to study the earliest member of the group

There are at least four good reasons for attempting to identify or reconstruct this earliest New World monkey (or the putative ancestor of any taxonomic group):

0047–2484/88/01/20155 + 38

(1) The most obvious reason is that a concept of what the ancestor looked like provides a more complete and better understanding of the evolutionary history of the group.

(2) Significant changes within the group since its founding can be identified and traced if a sound idea of the baseline or original member of the group is known. In fact, one could make a strong argument (as cladists have) that it is only by reconstructing the structure and morphology of the ancestor of a group that one can truly understand the sequence of evolutionary changes that have occurred within the group and their significance. Morphological adaptations to the environment and the behavioral repertoire of an animal do not occur in a void, but build upon the genetically determined structure that is inherited from an organism's ancestors.

(3) Comparisons of different groups as a whole, for example to determine how platyrrhines *as a group* solved basic problems of survival and reproduction differently from the approaches adopted by catarrhines or strepsirhines, are only appropriate when contrasting the basal adaptive strategies and associated morphologies of each group.

(4) Knowledge of the earliest New World monkey allows a better understanding of what is unique about platyrrhines as a group, what ways they have changed from their earlier primate ancestors and relatives, and what new and different strategies the earliest platyrrhines developed to cope with an arboreal life (this reason is in part related to reason 3). In this manner, it can be instructive to identify the ways in which platyrrhines, perhaps more than any other group, may best reflect the ancestral anthropoid or earliest euprimate.

There are two primary means of trying to reconstruct what the earliest member of a group may have looked like. The first, an approach used by paleontologists since fossils were first recognized, is to discover and study the earliest fossil representatives of the group, and through analogy to living animals with similar morphological features, to suggest the possible behavioral repertoire of the early members of the group. Although this is a common exercise in paleontology, it suffers from three serious problems: (1) because the fossil record is incomplete, we often do not have fossils from the earliest time periods of a group's existence. The platyrrhine fossil record is particularly poor (Rose & Fleagle, 1981; Fleagle & Bown, 1983; Ford, 1986a). (2) Complete specimens are extremely rare, again because of problems of preservation and the incomplete nature of the fossil record. Far more frequently, we have only a few or a single postcranial element, as in the case of the Gaiman astragalus (Fleagle & Bown, 1983; Reeser, 1984). Thus, the fossil record is quite limited in its ability to allow us to reconstruct the entire postcranium of the ancestor of the group. (3) Fossils may be highly derived in structure, in a direction away from the ancestral type and perhaps away from any living member of the group. Obviously, these specialized fossil species make a very poor and potentially misleading model for understanding the earliest members of a group. Yet it is sometimes difficult to determine how derived a particular fossil may be, especially when there are few known fossils and the fossils are very incomplete, precisely the situation for New World monkeys.

A subset of this first approach, selecting some extant form as the best model for the ancestral type, is subject to the same probability that it is in fact quite specialized in a number of ways from the common ancestor. Nonetheless, this last is an approach that has been used often in the study of platyrrhines to at least some degree (e.g., Gregory, 1920; Ribeiro, 1940; Hill, 1960; Orlosky, 1973; Rosenberger, 1979; Kay, 1980).

The second way to get a sense of what the earliest member of a group was like is to try to reconstruct this ancestor hypothetically through the examination of known members of the group (both fossil and extant). While cladistic analyses of any group have as their prime objective the elucidation of the phylogenetic relationships and identification of closest sister-taxa of known members of the group, these analyses also yield a characterization of hypothetical ancestral morphotypes. Lists can be produced of the shared derived morphologies or apomorphies of each branching point; these branching points represent hypothetical ancestors of all of the known taxa that branch from that point. By adding to the apomorphic features all of the primitive traits from earlier branching points that have not altered during the process of evolution up to that point, one can suggest a rather complete portrait of this hypothetical ancestor.

This second approach has several advantages. It avoids all three major problems inherent in the use of a particular fossil as the ancestor: there is no lack of hypothetical ancestors; they can be as complete as the number of characters one wishes to examine; and if the analysis is fairly accurate, this hypothetical ancestor should be the most primitive member of the group. An additional advantage is that one can then examine known fossils more critically. Are any known fossils very similar to or exactly like the hypothetical ancestor and therefore possibly represent the remains of that ancestor (Szalay, 1977)? If known fossil specimens appear derived, have they changed from the ancestor in the direction of any extant lineage or do they represent a unique evolutionary lineage within the group? Are any known fossils *more* primitive in some features than the hypothesized earliest common ancestor, and therefore should not be considered members of this monophyletic group? For platyrrhines, this would imply that either: (1) the pioneer primate stock was a more primitive anthropoid and "platyrrhines" developed within South America; or (2) more than one primate stock migrated to South America. And last, if all known South American fossil primates are clearly platyrrhines, can one assume they were "platyrrhines" when they first arrived on the continent, evolving elsewhere or while in transit from either Africa or North America? This last would suggest that we may find already distinct platyrrhines outside of South America.

While the preceding discussion is merely a summary of some of the basic rationale often given for performing cladistic analyses, the focus here is entirely on that hypothesized ancestral morphotype, which can be seen as a centerpiece for an investigation into the nature of the group as a unit and its origins. Certainly, this approach has problems as well, the primary one being that one could make mistakes and presume that certain features are characteristic of this hypothetical monkey, either as apomorphies or retained plesiomorphies, which in fact the earliest platyrrhine ancestor did not possess. But then, the alternative of assigning a particular fossil as ancestor falls prey to the same problem—one could in error assign the wrong fossil this lofty position (witness the ongoing controversies in hominid evolution).

There is a second problem in using a hypothetical morphotype reconstructed from cladistic analysis for functional analysis. A trait by trait analysis is a necessary, though in some ways unfortunate, byproduct of the method of deriving the morphotype. Clearly, in a real organism, anatomical features and regions interact in complex ways during locomotion and other activities. One could combine all aspects of the morphotype and analyze it as though it were a living organism, an approach that appeals to any functional morphologist. However, if errors were made in the reconstruction of a small number of the features, the entire analysis could be void. For the sake of accuracy and honesty, it is necessary to analyze discrete features separately for this hypothetical monkey; yet this limits the thoroughness of the reconstruction and the precision of the locomotor analysis.

However, as Novacek (1986, p. 100) has recently argued, the comparative approach of cladistics "provides a framework and a starting point for further inquiry."

In this study, this second approach is used, that of reconstructing the hypothetical common ancestor or hypothetical ancestral morphotype for platyrrhines. This model of the earliest platyrrhine will be the basis for suggesting aspects of locomotor behavior and substrate use in the earliest platyrrhines. Admittedly, much of this is speculative and may be seen at first glance as nothing more than an interesting flight of fancy. However, it is hoped that this study will demonstrate how important this particular use of "reasoned imagination" can be for evolutionary studies.

In the best of worlds, we would have complete information on the ways in which variations in each of these features affects locomotor and postural behavior and substrate preference in living primates. This is not yet the best of worlds, and the lack of, or conflicting information on, morphology/behavior links makes reconstruction of past behaviors more difficult and less precise than one might like. These problems are fully addressed by Dagosto (1986; see also Morbeck, 1983). Not only do we lack this information relative to normal behavioral patterns of animals, but no one has clearly addressed the relative effect on morphology of normal or common behaviors versus highly infrequent but crucial behavioral strategies in key survival situations. An example of the latter might be a posture adopted only during a brief but severe annual dry season, to reach a crucial food source for survival.

Although the comparative method of extant analogies is widely used in suggesting behavioral parameters for extinct forms (see Kay, 1984), this approach has been heavily criticized by Dagosto (1986). In addition, as Szalay & Dagosto (1980; Dagosto, 1986) have discussed, even if no living species has a particular attribute, careful consideration of biomechanics can sometimes allow one to suggest certain aspects of adaptive significance. This is particularly crucial for the many fossils which exhibit relatively unique traits or combinations of traits (e.g., Langdon, 1986).

Given the problems apparent in the comparative method or use of modern primate analogies, hypotheses of function and adaptive role will be formulated primarily from an understanding of bone mechanics (as in Dagosto, 1986 and elsewhere, and Gebo, 1986a). Analogy to structure and behavior in living primates will be used primarily to support or to test hypotheses of function based on mechanical principles. Locomotor categories will primarily follow those given in Fleagle & Mittermeier (1980).

Kay & Cartmill (1977, and discussed in Kay, 1984) have suggested that traits not be used if there is evidence that they evolved in a lineage before acquiring their present adaptive role. While in theory this would prevent misinterpreting the significance of features which are exaptations in the species under consideration, in practice it is often very difficult to determine if a particular feature served a different purpose in an ancestral form. One way of resolving this is not to consider any plesiomorphic traits (primitive retentions) in reconstructing probable adaptations, an approach that would severely handicap any study and nullify many that have been done in the past. More appropriate as an approach would be to utilize both primitive and derived features but to do so cautiously, on the assumption that features that were maladaptive for the biological role of a species would be eliminated or altered along the way. This of course still leaves the possible "noise" of traits which are non-functional or now serve an altered function in the species being studied. When a discrepancy occurs in interpreting features, and one of the features is derived and the other a primitive retention, the derived feature and its presumed adaptive role should take precedence in determining behavioral capabilities, since it is a newly occurring and thus immediately important attribute for the animal. The primitive feature in this case is "background noise", a feature evolved at some earlier point in time, whose current function is of uncertain relationship to its original function. This again highlights the importance of reconstructing and studying function and adaptive role within the context of phylogeny (see also Szalay & Dagosto, 1980).

Materials

The model of the earliest platyrrhine presented and discussed here is based on selected attributes of a number of joint surfaces. Although the entire skeleton will not be considered, the joints selected sample a large portion of the appendicular skeleton, including aspects of the shoulder and elbow joints of the forelimb, and the hip, knee, upper ankle (tibio-tarsal), subtalar, and transverse tarsal joints of the hindlimb. No features of the distal forelimb or the axial skeleton are included (there is not sufficient information on the distribution of features in these areas in all platyrrhines and many other primates to allow their inclusion at present).

The features comprising the ancestral platyrrhine morphotype are given in Table 1. They are drawn in the manner described above from a cladistic analysis of the skeleton of primates, especially platyrrhines. Figure 1 illustrates the cladogram of platyrrhine relationships that resulted from the studies that also generated the ancestral morphotype

Table 1	Features of the hypothetical ancestral platyrrhine[1]

Humerus

    1. entepicondylar foramen present

    2. entepicondylar foramen located medially, over medial epicondyle

    3. trochleocapitular ridge rounded, very distinct[2]

*   4. slight dorsal displacement of medial epicondyle[2]

    5. supinator crest wide[2]

**   6. dorsal epitrochlear fossa or at least concavity present[2]

**   7. epitrochlear notch present, of moderate depth in some

    8. ventral trochlea cone-shaped[2]

    9. dorsal trochlea spool-shaped

  10. radial fossa larger than coronoid fossa

  11. bicipital groove shallow and wide[2]

* 12. deltopectoral crest flattens superiorly[2]

  13. relative length of humerus moderate, index (L/ML of femur × 100) between 72–83

* 14. relative width of lateral epicondyle reduced, mean (DLEW/BEW × 100) less than 25[2]

  15. humeroradial aspect of distal joint transmits slightly more weight than humeroulnar, index (VCW/DTW × 100) between 105–120[2]

  16. relative breadth of distal epiphysis moderate, index (BEW/L × 100) between 18–20

  17. relative breadth of distal articular surface moderate, index (ASW/L × 100) between 13–16

  18. capitulum not spherical, index (CD/VCW × 100) between 53–70

  19. trochlea not strongly cylindrical, index (MinTD/MaxTD × 100) between 72–82[2]

Table 1 *continued*

Femur
    1. slight to no extension of facet of head on posterior neck[2]
    2. fovea capitis offset posteriorly
a   3. deepest point of notch between head and greater trochanter is midway, bisects distance
*? 4. at least low rounded eminence (mound) on posterior proximal femur (posterior neck)[2]
*   5. intertrochanteric crest extends distally below fossa in at least some (vs. no crest)
    6. lesser trochanter extends postero-medially[2]
    7. area of insertion of iliopsoas tendon is round
    8. straight shaft, no anterior bow
*   9. remnant only of third trochanter—slightly raised or roughened area
   10. medial condyle extends farther inferiorly[2]
   11. condyles subequal extension posteriorly[2]
   12. patellar groove of moderate depth[2]
   13. lateral margin of patellar groove rounded
   14. facet of patellar groove curls over lateral margin
   15. medial margin of patellar groove sharp
   16. lateral margin of patellar groove extends farther proximally
   17. neck wide, index (ND/PSTD × 100) between 102–120
   18. relative transverse trochanter breadth moderate, index (TTB/LL × 10) between 18–21
   19. relative length of third trochanter (or remnant area) short, index (3TL/LL × 100) between 4–7
   20. location of third trochanter mid-range (approximately opposite lesser trochanter), index (3TGTL/LL × 100) between 11–14
   21. patellar groove moderate to moderately narrow, index (SPW/BEB × 100) less than 46[2]
a 22. distal epiphysis wider than deep, index (DED/BEB × 100) between 76–83
   23. head and greater trochanter subequal proximal extension, index (HP/GTP × 100) between 88–114[2]
a 24. lesser trochanter small to moderate extension, index ((LTD-BLTD)/PSTD × 100) less than 75[2]
   25. medial condyle slightly larger than lateral, index ((MCW X MCH)/(LCW X LCH) × 100) between 105–130[2]
   26. medial condyle a little less than twice as tall as wide, index (MCW/MCH × 100) between 52–64

Tibia (distal only)
    1. distal shaft triangular in cross-section
*   2. ridge anterior to fibular facet absent in at least some
** 3. fibular facet moderately long to long[2]
    4. fibular facet shallow
** 5. fibular facet posteriorly displaced to central or more posterior location[2]
    6. anterior trochlear (lateral tibio-astragalar facet) border extends farther inferiorly
*   7. posterior endpoint of median antero-posterior trochlear crest more medial in some or all
    8. posterior border of trochlea flat
    9. anterior border of trochlea sharp
   10. trochlear facet has moderate extension onto anterior face of shaft[2]

Table 1 *continued*

    *   11. posterior malleolar groove extends straight down over tip of malleolus

         12. horizontal posterior malleolar groove present

         13. lateral flange on posterior malleolar groove present

         14. trochlear facet moderately rectanguar, index (FW/FL × 100) between 110–130

         15. distal shaft fairly equal dimensions, index (SMLD/SAPD × 100) between 95–120

         16. relative height of medial malleolus moderate, index (MMH/FL × 100) between 57–70

         17. relative width of medial malleolus moderate, index (MMW/FL × 100) between 75–95

         18. medial malleolus slightly taller than wide, index (MMH/MMW × 100) between 72–90

         19. distal tibia and fibula unfused[2]

Calcaneus

a   1. peroneal tubercle of moderate size

a   2. peroneal tubercle approximately central on long axis of bone (vs. anterior)

      3. sustentaculum astragali with "waist"

*   4. anterior articular surface (AAS, anterior calcaneo-astragalar facet) reaches distal end of bone only in small point or not at all in some individuals[2]

a   5. AAS and medial articular surface (MAS, medial calcaneo-astragalar facet) confluent in at least some individuals[2]

*   6. AAS large oval, slightly reduced from round (change from Ford, 1980*a*, 1986*a*).

*   7. MAS oval (not round)

a   8. cuboid articular surface (CAS, calcaneo-cuboid facet) slightly concave (not flat or with deep "cuboid pivot")[2]

*   9. pit for inferior calcaneo-cuboid ligament absent in at least some individuals

*   10. CAS fat "L" or "U" shape (to thin, vs. "D")

      11. Sustentaculum forms angle of 10–40° to flat latero-plantar surface of bone

*   12. both ends of calcaneus oblique to long axis of bone in at least some

*   13. navicular articulation (calcaneo-navicular facet) present in at least some[2]

*   14. sulcus for tendon *m. flexor digitorum fibularis* moderately narrow and moderately deep

      15. lateral surface of heel lacks concavity

      16. tuber calcanei small and indistinct

      17. groove for tendo calcanei flat or shallow

      18. angle of posterior articular surface (∠PAS, posterior calcaneo-astragalar facet) to long axis of bone between 10–20°[2]

      19. relative length of peroneal tubercle index (PTL/L × 100) between 30–50 (highly variable, poor trait)

      20. CAS fairly perpendicular to bone, index (3PtL/L × 100) between 97–99 (vs. mediolaterally oblique)[2]

      21. relative length of heel short to medium, index (PASGT/L × 100) between 22–32[2]

      22. relative length of PAS short to moderate, index (PASL/L × 100) less than 38[2]

      23. relative width of PAS moderate, index (PASW/L × 100) between 17–24[2]

      24. shape of PAS rectangular, length approx. 2X width, index (PASW/PASL × 100) between 45–59[2]

Table 1 *continued*

25. calcaneus of moderate relative length and width,
    index (W/L × 100) between 46–56
26. dorsoventral height of bone moderate to low, index
    (DVH/L × 100) between 36–44

[1] Characters are listed by bone, not by joint, as some features are not directly associated with a particular joint complex. Abbreviations for measurements are explained and measurements illustrated in Figures 2–5. Notations in the left-hand column are as follows:

a   is a shared derived trait of the Anthropoidea, but a primitive retention in the earliest platyrrhine;
*   is a derived trait first appearing in the hypothetical ancestral platyrrhine, albeit one of low value (that is, it also develops independently in many other primates and/or the derived condition is subsequently lost by a number of platyrrhines);
**   is a derived trait of platyrrhines considered of high weight, completely or nearly unique to platyrrhines;
—   no symbol indicates that the feature is interpreted as a retained generalized mammalian or early primate condition.

[2] These features are discussed in the text. For all other features, at present there is not sufficient information to discuss in detail the implications of differing morphologies for behavior and adaptive role.

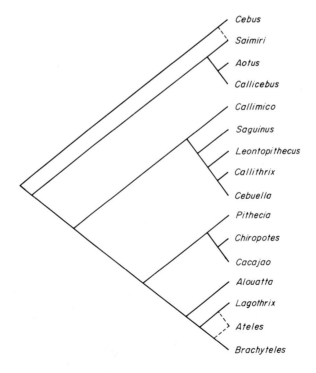

Figure 1. Cladogram of platyrrhine relationships resuting from the same cladistic analysis that generated the hypothetical ancestral platyrrhine morphotype discussed in text (from Ford, 1986a).

discussed here, and represents the most parsimonious arrangement based on a combined analysis of postcranial and cranial anatomy, dentition, karyotypic and immunodiffusion data, and other miscellaneous traits (see Ford, 1986a). Analysis of the overall relationships within the Platyrrhini is described fully elsewhere (Ford, 1980a, 1986a), including detailed explanations of determination of character states, polarities and morphoclines, recognition of homology or homoplasy, and weighting of traits, as well as descriptions of the various character states and their distributions. These topics will be dealt with only briefly here.

In reconstructing the ancestral morphotype, determination of the polarity of morphoclines of change for traits is crucial. While several criteria were employed, it is important to note here that heaviest emphasis was placed on the out-group criterion in the determination of polarity for individual characters (e.g., Eldredge & Cracraft, 1980; Wiley, 1981). Thus, presence and distribution of traits within platyrrhines were relatively insignificant in determining polarity. Instead, presence of a particular character state in a large number of members of the "out-group" was used to determine the primitive condition for the morphocline. In the past there has been considerable controversy over the monophyletic nature of the Anthropoidea and over which Paleogene group of primates is most likely to have included the direct ancestors of the Anthropoidea and, by extension, the Platyrrhini (e.g., Ford, 1980b, 1986a; Rosenberger & Szalay, 1980). Therefore representatives of all major groups of primates, extinct and extant, were included in the out-group as well as a number of generalized and/or early mammals (Table 2). For those features in which each major primate group exhibited a different morphology, presence of a character state in early or generalized mammals (a broader out-group) or other, less satisfactory criteria were used to determine polarity.

Early analyses were done using a rooted, parsimony-based program for cladistic analysis (Wagner79), with the root being the primitive state for each trait as determined above. Thus, the method of analysis essentially coded for the primitive primate state, but not for how or how much each lineage branching from that had changed. These analyses included all extant primates listed in Table 2 (because of incomplete data sets, fossils could not be included in the computer runs). All platyrrhines grouped together, as did all catarrhines included, and these formed a sister-group to all strepsirhines. Later modifications based on more complete data sets for platyrrhines were based on the same principles, but did not use a computer-based algorithm because there were not complete distributions for the additional data (Ford, 1986a).

Characters were not formally weighted prior to analysis. However, some characters are clearly of greater value than others. Characters that were more variable (although highly variable features were excluded entirely) or that exhibited large amounts of homoplasy (either repeated convergent development or reversals in lineages) were considered of less importance in deciding branching points that were not supported by a large number of shared, derived traits. These are discussed below where appropriate. Traits exhibiting some degree of intrageneric variability were included if some taxa appeared non-variable. That is, if a character state was present in all members of one or more taxa, absent in all members of other taxa, but variably present in yet some other taxa, it was included in the study in the belief that this reflects meaningful differences in the evolution of the taxa involved (Cartmill, 1978). In these cases, it is recognized that taxa for which only one or a very few specimens were examined could be miscoded by one character state (as having or lacking the feature rather than exhibiting variability); this does not pose a major problem in a multi-state character for the study as a whole. Indices were used for a number of

Table 2   **List of taxa examined for polarity determination***

Species	Number
Platyrrhini:	
Alouatta belzebul	2
A. caraya	1
A. palliata	1
A. seniculus	2
Aotus trivirgatus	17
Ateles belzebuth	1
A. geoffroyi	3
A. paniscus	2
Brachyteles arachnoides	1
Cacajao calvus	1
C. rubicundus	3
Callicebus moloch	4
C. pallescens	1
Callimico goeldii	5
Callithrix argentata	1
C. jacchus	6
Cebupithecia sarmientoi	1
Ceboid H	1
Ceboid M	1
Cebuella pygmaea	4
Cebus albifrons	1
Cebus apella	2
C. capucinus	1
C. nigrivittatus	2
Chiropotes satanas	4
C. sp.	2
Lagothrix lagothriche	5
Leontopithecus rosalia	5
Pithecia pithecia	5
Saguinus fuscicollis	1
S. imperator	2
S. leucopus	1
S. midas	1
S. oedipus geoffroyi	1
S. sp.	1
Saimiri sciureus	6
OUT-GROUPS	
Catarrhini	
Cercopithecus aethiops	1
C. mitis	1
Macaca fuscata	1
M. mulatta	1
M. thibetana	1
Nasalis larvatus	1
Presbytis melalophos	1
Rhinopithecus roxellanae	1
Hylobates lar	1
Pan paniscus	1
Pan troglodytes	1
Tarsiidae	
Tarsius spectrum	1
T. syrichta	1
Strepsirhini	
Lemur catta	1
L. macaco	1

Table 2 *continued*

L. mongoz	1
Avahi laniger	1
Daubentonia madagascariensis	1
Galago crassicaudatus	1
(portions of some lorisoids examined)	
Fossil primates	
Hemiacodon gracilis	1
Adapis parisiensis	1
Leptadapis magnus	1
Notharctus tenebrosus	4
N. sp.	1
Smilodectes gracilis	1
Pliopithecus vindobonensis (cast)	1
Aegyptopithecus zeuxis	1
parapithecids, unassigned	5
unassigned, Fayum specimens	9

* This list includes those specimens examined by the author; all were wild-caught adults, execpt some platyrrhines where wild-caught specimens were not available in collections. The out-group taxa were chosen as a sample, avoiding obviously highly specialized members of groups where possible. Descriptions available in the literature for other primates and many mammals (fossil and extant) were also utilized. (Reader is referred to Ford, 1980*a*, for further discussion and references.)

quantitatively defined features, following standard practice for many of these features. Despite the problems that can be inherent in the use of ratios, it was felt they were an acceptable way to describe aspects of shape considered here (see discussion in Ford, 1980*a*).

Table 1 presents a list of the postcranial features that are here hypothesized as present in the earliest platyrrhine. (Measurements taken are illustrated in Figures 2–5.) From these features, one can get a crude "picture" of how this animal may have looked, at least in terms of its appendicular skeleton. This list includes both apomorphic features and primitive retentions, as it is a composite of both that comprises the total morphological pattern of the animal. However, it is the new derivations that set this monkey apart from its immediate, non-platyrrhine ancestor, and derived traits are therefore indicated as such in the table. In fact, at present, reasonable speculation of the implications for behavior can only be made for a small number of the features listed in Table 1. These traits will be discussed here; the remainder are presented in Table 1 in the hope that in the future they can also be interpreted to shed light on the adaptations of this hypothesized platyrrhine ancestor.

Morphological reconstruction of the earliest platyrrhine common ancestor: description and analysis

Abbreviations for measurements and indices are used throughout this section. For illustration and explanation, see Figures 2–5.

Humerus
The bicipital groove on the proximal humerus is shallow and wide, a retained primitive trait. This groove carries the tendon for *m. biceps brachii*, and a narrow and deep groove would maintain the tendon within the proper track through a wider range of movements of

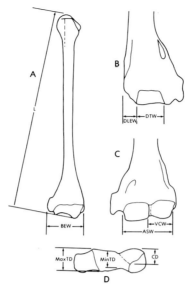

Figure 2. Measurements taken on the humerus. A, ventral view, left humerus; B, dorsal view, distal left humerus; C, ventral view; D, inferior view. L, maximum length; BEW, biepicondylar width; ASW, width of articular surface, ventrally; DLEW, dorsal lateral epicondylar width; VTW, ventral trochlear width; DTW = dorsal trochlear width; VCW, ventral width of capitulum; CD, anteroposterior diameter of capitulum; MaxTD, maximum anteroposterior diameter of trochlea; MinTD, minimum anteroposterior diameter of trochlea (fuller descriptions can be found in Ford, 1980a).

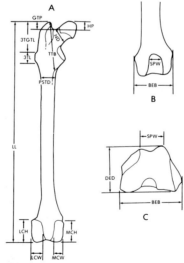

Figure 3. Measurements taken on the femur. A, posterior view, left femur; B, anterior view; C, inferior view. LL, maximum lateral length; ND, neck diameter, superior–inferior; HP, superior projection of head; GTP, superior projection of greater trochanter; TTB, transverse trochanter breadth; 3TL, length of third trochanter; 3TGTL, distance of top of third trochanter from superior point of greater trochanter; PSTD, proximal shaft transverse diameter, taken just below lesser trochanter; BEB, biepicondylar breadth; DED, distal epiphysis depth, anteroposterior; SPW, superior patellar width; MCW, width of medial condyle; MCH, height of medial condyle; LCW, width of lateral condyle; LCH, height of lateral condyle. Not illustrated: LTD, diameter at lesser trochanter, measured in axis of lesser trochanter, not including third trochanter or ridge, if present; BLTD, shaft diameter below lesser trochanter (same level as PSTD), in same axis as LTD (fuller descriptions can be found in Ford, 1980a).

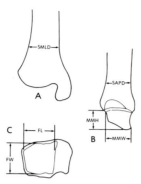

Figure 4. Measurements taken on the tibia. A, anterior view, distal right tibia; B, lateral view; C, inferior view. SMLD, mediolateral diameter of distal shaft; SAPD, anteroposterior diameter of distal shaft; FW, width, anteroposterior, of trochlear facet for astragalus; FL, length, mediolateral, of trochlear facet for astragalus; MMH, medial malleolus height; MMW, medial malleolus width (fuller descriptions can be found in Ford, 1980a).

the shoulder (Ford, 1980a). This narrow groove is clearly advantageous for suspensory and brachiating primates exhibiting a great degree of circumduction and occurs in hominoids, colobines, and atelines (but not *Alouatta*) (Ford, 1980a; Fleagle & Simons, 1982). The shallow, wide groove in the ancestral platyrrhine would indicate a lack of anatomical specialization for suspensory postures.

The deltopectoral crest descending from the greater tuberosity is derived from the primitive condition of a well-defined crest with moderate flare, so that at the superior portion of the crest in the ancestral platyrrhine, the crest flattens out and almost disappears. This is not the same "flattening" described by Fleagle (1983; also Fleagle & Simons, 1982) for Fayum humeri, where a flat plane between the deltopectoral and deltotriceps crests is being referred to. While a rounded crest appears to be typical of brachiators (Le Gros Clark & Thomas, 1951; Napier & Davis, 1959; Delson & Andrews, 1975), the possible significance of the flattening seen in platyrrhines is unclear.

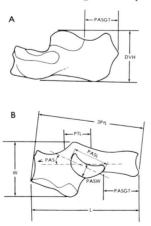

Figure 5. Measurements taken on the calcaneus. A, medial view; B, superior view. L, maximum length, parallel to long axis of bone; 3PtL, length using three points, perpendicular to plane of cuboid facet; W, maximum width; DVH, dorsoventral height; PTL, peroneal tubercle length; PASGT, length of heel; PASL, length of posterior articular surface; PASW, width of posterior articular surface; ∠PAS, angle of posterior articular surface to long axis of bone (fuller descriptions can be found in Ford, 1980a).

At the distal end of the humerus, the medial epicondyle of the ancestral platyrrhine exhibits a very slight dorsal displacement, derived from the primitive condition of a straight medial extension (Figure 6). The primitive medial projection increases the medial rotatory torques of the digital and carpal flexors and pronators about the elbow joint (Jenkins, 1973; Fleagle & Simons, 1982) and has been suggested as indicative of arboreal climbing and suspensory behavior (Fleagle, 1983), although it occurs in a wide variety of primates (such as *Plesiadapis*, many strepsirhines and adapids, *Tarsius*, many Fayum humeri, *Pliopithecus*, and hominoids [Ford, 1980a]). A strong posterior projection would reduce this torque and enhance action of the flexor and pronator muscles, as in pronograde quadrupedalism (Szalay & Dagosto, 1980; Fleagle & Simons, 1982). This morphology has been particularly linked with terrestrial locomotion (Jolly, 1967; Conroy, 1976). However, a large posterior displacement also occurs in a number of arboreal, non-habitual pronators,

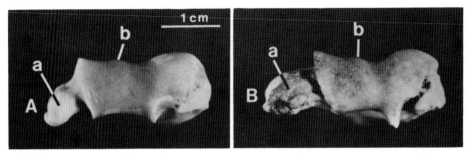

Figure 6. Inferior view of distal left humeri of: A, *Ateles geoffroyi*; B, *Lagothrix lagotricha*. Note (a) large degree of dorsal displacement of medial condyle in *Ateles* versus moderate to slight dorsal displacement in *Lagothrix*; (b) rounded but distinct trochleocapitular ridge in both.

including many platyrrhines (especially *Cebus* and *Chiropotes*, and some *Saimiri*, *Lagothrix*, and the other pitheciines) and colobines (Ford, 1980a). The intermediate position, characteristic of the earliest platyrrhine, is considered by Fleagle & Simons (1982, p. 185) as "an adaptation to arboreal quadrupedalism involving a semipronated, grasping forelimb", and indicates a lack of specializations for suspensory behavior.

The presence of a dorsal epitrochlear fossa, with at least some individuals having a distinct pit and the others a marked concavity, seems a strong derived trait in the platyrrhine common ancestor, unless it is retained from the common anthropoid ancestor (less certain). A fossa or pit is variably present in a number of Miocene hominoids, cercopithecoids, and gibbons as well. A distinct pit also occurs in several humeri from the Fayum (Conroy, 1976; Fleagle & Simons, 1978, 1982; Fleagle, 1983), and in some or all *Plesiadapis* (Szalay, Tattersall and Decker, 1975), presumably as independent derivations, and clearly independently in talpids (Campbell, 1939; Ford, 1980a). A shallow fossa has been described for *Cantius* (Rose & Walker, 1985), which may be similar to the shallow concavity described for *Notharctus* and *Adapis* by Ford (1980a). A fossa is also found in omomyids (Szalay & Dagosto, 1980), but not in extant strepsirhines or *Tarsius* (Ford, 1980a). The ulnar collateral ligament attaches in this area. Conroy (1976) suggested that the pit may be associated with a larger medial epicondyle and associated muscles; however, Ford (1980a) found there is no correlation between size of the medial epicondyle and presence of the pit. Thus, the functional significance of this feature is at present unclear.

The brachialis flange is relatively wide, a primitive retention. Although there are some discrepancies in descriptions of the width of the crest in different taxa (see discussion in Ford, 1980a), it is clear that a wide crest reflects well-developed forearm flexors (Ford, 1980a; Fleagle & Simons, 1982; Rose, 1983; Rose & Walker, 1985; Rose, 1987) and is indicative of arboreal quadrupedalism (Dagosto, 1983; Fleagle & Simons, 1982, but see Conroy, 1976, for alternate view).

The lateral epicondyle is somewhat reduced in size, a shared derived trait of platyrrhines, although not of strong weight. This is measured as the dorsal width of the lateral epicondyle relative to biepicondylar width: DLEW/BEW × 100. The relative size of the lateral epicondyle does not correlate strongly with the size of the supinator crest (Ford, 1980a). The lateral epicondyle is the area of attachment of the carpal extensors and supinator muscles. Conroy (1976) felt that Erikson's (1963) "climber" class had larger epicondyles than those in the "springer" class; however, this author (1980a) could not make the same distinction because of the large amount of variability found in *Cebus* and *Saimiri*. Conroy suggested a large lateral epicondyle would characterize vertical clingers and leapers because of the relative importance for them of carpal extension and, to a degree, supination. This appears to be true for indriids but not for *Galago*, supporting the independent development of vertical clinging and leaping in these two groups (see also Gebo & Dagosto, 1988). The reduction of the lateral epicondyle in platyrrhines suggests a reduction in the importance of carpal extension and/or supination.

The relative size and shape of the distal articular surfaces are examined through several features. The trochlea is cone-shaped with a slight anteromedial flare (not spiraled), with distinct but not extremely high borders (measured by the index MinTD/MaxTD), and a rounded, distinct trochleocapitular ridge, all here considered primitive retentions (see Ford, 1980a; Szalay & Dagosto, 1980, suggest a cone-shaped trochlea may be primitive for haplorhines). This again seems to suggest an unspecialized arboreal quadruped, with the elbow often flexed and the forearm semi-pronated (Grand, 1968; Fleagle & Simons, 1978, 1982). This is in distinct contrast to a narrow trochlea with high margins, constraining conjunct movement and providing a stable joint for strong flexion/extension in hominoids and cercopithecoids (Ford, 1980a; Rose, 1983, 1988), and to very low margins in more suspensory forms (Fleagle, 1983). The moderate development of a trochleocapitular ridge (see Figure 6), not present in early eutherians, would limit mediolateral sliding of the humerus over the distal arm bones (Jenkins, 1973) and transfer the stabilizing role to the ulna, freeing the radius for increased rotation (Morbeck, 1972). Its presence in most primates leads to the conclusion that it is primitive for the order and suggests that early primates were marked by the development of at least some capability for pronation/supination (Ford, 1980a). The trochleocapitular ridge is absent in *Plesiadapis*, suggesting either its loss in a specialized *Plesiadapis* or its presence as a derived euprimate, rather than primate, feature. It is secondarily lost in *Adapis* and to a degree in lorisines (Szalay & Dagosto, 1980; Dagosto, 1983).

The capitulum is slightly larger than the trochlea. This is measured as the ratio of ventral capitular width to dorsal trochlear width (VCW/DTW × 100). (Others have used a similar index but using ventral trochlear width rather than dorsal trochlear width. Here, ventral trochlear width did not yield results that fell into discrete groups, but results using DTW are consistent with those of others [Ford, 1980a]). The wider capitulum suggests that slightly more weight is being transmitted through the humeroradial than the humeroulnar aspect of the elbow joint. Szalay & Dagosto (1980, p. 42) felt their index was "not clearly indicative of locomotor specializations", but that a slightly larger capitulum (as here) suggested grasp leaping and quadrupedal tendencies. This is a retained primitive condition, also found in most strepsirhines, notharctines, and omomyids. Two separate directions of change from this condition are seen in primates: (1) the specialized wide trochlea of hominoids and some of the Fayum humeri, as well as *Galago* (which is clearly derived—see Ford, 1980*a*, pp. 298–301; Jenkins, 1973; and contra Fleagle & Simons, 1978); and (2) an extremely narrow trochlea, in specialized pronograde Old World monkeys, as well as atelines (not *Alouatta*), *Adapis*, and *Avahi* (also derived, contra Conroy, 1976—see Ford, 1980*a*).

Femur

The articular surface of the femoral head extends only slightly or not at all onto the posterior surface of the neck, a retained primitive condition (Figure 7). A large extension on the posterior surface has been found in leapers and vertical clingers, such as *Tarsius*, *Galago*, *Avahi*, *Indri*, *Lepilemur*, *Presbytis melalophos*, *Cebuella*, and *Callithrix* (Napier & Walker, 1967; Walker, 1974; Jungers, 1977; Fleagle, 1977; Ford, 1980*a,b*). The literature is not consistent on the biomechanical correlates of a posterior extension. While Jungers (1977) suggested it allows increased flexion, abduction, and lateral rotation in vertical support postures, Fleagle (1977) and Rose (1983) have suggested that the same excursion (flexion, abduction, and lateral rotation) characterizes non-cursorial, quadrupedal mammals, which lack a posterior extension of the femoral head. Jenkins & Parrington (1976) and Jenkins & Camazine (1977) suggested a posterior extension is characteristic of generalized, non-cursorial mammals, allowing a wide range of femoral postures, especially abduction and extension during climbing. Although the lack of a posterior extension is considered primitive for platyrrhines, many New World monkeys do have some

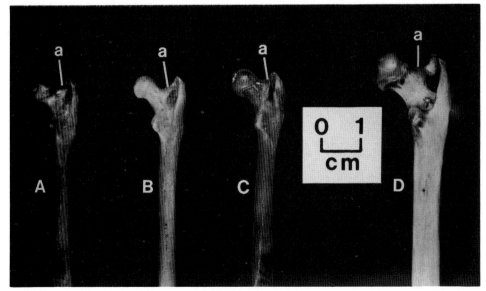

Figure 7. Posterior view of proximal right femora of: A, *Saguinus oedipus geoffroyi*; B, *Saimiri sciureus*; C, *Aotus trivirgatus*; D, *Cebus apella*. Note presence of eminence of bone (a) on A and B and of a ridge on posterior neck of C (*Aotus*), but a smooth surface in *Cebus*. All show a lack of any posterior extension of the femoral head, as in the ancestral platyrrhine.

development of the facet onto the neck, and these exhibit a wide variety of locomotor and postural behaviors (Ford, 1980a). None have the almost cylindrical, hinge-like head of galagos (Jouffroy & Gunther, 1985). It appears that an extension of the head onto the posterior neck facilitates lateral rotation (enhanced in the adducted femur of habitual leapers [Davis, 1987]); it may also facilitate abduction, whether in a flexed (for vertical clinging) or extended (for quadrupedal climbing) posture. The lack of an extension, as in the hypothesized ancestral platyrrhine, still leaves a broad, almost spherical articular surface allowing for a variety of femoral movements and postures, but with somewhat less mobility (Dagosto, 1983).

There is no ridge on the posterior proximal femur, at the lateral end of the neck near the trochanteric fossa. The bone is smooth here in most primates and mammals, as well. However, this is only true for *Cebus* among platyrrhines; all other platyrrhines have at least a low "mound" or rounded eminence of bone, and often a distinct, proximo-distally elongated ridge is present (Figure 7). As discussed in Ford (1980a, 1986a), currently it is the best hypothesis of relationships to assume that *Cebus* is the sister-group of all other platyrrhines, and thus presence of a "mound" or ridge of bone is a shared derived feature of all other platyrrhines. Clearly, though, an argument could be made for the presence of a mound of bone as a shared derived platyrrhine trait, with subsequent loss in *Cebus*. A rounded eminence of bone in this position also occurs in *Daubentonia* and some macaques (Ford, 1980a), in several Miocene hominoids (Rose, 1983), and in *Apidium*. From Hill's (1959) description of the morphology of this region in *Callimico*, it appears likely that primates with a bony mound or ridge have a thicker, stronger, and more prominent ischio-femoral ligament, which could limit the degree of flexion of the femur on the trunk and, to some degree, internal (medial) rotation (Ford, 1980a). The adaptive value and related constraints on posture of this morphology are unclear, given the extremely varied locomotor habits of the platyrrhines that exhibit this trait (all but *Cebus*), including some engaging in large amounts of vertical leaping, such as *Callimico*. However, a ridge or eminence may favor external rotation and extension as correlates of arboreal climbing, which all platyrrhines do to some extent (Moynihan, 1976; Kinzey, 1986). This would agree with Walker's suggestion (in Napier, 1964) that this ridge would increase the angle of insertion and thus the leverage of obturator internus, an external rotator of the femur (see also Napier & Walker, 1967; Davis, 1987). The significance of the lack of a ridge or "mound", as likely in the hypothesized ancestral platyrrhine, is obscure, since this is the common condition and widespread among primates of varying habits.

The head and greater trochanter are subequal in height (degree of proximal extension), which is the primitive condition (index of HP/GTP, see Figure 3). The head extending farther proximally has been linked to increased freedom of movement at the hip, especially for climbing or suspensory locomotion (Walker, 1974; Dagosto, 1983), and indeed this situation occurs in the platyrrhines exhibiting suspensory postures, in *Hylobates*, and in other hominoids (Ford, 1980a). The angle of the femoral neck with the femoral shaft may influence this to some degree; however, this angle is fairly variable in a number of primates, especially *Ateles*. This relationship needs to be examined in greater detail. The reverse situation of a more proximally extending greater trochanter suggests powerful gluteal muscles, important in leaping and cursorial locomotion (Howell, 1944), and is found in cercopithecoids, *Cheirogaleus*, indriids, and the vertical clinging small callitrichids (Ford, 1980a). However, neither *Tarsius* nor *Galago* have a high greater trochanter; in *Tarsius* it is subequal to the head (Grand & Lorenz, 1968), and in *Galago* it has been described as subequal (Jouffroy & Gunther, 1985) or with the head more proximal (Ford, 1980a,b). Jouffroy & Gunther suggest that the subequal condition they report in *Galago* favors extension and flexion at the expense of abduction and adduction, but this would appear to be an interpretation based on the behavior of *Galago* and not suited to the wide variety of primates with subequal extensions. *Galago* and *Tarsius* appear to have adapted the proximal femur to the demands of leaping in different ways from those of other saltatorial primates, in particular in extreme modifications of the shape of the articular head. It may be more appropriate to interpret a subequal condition as one *not* favoring cursorial locomotion, leaping (except in highly specialized cases), or suspensory movements.

The lesser trochanter extends postero-medially, the primitive condition for primates. A more medial displacement suggests increased power for iliopsoas as a femoral flexor (Rose, 1987) and adductor (Jungers, 1977), as seen in both vertical clinging and suspensory primates (Ford, 1980a). An extremely posterior location seems to be associated with more cursorial and/or semi-terrestrial habits (Ford, 1980a), and involves greater torsion in the action of iliopsoas. Thus, the condition suggested for the earliest platyrrhine would suggest a generalized arboreal quadruped.

The lesser trochanter is moderate to reduced in size (measured by femoral index [LTD-BLTD/PSTD]), as in the ancestral anthropoid. A reduced lesser trochanter also occurs in many other mammalian groups, but early primates, strepsirhines, and many fossil and generalized mammals have a large lesser trochanter, which is considered the primitive condition (Ford, 1980a,b; Novacek, 1980). The reduced size suggests a reduced importance of the iliopsoas muscle as a flexor of the thigh, perhaps in association with increased quadrupedal movements, requiring less habitual hip flexion than leaping. The fact that the lesser trochanter is secondarily increased in size in *Cebuella* and *Callithrix*, the New World monkeys that most frequently engage in vertical clinging with strongly flexed thighs (Kinzey, Rosenberger and Ramirez, 1975), supports this interpretation.

In the distal femur, the ancestral platyrrhine has a medial condyle which extends slightly further distally than does the lateral. However, the condyles are relatively equal in their distal extension in many extant platyrrhines (see Figure 8). The two condyles are subequal in their extension posteriorly, and the medial condyle is subequal to or slightly larger than the lateral condyle (in area as measured in the index [MCW × MCH/LCW × LCH], see Figure 3). All of these traits are retained primitive conditions. The medial condyle extending further distally, when the corresponding facets on the proximal tibia are level and perpendicular to the tibial shaft, creates a "knock-kneed" stance with an inverted knee. This condition is quite common in mammals (Lessertisseur & Saban, 1967). The opposite condition of an everted knee with a longer lateral condyle is quite rare, occurring in bats (Lessertisseur & Saban, 1967) and, among primates, in *Nasalis* and *Notharctus* (Ford, 1980a). Although Lessertisseur & Saban (1967) reported it for *Megaladapis* as well, Jungers (1977) has shown that it is at least partly due to an elevation of the lateral condyle of the tibia. The femoral condyles are subequal in distal extension in many hominoids (but not humans), some Old World monkeys, and many New World monkeys, including *Callicebus*, *Aotus*, and *Saimiri* (Ford, 1980a). Rose (1983) suggests that this subequal condylar extension in early hominoids, along with other features (condyles subequal in size, wide patellar groove, and narrow articular region; see below), places the knees directly below the hip in quadrupedal posture, allowing generalized hip mobility with habitual hip flexion, abduction, and lateral rotation coupled with the semiflexed knee pointing antero-laterally, as in an arboreal quadruped.

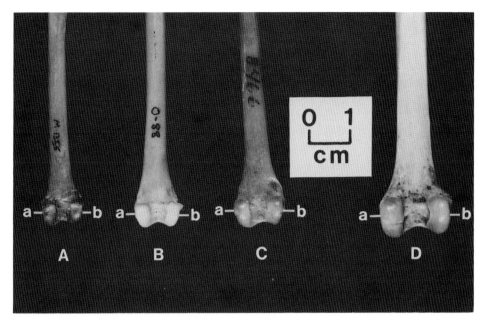

Figure 8. Posterior view of distal right femora of: A, *Saguinus oedipus geoffroyi*; B, *Saimiri sciureus*; C, *Aotus trivirgatus*; D, *Cebus apella*. The medial condyle (a) extends farther distally (least so in *Saimiri*) and is slightly larger than the lateral condyle (b), as in the ancestral platyrrhine.

It might be expected that the relative posterior extension of the condyles would mirror the distribution of the distal extension, influencing knee inversion/eversion in the flexed rather than extended posture. To some degree this is true (e.g., the lateral condyle of *Notharctus* extends further posteriorly as well as distally). In these cases, there would be increased curvilinear length of one of the condyles, increasing rotation at the knee during flexion/extension (Tardieu, 1981). However, these two features do not correlate precisely (Ford, 1980a). The reasons for this are as yet unclear.

The medial condyle being just slightly larger (width times length in posterior aspect) in the ancestral platyrrhine morphotype suggests that weight is being transmitted almost equally but with a slim majority passing through the inner aspect of the knee in a flexed position (Ford, 1980a). Tardieu (1981) found most primates to have condyles of roughly equal curvilinear length (she did not measure relative area, the comparison used here). Jouffroy & Gunther (1985) suggest that condyles of roughly equal size may relate to strong flexion, as they found this condition in *Galago*. Ford (1980a) also found *Galago* to have more equal-sized condyles than most other primates, or even a larger lateral condyle, which was also true of *Tarsius* and *Hemiacodon*, but not indriids. The majority of primates and the ancestral platyrrhine seem to have the medial condyle slightly larger rather than equal, and it is this morphology (rather than subequal condyles) that may be a more appropriate part of Rose's (1983) picture of distal femoral morphology for generalized arboreal quadrupeds.

The patellar groove is moderately narrow, rather than wide or very narrow (reflected in the index SPW/BEB), and of moderate depth, both primitive retentions. Distribution of morphologies for these traits and proposed interpretations in the literature are difficult to interpret. There are several problems in interpreting these features. These two aspects of patellar groove morphology vary somewhat independently (Ford, 1980a) and so should be examined as separate features, but most authors deal with both together, describing patellar grooves as either "wide and shallow" or "deep and narrow". These features are seldom quantified, particularly patellar groove depth, which is difficult to measure. Descriptions become highly subjective, depending on the reference point and comparative sample of the examiner. Someone examining primarily strepsirhine femora, which are generally both deep and narrow, would have a very different sense of the words "relatively shallow" from someone examining platyrrhine femora, which range from among the deepest and narrowest in primates to the shallowest and widest, approaching some early eutherians (Ford, 1980a). Patellar width is always assessed relative to biepicondylar breadth (whether measured or qualitatively assessed), which itself can vary depending on condylar breadths.

Given these problems of interpretation, patellar groove shape is difficult to analyze. Many have suggested that a deep, narrow groove provides extreme stability, preventing dislocation of the patella during extension/flexion and relating to the power of the quadriceps during extension, especially in leaping or vertical clinging and leaping (Walker, 1974; Fleagle, 1977; Tardieu, 1981; Tattersall, 1982; Jouffroy & Gunther, 1985; Rose & Walker, 1985). A deep patellar groove is found in all leaping primates (e.g., *Tarsius*, *Galago*, *Avahi*, and slightly less so in *Callithrix* and *Cebuella*), as well as a number of fossils that have been suggested to exhibit leaping tendencies (*Cantius*, *Notharctus*, *Smilodectes*, and *Hemiacodon*—Gregory, 1920; Napier & Walker, 1967; Walker, 1974; Szalay, 1975; Szalay & Dagosto, 1980; Rose & Walker, 1985; Dagosto, 1986; also *Cebupithecia*—Davis, 1987). However, it should be noted that the groove is equally deep in *Lemur catta* and *Daubentonia* (Ford, 1980a), which are generally considered quadrupeds (Jolly, 1966; Walker, 1974; Tattersall, 1982). Narrowness of the patellar groove, as opposed to depth, does not appear to be important for leapers. When width is assessed metrically (SPW/BEB), leaping primates have *roughly* the same widths as *Daubentonia*, cercopithecoids, and many platyrrhines, including *Cebus* and *Aotus* which are generally considered more quadrupedal (Fleagle & Mittermeier, 1980; Wright, 1981; Fleagle, Mittermeier and Skopec, 1981; Terborgh, 1983).

A wide patellar groove, along with other features, has been suggested to be characteristic of arboreal quadrupeds (Rose, 1983), and a broad and shallow groove as indicative of generalized climbers (Rose, 1987). Yet a wide groove is found not only in quadrupedal primates but also in *Avahi*, generally considered a leaper (Tattersall, 1982). It is notably wide in platyrrhines which frequently exhibit suspensory behavior (*Ateles*, *Brachyteles*, *Alouatta*, and *Lagothrix*—Moynihan, 1976; Fleagle & Mittermeier, 1980; Terborgh, 1983), and the widest of all in *Hylobates* (see Ford, 1980a on morphologies). The wider groove in suspensory primates may reflect the lack of a need for great stability in the hindlimb. A shallow groove as well as one of moderate depth each occur in a variety of animals with a broad range of locomotor and postural specializations (see Ford, 1980a).

Thus, while one can in general say leapers tend to have deeper (but not necessarily narrower) patellar grooves and primates with suspensory habits tend to wider ones, quadrupedal primates vary from one end of the spectrum to the other. Of the ancestral platyrrhine morphotype, it can be suggested that the moderately narrow patellar groove suggests it was not suspensory, and the moderate rather than extreme depth of the groove suggests it did not do much leaping.

Tibia

Form/function relationships have not been investigated or are only poorly understood for most of the features of the distal tibia, although Dagosto (1985, 1986) gives an excellent discussion of the interactions within the upper ankle joint. This is especially unfortunate as it is on this bone that a number of key shared platyrrhine features are located.

The fibular facet is particularly distinctive, being both unusually long antero-posteriorly and more posterior in its location on the shaft (see Figure 9). Yet the nature of the tibiofibular articulation is not well understood beyond the simple issue of whether or not the bones are fused. Barnett & Napier (1952, 1953) described three types of tibio-fibular articulation: (1) whole or partial fusion allowing no movement at the joint; (2) a mobile synovial joint allowing conjunct rotation and flexion/extension at the upper ankle joint, with a semi-lunar facet on the distal tibia for articulation with the fibula usually present; and (3) an intermediate joint that is unfused distally but in which the bones are closely apposed and joined by a syndesmosis, severely limiting or excluding any rotational movements.

Almost all primates exhibit the second type of joint, with a semilunar facet which allows some mobility. This would appear to be the primitive condition for primates as an order. It may be retained from and represent the primitive condition for eutherian mammals in general, as argued by Matthew (1937) and suggested by Carleton (1941, who derives it from a pre-mammalian joint marked by no contact between the distal tibia and fibula). Or, as argued by Barnett & Napier (1953), it may represent a basal primate innovation associated with increased body size in an arboreal milieu. This basal primate would have been of very small body size, derived from an early eutherian which had a syndesmosis at this joint with little to no movement.

Tarsius and some Eocene tarsioids from Europe are the only primates with consistent fusion of the distal tibia and fibula (Schlosser, 1907; Simons, 1972; Schmid, 1979; Dagosto, 1985). Recently, Fleagle & Simons (1983) described the third type of joint, with close apposition of the bone distally, well-developed tibio-fibular ligament crests, and limited freedom of movement at the joint, in the Oligocene anthropoid *Apidium* as well as in *Microcebus* and a number of smaller platyrrhines, all of which exhibit frequent leaping behavior. A similar but not identical condition occurs in *Galago* and is reported for the omomyid *Hemiacodon* (Dagosto, 1985), and a similar morphology occurs in *Necrolemur* (Fleagle, personal communication).

Both Fleagle & Simons (1983) and, especially, Dagosto (1985) suggest that this condition of a syndesmosis with closely apposed bones may represent the primitive haplorhine morphology, derived from that of most strepsirhines and notharctines. (If Barnett & Napier are correct, and this is the primitive eutherian condition, it would be secondarily derived in haplorhines). However, although it occurs in omomyids, one early anthropoid, and some platyrrhines, all of these primates share (or are presumed to share) a distinctive behavior, frequent leaping. Leaping also characterizes most of the other eutherians which exhibit a syndesmosis at this joint (Carleton, 1941; Barnett & Napier, 1953). This suggests it is highly possible that a syndesmotic joint is a derived, specialized feature in primates, developed convergently a number of times in haplorhines, as is patently the case for tibio-fibular fusion in mammals (see review in Ford, 1980a).

In support of this conclusion is the fact that the shape and structure of the fibular facet on the tibia is quite different in each of the primates exhibiting syndesmosis, and in each case the facet is much more similar to that of other closely related primates than to those with a similarly closely appressed fibula. The fibular facet is long and either central or posteriorly offset in the platyrrhines, short and anterior in the strepsirhines, and lacking in *Hemiacodon* (from illustration in Dagosto, 1985) (see discussion below and Figure 9). This suggests that the syndesmotic nature of the joint in leaping primates is a derived adaptation to the constraints of leaping, which is superimposed, in each case, on whatever other morphology of the tibiofibular joint is inherited from each leaper's most recent, non-leaping ancestor, rather than it being a retained homologous feature from a common haplorhine ancestor.

A more intriguing possibility, raised by Fleagle & Simons (1983), is that the "incipient fusion" morphology (syndesmosis) is not convergent but homologous in the Fayum forms and platyrrhines. This would require one of the following two scenarios to be true. Either (1) this specialized feature and a propensity for leaping was ancestral only for all platyrrhines and for the Fayum primates. This conclusion seems to find no support in any other aspect of the postcranial morphology of the presumed ancestral platyrrhine. Or (2) *Apidium* is more closely related to a small number of platyrrhines (who themselves do not form a monophyletic group within Platyrrhini) than to any other primates, which seems highly unlikely given our current understanding of primate relationships and biogeography. While neither of these hypotheses seem at all reasonable, there are some extremely interesting similarities between the tibiae of platyrrhines and some other Fayum specimens (see below).

Looking more closely at the shape and location of the tibial articular surface for the fibula, it becomes clear that there are four distinct morphologies, each primarily restricted to and characterizing a different phylogenetic group of primates (Ford, 1980a). These are (Figure 9): (1) fusion, found in two tarsioids; (2) a short, anteriorly located facet, found in strepsirhines and adapids; (3) a tiny, anteriorly located facet or no distinct facetal surface, found in all cercopithecoids and *Pliopithecus*; and (4) a moderately to very long facet, centrally or posteriorly located, found in all platyrrhines, extant and extinct. Extant hominoids exhibit some degree of variability, with most ranging from the typical cercopithecoid condition of a small anterior facet (Ford, 1980a) to a somewhat longer, central or anteriorly located facet, tallest supero-inferiorly at the anterior end (E. Strasser, personal communication). *Hylobates* is particularly interesting, exhibiting a great deal of variation, but none exactly like that seen in the platyrrhines (E. Strasser, personal communication).

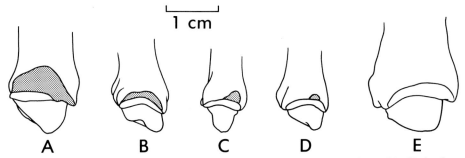

Figure 9. Lateral view of distal right tibiae of selected primates, illustrating the form of the fibular facet (stipple). A, *Ateles geoffroyi*, with a long facet posteriorly displaced, and angled somewhat. B, *Cebus apella*, with a long facet extending the entire length of the lateral surface, as in many platyrrhines. C, *Lemur macaco*, with a small, anteriorly placed facet, typical of strepsirhines. D, *Cercopithecus cephus*, with a tiny, anteriorly placed facet as found in many catarrhines. E, *Macaca mulatta*, lacking any distinct fibular facet, as in many cercopithecoids.

As discussed above, tibio-fibular fusion clearly seems to be a derived condition in those that exhibit it. The cercopithecoid condition of a tiny or absent facet is also best considered derived, as most mammals have a clear facetal contact (Lessertisseur & Saban, 1967; see also Carleton, 1941; Barnett & Napier, 1953; Ford, 1980*a*; Strasser, 1988). Since it also characterizes Miocene catarrhines and some hominoids, this may be the primitive condition for catarrhines. This tiny facet would suggest limited movement between the tibia and fibula, even though there are no indications of syndesmosis. It is particularly interesting and enigmatic that the platyrrhines which have developed the closely appressed, tightly bonded tibio-fibular joint, such as *Saimiri*, which would appear to severely limit movement at this joint (Fleagle & Simons, 1983; Dagosto, 1985), still retain a moderately long to very long facet (with some degree of variability in length). Indications from upper ankle joint morphology on mobility and movement are often in conflict with tibio-fibular joint morphology in strepsirhines, platyrrhines, and catarrhines.

Ford (1980*a*) examined three tibial fragments from the Fayum, and found that one (DPC 1086) exhibits the catarrhine morphology of a very small, anteriorly located facet. Of the other two, one (YPM 25967) has the platyrrhine-like very long facet which appears to be offset posteriorly, although the specimen is broken in this region. The last (YPM 25968) is intermediate, with a moderately long facet that is located anteriorly. Ford (1980*a*, 1986*a*) interpreted the widespread anterior location of the facet and the strepsirhine short facet as primitive for the order, with the platyrrhine condition being derived, suggesting that the longer facet provides greater capacities for movement associated with the development of more varied locomotor capacities. If this morphocline is correct, then within the Fayum there is one tibial fragment that clearly belongs to a derived catarrhine (DPC 1086), but also one that if not for geography must be considered a platyrrhine (YPM 25967), as it exhibits a derived morphology seen in no other known primates. A third (YPM 25968) would be most reasonably interpreted as intermediate, still retaining a non- or pre-platyrrhine anterior location of the fibular facet, but having acquired a platyrrhine longer facet. This third specimen then could represent a sister-group to all other platyrrhines, including YPM 25968, or possibly a derived hominoid.

But there is another possible explanation. Gebo (1986*a,b*) has suggested that prosimians are marked by the use of vertical supports and vertical climbing, as opposed to horizontal climbing and quadrupedalism in anthropoids. Gebo interprets the anthropoid adaptation to horizontal supports as derived. However, if, instead, the strepsirhine adaptation to vertical supports is seen as derived (see discussion below as well), the shorter fibular facet of strepsirhines may be a derived condition. Dagosto (1985) also suggested that it is just as reasonable, at present, to assume aspects of the distal tibial morphology in anthropoids represent the primitive eutherian condition as that of strepsirhines. The platyrrhine longer facet could represent the primitive, generalized euprimate condition, retained in two of the Fayum tibiae as well. This otherwise elegant solution to the sticky problem of the Fayum specimens still leaves a few unresolved problems, such as (1) why have the catarrhines converged (and surpassed) the strepsirhine condition of a shorter facet; (2) do most hominoids retain the primitive condition or have they secondarily reverted to it; and (3) why does one of the Fayum tibiae still have a derived platyrrhine trait—a posteriorly displaced facet? This scenario would also remove one of the most compelling features supporting a monophyletic Platyrrhini (Ford, 1986*a*).

The last tibial trait to be mentioned is that the trochlear facet for articulation with the astragalus curls around onto the anterior face of the shaft of the tibia to a moderate degree.

One could expect that this facet would meet a cup-like extension of the astragalar trochlea onto the superior surface of the neck of the astragalus ("tibial stop") during extreme dorsiflexion, the close-packed position (Lewis, 1980; Dagosto, 1986; Gebo, 1986a). The corresponding facets should form a complex, and their presence or absence covary precisely within primates. However, this does not appear to be the case; frequently, individual specimens exhibit one facet but not the other, or facets of widely differing size occur on the two bones (Ford, 1980a). The reasons for this are unclear.

Calcaneus

The bones of the tarsus are currently probably the best studied of the entire postcranial skeleton, thanks in large part to the work of Szalay and his coworkers and students (e.g., Szalay, 1966; Decker & Szalay, 1974; Szalay & Decker, 1974; Szalay *et al.*, 1975; Szalay & Drawhorn, 1980; Dagosto, 1983, 1986, 1988; Strasser, 1988), as well as others (e.g., Grand, 1967; Barnett, 1970; Conroy, 1976; Lewis, 1980; Ford, 1980a; Langdon, 1986; Gebo, 1986a; Gebo & Simons, 1987). The reader is particularly addressed to Dagosto (1986) for a discussion of joint movements in the foot involving the calcaneus. Despite the attention devoted to this region, the relationships of specific morphologies to and implications for behavior often remain unclear; the features discussed here illustrate this problem.

The articulations of the calcaneus with the astragalus are particularly important in the adaptive role and function of the foot, as it is at the subtalar joint that the foot articulates with the rest of the leg. This is a complex joint, comprising three facets (see Figures 5 and 10): the posterior articular facet (PAS, equivalent to Dagosto's pAC[C]) and the medial and anterior articular facets, which are often merged into one (MAS and AAS, respectively, both equivalent to Dagosto's aAC[C]). The PAS articulates with the posterior articular facet of the astragalus, and both the AAS and MAS articulate with the inferior surface of the astragalar head and neck. The PAS and combined AAS/MAS will be discussed separately.

PAS. The posterior articular facet (PAS) of the calcaneus, which is roughly a section of a cone (Dagosto, 1986; Langdon, 1986), is of short to moderate length in the ancestral platyrrhine, of moderate width, and is a rectangle (curved) approximately twice as long as wide. All of these represent retained primitive aspects of PAS shape (Ford, 1980a). These aspects of PAS size and shape are measured through a comparison of the indices PASL/L, PASW/L, and PASW/PASL, as well as W/L (Figure 5), which, when considered together, allow a sorting out of the variables which appear to be changing in any given taxon (see Ford, 1980a). Thus, if PASL/L yields a value different from that of most primates, but calcaneal length (L) retains a relationship to calcaneal width (W) and PAS width consistent with that seen in other primates, one can assume it is PAS length that is changing, and not calcaneal length. Similar reasoning is applied to interpretation of the other traits. The angle of the PAS to the long axis of the calcaneus ($\angle$PAS) is between 10° and 20°, which is primitive for euprimates and found in all platyrrhines (Ford, 1980a). This relatively low angle appears to be a shared derived euprimate trait. The angle is much larger, up to 40°, in early mammals and plesiadapiforms (Szalay, 1966; Barnett, 1970; Szalay & Decker, 1974; Novacek, 1980).

While a PAS of short-to-moderate length characterizes the ancestral platyrrhine morphotype, a longer PAS is found in indris, and especially, in the large, prehensile-tailed platyrrhines (*Ateles, Brachyteles, Lagothrix,* and *Alouatta*—see Figure 10), the Miocene fossil *Pliopithecus* (Ford, 1980a), and extant great apes (Langdon, 1986). All of the anthropoids with a relatively long PAS exhibit some degree of suspensory or forelimb-oriented climbing locomotion (*Alouatta* is less forelimb-oriented) (MacKinnon, 1974; Moynihan, 1976; Prost, 1976; Fleagle & Mittermeier, 1980; Susman, Badrian and Badrian, 1980; Maple & Hoff, 1982). In the indris, thigh-powered hindlimb leapers, the PAS forms a very low angle to the long axis of the calcaneus (Ford, 1980a; Dagosto, 1986). While allowing a greater range of motion (Dagosto, 1986), a long PAS in these primates would allow primarily greater flexion/extension at the subtalar joint. In the more suspensory hominoids, the long PAS forms a very high angle to the calcaneus (approaching 30–40° in the great apes, and slightly less but still high *Hylobates*—Langdon, 1986; see also Ford, 1980a). This allows more complex motions at the subtalar joint; Langdon points out that a greater angle relates to a great advance of the astragalar head into the midtarsal joint during inversion and a tighter packing of the joint.

Dagosto (1986) found a shortened PAS in lorisines, which she described as characteristic of climbers, allowing a greater range of inversion/eversion but with decreased stability, especially in eversion, due to the low degree of overlap of the astragalar facet on the calcaneal facet. Ford (1980a) also found an extremely shortened PAS in *Cercopithecus* (Figure 10; see also Strasser, 1988) and one Fayum calcaneus, YPM 25804a (neither of

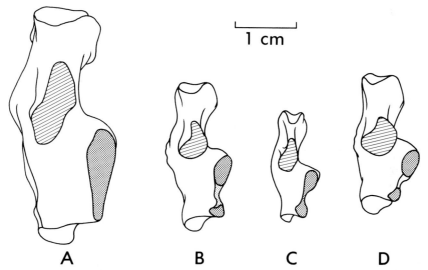

Figure 10. Superior view of right calcanei of: A, *Ateles geoffroyi*; B, *Cebus apella*; C, *Aotus trivirgatus*; and D, *Cercopithecus cephus*. Note moderate posterior articular surface (parallel lines) of (B) and (C), and the noticeably longer facet of *Ateles* and short facet of *Cercopithecus*. Also, the confluent anterior and medial articular surfaces (stippled) of *Ateles* and *Aotus*, and the separate facets of *Cercopithecus*. The middle portion of the facets on *Cebus* would contact the astragalar neck only in certain positions.

which have elongated calcanei that might unduly influence the index of relative PAS length). The implications of this for the Fayum specimen are unclear, since the condition is not confined to the slow-climbing lorisines within primates, but it would seem to suggest an emphasis on climbing. Langdon (1986) relates the short PAS of cercopithecines to the greater depth of curvature of the facet in these forms.

The PAS of lorisines also forms a high angle to the long axis of the calcaneus, approaching 20° (Dagosto, 1986), which Dagosto related to their habitually inverted foot.

This value is at the upper limit but within the ancestral range of Ford (1980a), within which range the data did not allow significant discrimination of groups. Values in the range of 10°–20°, as in many euprimates, allow the astragalus to advance more straight forward rather than with a strong component of rotation with inversion. Langdon (1986) suggests this more constrained movement would relieve strain during suspensory postures and increase plantar- and dorsiflexion. However, it must be noted that Langdon only examined atelines and *Cebus* among non-catarrhine primates, which he contrasted to the relatively high angle found in most catarrhines. The angle seen in these platyrrhines is shared by many other euprimates that do not engage in significant amounts of suspensory behavior and is better considered the ancestral primate condition. The higher angle of most catarrhines is derived (*Pongo* has apparently retained or reverted to the ancestral condition—see Langdon, 1986).

A wide PAS is found only in hominoids and colobines (Ford, 1980a; see also Langdon, 1986; Strasser, 1988); all other primates, regardless of locomotor and postural behavior, exhibit PAS widths that are relatively the same. The Miocene fossil *Pliopithecus* appears to have increased both the relative length and, to a lesser extent, the relative width of the PAS, resulting in an increased surface area for weight-bearing, but one that has retained roughly the same shape as in ancestral and related forms (Ford, 1980a).

Thus, the condition of the PAS seen in the ancestral platyrrhine would seem to indicate the lack of extremely specialized locomotor modes. It lacks the long PAS of low angle found in the thigh-powered leaping indriids, the long PAS (often of high angle) of suspensory or forelimb oriented climbers, and the short PAS set at a high angle found in slow-climbers.

AAS/MAS. The anterior and medial astragalo-calcaneal facets (AAS and MAS) are closely approximated and variably confluent, as may be primitive for anthropoids or else a weak derived trait of platyrrhines. Both the AAS and MAS are oval in shape (rather than round), and the AAS only reaches the anterior end of the calcaneus in a small point or not at all, all of which are derived features weakly uniting platyrrhines.

The AAS and MAS facets are separate in many strepsirhines (e.g., *Galago*, and Miocene lorisoids), *Notharctus, Tarsius, Apidium*, and cercopithecoids. They are confluent in extant hominoids, some omomyids and adapids, *Plesiadapis*, and many platyrrhines (and in at least some specimens of all platyrrhines—see Figure 10) (Ford, 1980a; Gebo, 1986c; Gebo & Simons, 1987; Langdon, 1986). Gebo & Simons (1987) suggest that the separate facets of *Apidium* may indicate less subtalar mobility and derived affinities to Old World monkeys. It should be noted, however, that there is substantial variability in the degree of separateness of these facets in most primates, especially among platyrrhines. In many taxa, some individuals have totally separate facets while others have a single, confluent facetal surface; in yet other taxa or individuals, the facets are continuous but the midportion only contacts the astragalus when the joint is in certain positions. Variable taxa include some strepsirhines (e.g., *Lemur, Daubentonia*), some cercopithecoids (e.g., *Macaca, Rhinopithecus*), and many platyrrhines (Ford, 1980a). The distributions of the various character states crosscut both locomotor specializations and taxonomic groups, making this a difficult and problematic feature for phylogenetic or functional inference.

The AAS *not* reaching the distal (anterior) end of the calcaneus with a broad area of contact reflects an alternating tarsus, in which the astragalo-navicular and calcaneo-cuboid elements of the transverse tarsal joint are staggered. Dagosto (1986) suggests movement potential would be limited, especially for eversion and plantar flexion, but Ford (1980a) has suggested this structure might actually enhance plantar flexion of an inverted foot, important for quadrupedal climbing or walking using grasping in an arboreal environment. While one might expect calcaneo-navicular contact to limit inversion potential in an alternating tarsus, in fact, Ford (1980a) did not find concordant distributions for the presence of a navicular facet on the calcaneus and an AAS which does not reach the anterior end of the bone. In a serial tarsus, the two elements of the transverse tarsal joint are approximately level, evidenced by the AAS broadly reaching the distal end of the calcaneus. Dagosto (1986) suggests this would allow more supination, expected in both climbers and primates of large body size (she also found this condition in indriines, interpreting it as signifying a mixed adaptation in the foot of these strepsirhines).

Some members of each major group of primates exhibit each of the two conditions for the transverse tarsal joint (alternating vs. serial), making determination of polarity for this feature quite difficult. However, the majority of platyrrhines with an alternating tarsus are arboreal quadrupeds exhibiting varying degrees of leaping and branch running, while those with a serial condition are either small-bodied forms with claws, exhibiting some vertical leaping (especially *Callithrix* and *Cebuella*), or large-bodied with some suspensory behavior (*Ateles* and *Brachyteles*) (Ford, 1980a, 1986a). The linking of a serial joint with specialized, non-quadrupedal behaviors is further supported by the presence of a serial tarsus in the partially suspensory extant hominoids and colobines (Ford, 1980a), and the leaping indriines (Dagosto, 1986). However, an alternating tarsus is found in specialized leapers with elongated tarsals (e.g., *Tarsius*, *Galago*). And several platyrrhine genera as well as *Macaca* exhibit considerable variability for this trait (Ford, 1980a).

Recognizing the above problems, it is suggested here that an alternating tarsus, reflecting generalized arboreal quadrupedalism, is the primitive euprimate condition which is retained in the earliest playtrrhines. A serial tarsus has then developed several times within primates, associated with a number of derived locomotor modes, including thigh-powered vertical leaping (but not in those with elongated tarsals) and suspensory locomotion, and in both large-bodied and small, clawed forms. The primitive platyrrhine condition of either an alternating tarsus or, variably, a very small extension of the AAS to the anterior end (as seen in *Cebus*, some *Aotus*, and one specimen from the Fayum, YPM 25804a), suggests an arboreal quadruped with locomotion similar to that of *Aotus*. The purpose of the slight extension of the facet to the anterior end (as opposed to a broad area of contact) is unclear.

The heel of the calcaneus is of moderate length, retaining the primitive primate condition (measured by the index PASGT/L). The heel acts as a lever or power arm for the load arm or distal portion of the foot (some researchers include the length of the PAS, which acts as a fulcrum, in their measure of the power arm). A short heel suggests more speed at the expense of power, as in vertical clinger and leapers such as *Galago* and *Tarsius* (Hall-Craggs, 1965; Walker, 1974; Berge & Jouffroy, 1986; Davis, 1987). A longer heel indicates more powerful action by the triceps surae during extension of the foot (Reed, 1951), and has been associated with cursoriality in Old World monkeys, and with a general increase in body size, especially in hominoids (Langdon, 1986). A long heel also occurs in larger-bodied leapers, such as the indriids (Ford, 1980a; Dagosto, 1983, 1986), suggesting the need for increased torque at the expense of speed (Davis, 1987). However, several individuals have noted problems in interpreting this feature (e.g., Ford, 1980a; Dagosto, 1983; Langdon, 1986). At present, the intermediate condition hypothesized for the ancestral platyrrhine can only be interpreted as not suggestive of extreme ricochetal leaping.

The cuboid articular surface (CAS) at the distal end of the calcaneus is slightly concave, as in the earliest anthropoid, and shaped like a fat "L" or "U", rather than a "D", this being a weak platyrrhine apomorphy (there is a great deal of variability in expression of this aspect of CAS morphology). There is quite a bit of diversity within primates in the shape and nature of the slope or concavity of the CAS (see discussion in Ford, 1980a). Szalay & Decker (1974; also Decker & Szalay, 1974) suggested that a facet shaped in a half-circle ("D"-shaped), with an essentially flat perimeter and a deep concavity on the straight edge, a cuboid pivot, was primitive for euprimates and allows greater mobility at the expense of stability at this joint. However, a deep pivot is actually found in only a small number of primates, including some Eocene adapids, galagos, and hominoids (Ford, 1980a), and seems more clearly derived. Dagosto (1986) came to the same conclusion for the very deep pivot of galagos and lorises, suggesting that it allows more conjunct rotation and a dominance of pronation/supination over sliding movements, such as abduction/adduction; this would be a very stable morphology, limiting flexion/extension at the midtarsal joint (which is staggered in galagos, with their extremely elongated tarsal elements). Rose (1983) pointed out that this deep-pivot joint in hominoids closepacks in supination. The flatter CAS of indriines would allow greater eversion and easier sliding movements. This slightly concave facet, not flat as in early eutherians but still lacking a deep pivot, is found in many primates, especially many haplorhines (including omomyids [Savage & Waters, 1978], Fayum calcanei, and many platyrrhines and catarrhines [Ford, 1980a]). This morphology, present in the hypothetical ancestral platyrrhine, does not appear to imply any specific locomotor modes or special behaviors.

The cuboid articular surface, and distal end in general, is perpendicular to the long axis of the calcaneus, rather than oblique medio-laterally, as is primitive for primates (reflected in the index 3PtL/L). An oblique orientation typifies many mammals and is primitive for eutherians (see Ford, 1980a, and references therein), and Szalay & Decker (1974) argue that the oblique orientation limits movement to a basic hinge-joint. A more perpendicular orientation is found in primates, allowing greater mobility at the CAS. While all primates have values close to 100, indicating an orientation close to perpendicular, ranges of variation are very narrow for genera and some have values slightly lower than others (closer to 95) (see Ford, 1980a). Those with lower values and a slightly more oblique orientation of the CAS include many vertical clingers and leapers (Callithrix, Cebuella, Avahi, and Galago, but not Tarsius), those exhibiting suspensory behavior (Hylobates, Brachyteles, Ateles), and the semi-terrestrial Macaca. As for many other features discussed here, similar, convergent adaptations in the postcranium have developed in primates which hold their torso vertically in locomotion, whether they do so for vertical leaping or forelimb suspension: in this case, an oblique CAS which limits mobility in the midtarsal joint to a degree. Increased obliqueness of the CAS is also seen in Aotus and a few other primates in which it is less easily explained. The perpendicular CAS seen in the ancestral platyrrhine suggests mobility in the joint and the absence of significant amounts of vertical support postures (either for leaping or forelimb suspension).

Discussion

Locomotor and postural capabilities of the earliest "hypothetical platyrrhine"
The portrait we are able to draw of this earliest platyrrhine, from the analysis above, is an extremely consistent one. The consistency is all the more striking since the list of

morphological features considered ancestral was based primarily on out-group comparisons and not on prior detailed functional analyses, although some consideration of function (as well as other criteria) was used (see discussion of polarity determination in Ford, 1980a). This, combined with the sheer number of features considered (see Table 1) made it virtually impossible to "fudge" the ancestral morphotype towards features favoring a particular locomotor mode.

What emerges is a pattern indicating an arboreal, quadrupedal monkey. Features supporting this include: on the humerus—a shallow, wide bicipital groove, slight dorsal displacement of the medial epicondyle, cone-shaped trochlea, moderate height for the trochlear margins, distinct but rounded trochleocapitular ridge, and a wide supinator crest; on the femur—no posterior extension of the articular head, small lesser trochanter, medial condyle just slightly larger than lateral and extends slightly further distally; on the tibia a long and centrally located fibular facet; and on the calcaneus an alternating tarsus with the AAS not reaching or just narrowly reaching the anterior end of the bone. This monkey would exhibit a large component of branch walking and running using a semipronated, grasping limb (perhaps flat-palmed on very large supports).

The arboreal quadrupedalism of the earliest platyrrhine almost certainly was one utilizing primarily horizontal branches and without any significant leaping behavior. This is clear since the morphotype lacks any of the distinctive features analysed here that suggest leaping or vertical postures, whether ricochetal, thigh-powered, or foot-powered. Instead, it has a femoral head and greater trochanter subequal in height (rather than a higher greater trochanter); a postero-medial rather than medial extension of the lesser trochanter, which is small in size; a patellar groove of moderate rather than extreme depth; the lack of a synostosis or syndesmosis at the distal tibio-fibular articulation; a calcaneal PAS angled between 10–20°, not lower; an alternating midtarsal joint rather than a serial one; a moderate lever arm with a relatively short calcaneal heel; and a distal calcaneal end and CAS which is perpendicular to the long axis of the calcaneus medio-laterally. While having slightly more weight transferred through the humero-radial rather than the humero-ulnar aspect of the elbow joint has been said to suggest quadrupedal grasp-leaping (Szalay & Dagosto, 1980), this forelimb trait does not reflect leaping so much as grasping, and could as easily allow grasp-clinging or branch-walking with a grasping forelimb without significant amounts of leaping. While none of the traits listed here preclude some leaping, and in fact many occur in at least one primate that exhibits high frequencies of leaping behavior as well as in many quadrupedal primates, the combination of these traits is best associated with marked propensities for arboreal quadrupedalism. This conclusion is reinforced by the absence of any features specifically identified as biomechanical correlates of leaping behavior.

The ancestral platyrrhine morphotype presented here also lacks any of the features that characterize primates with high percentages of suspensory behavior in their locomotor repertoires. These characters in the ancestral platyrrhine include: the absence of a rounded deltopectoral crest; the reduced rather than enlarged lateral epicondyle; the presence of a dorsal epitrochlear fossa; subequal femoral head and greater trochanter rather than a higher head; a postero-medial rather than medial placement of the lesser trochanter; the patellar groove moderately narrow rather than wide; a calcaneal PAS of moderate rather than extreme length and not located at a high angle; an alternating midtarsal joint; and a CAS oriented perpendicular to the long axis of the calcaneus. The perpendicular CAS as well as a postero-medial rather than posterior location of the lesser trochanter and subequal extension of the femoral head and greater trochanter also mediate against cursorial, semi-terrestrial locomotor habits. And the ancestral platyrrhine lacks specializations for slow-climbing as seen in lorises, lacking a shortened PAS on the calcaneus at a high angle.

Thus, the ancestral platyrrhine emerges as a quadrupedal arboreal branch runner and walker, with significantly limited or non-existent propensities for large amounts of leaping, suspensory behavior, cursoriality, or slow-climbing. This generalized structure would allow this monkey to engage infrequently in a wide variety of movements, including some leaping, and indeed primarily quadrupedal platyrrhines such as *Cebus, Chiropotes, Aotus, Callicebus,* and *Saguinus* do leap between 10% and 25% of the time during locomotion (Fleagle & Mittermeier, 1980, 1981; Wright, 1981; Terborgh, 1983).

An important aspect of this study is an examination of the features present in this hypothetical ancestor that are unique platyrrhine traits and their functional significance. While a number of postcranial features are considered shared derived platyrrhine features (Table 1), only a small number of these can be discussed at present in terms of implications for behavior and adaptations. These include: a humeral medial epicondyle which is slightly displaced dorsally; a reduced lateral epicondyle; a rounded bony eminence (mound) on the posterior neck of the femur (which may represent the ancestral platyrrhine condition, see above); a long, centrally located fibular facet on the distal tibia; and an alternating midtarsal joint where the AAS just barely reaches the anterior end of the calcaneus or not at all. As discussed above, all of these simply reinforce the adaptation for arboreal quadrupedalism and grasping. In addition, the alternating midtarsal joint without a narrow anterior extension is primitive for euprimates, and as discussed above, the long fibular facet of platyrrhines may, in fact, be best interpreted as the primitive euprimate condition rather than a specifically platyrrhine trait.

Implications for primate evolution

What does this model of the earliest platyrrhine allow us to say about the known platyrrhine fossil record? We can use it to weigh the various possibilities presented at the beginning of this paper.

(1) Are any fossils sufficiently similar to this morphotype to represent the remains of members of this ancestral population? Most of the known fossils from South America and the Antilles for which there are postcrania are already specialized beyond the level suggested for the ancestral platyrrhine (Ford, 1986a,b; Ford & Morgan, 1986; Davis, 1987), and none of these are a good model or represent potential "hard evidence" for the place and age of the earliest platyrrhine. The single possible exception at present is the astragalus from Gaiman, Argentina, assigned to *Dolichocebus gaimanensis* (Fleagle & Bown, 1983). This astragalus appears to be very similar to that of *Aotus* and *Callicebus* (Reeser, 1984), platyrrhines suggested as being little changed from the ancestral platyrrhine by Ford (1980a, 1986a). However, Gebo & Simons (1987) suggest there may be features indicating greater leaping propensities in this astragalus.

(2) Are any known fossils from South America *more* primitive than this hypothesized platyrrhine ancestor, which would lead to several possible hypotheses? None of the known South American primate postcrania are more primitive than the proposed common platyrrhine ancestor, and thus there is no evidence from the skeleton for non-platyrrhine primates having reached the continent.

(3) If, as indicated in answer to (2), all known South American primates are

platyrrhines, did the first derived platyrrhine evolve while in transit to South America or prior to movement out of the geographic area in which this branch separated from other anthropoids? At least one Fayum specimen (YPM 25967, a distal tibial fragment) does exhibit a derived platyrrhine morphology, the posterior displacement of a long fibular facet. Either this is a rare convergent development for this trait within primates, or it is evidence that platyrrhines diverged from other anthropoids in Africa, and the stock developed at least some of their unique platyrrhine attributes prior to introduction to South America. In this latter interpretation, the Fayum tibia would represent a remnant of this early African platyrrhine root. Hoffstetter (1977) has suggested that the parapithecids from the Fayum are platyrrhines. However, aspects of the postcranial morphology of *Apidium* suggest several departures from the ancestral platyrrhine morphotype constructed here and indicate more leaping specializations in these particular Fayum primates (Fleagle and Simons, 1983; Gebo & Simons, 1987—the latter have also suggested derived cercopithecoid affinities for this group). Only further fossil discoveries in Africa, including better preserved and more complete specimens similar to YPM 25967, will allow a clear resolution of this problem.

This hypothesized structure for the common platyrrhine ancestor and the presence of some similar fossils in Africa alongside other early anthropoids allows a consideration of early anthropoid adaptation and of ways in which the earliest platyrrhines and anthropoids may have been like or unlike the earliest euprimates. Several other workers have also addressed some of these issues recently, primarily from the perspective of identifying discrete strepsirhine/adapid features. Consideration of alternate views on hypothesized early anthropoid morphology therefore requires some discussion of models of early euprimate and strepsirhine adaptations.

Dagosto (1986, 1988) has demonstrated clearly that the strepsirhines share a suite of postcranial features which are derived within primates, and which are shared with adapid primates. For these features, Dagosto has found that *Tarsius*, the omomyids, and the anthropoids share a retention of the primitive eutherian condition. Dagosto interprets the shared derived features of strepsirhines as adaptations for grasping and climbing. Gebo (1986b) interprets strepsirhine pedal morphology as one adapted for the use of vertical supports, primarily vertical climbing rather than leaping. These two models of strepsirhine adaptation are therefore quite similar, except in the nature of the ancestral, pre-strepsirhine state that each proposes.

The ancestral euprimate proposed by Dagosto (1988) would have been small (less than 500 g) and exhibited more leaping ("grasp-leaping"), perhaps similar to the condition seen in many omomyids. This model is similar to the hindlimb dominated locomotion predicted by Martin (1972) in the common primate ancestor. Dagosto then suggests that the earliest strepsirhines developed arboreal climbing and decreased leaping in association with increased body size. Gebo (1986b), in distinct contrast, suggests that the vertical climbing of strepsirhines is related to *small* body size and is ancestral to the condition seen in anthropoids (thus representing the ancestral euprimate morphotype), and that anthropoids diverged by increased use of horizontal supports for climbing, in association with increased body size. However, he notes flaws in his model, in particular the similarity of omomyids and *Tarsius* to anthropoids in many features and, more crucial, the apparent very small body size of at least one early anthropoid, *Qatrania* (Simons & Kay, 1983).

In both Dagosto's and Gebo's models, the ancestral euprimate is small in body size, as supported by much of the Eocene fossil evidence, and I find no reason to dispute that assumption. However, the vertical support orientation of strepsirhines, tied with other aspects of their postcranial and behavioral makeup, seems best interpreted as a derived adaptation within primates, rather than as the ancestral euprimate type; the same would be true of the vertical support orientation of *Tarsius* and omomyids.

Most features cited by Dagosto (1986, 1988; also Szalay & Dagosto, 1980) as characterizing the ancestral euprimate morphotype are adaptive for the complex joint movements associated with climbing, walking, and grasping in an arboreal milieu. Very few features suggest significant (as opposed to occasional) amounts of leaping with associated morphological adaptations. Of those that do, at least one (increased length of tarsal bones) is absent in most anthropoids and can as easily be interpreted as a convergent development for leaping in many strepsirhines and the omomyids and *Tarsius* (as opposed to their hypothesized reversal in anthropoids). This, when combined with the reconstructed platyrrhine morphotype offered here that appears but little changed from its earlier, non-platyrrhine primate ancestors, suggests that it is more likely that the ancestral euprimate was an arboreal quadruped emphasizing grasping and walking on horizontal supports. This model agrees with that suggested by Godinot & Jouffroy (1984), who found the wrist of *Adapis parisiensis* is very similar to that of many platyrrhines, and lacks any specializations characteristic of either leaping or slow-climbing primates (although Dagosto [1986] has shown that the hindlimb of *Adapis* exhibits a highly derived structure, not associated with arboreal quadrupedalism). Godinot & Jouffroy suggest a locomotor mode marked by quadrupedal branch running and walking, and further suggest that similarities between the hands of adapids and platyrrhines are perhaps better interpreted as retentions from their common euprimate ancestor than as derived (either homologous or convergent) features (but see also Beard & Godinot, 1988).

The locomotor repertoire of this early euprimate would be marked by movement using a grasping forelimb and hindlimb in its quadrupedal and other, far less frequent, movements, especially on medium- to small-sized supports, as a distinct derivation from early eutherian mammals. In this model, similarities between the hindlimbs of adapids and omomyids that suggest specialized leaping abilities would be independently derived from some earlier (Paleocene) common euprimate ancestor, as yet undiscovered (and thus probably not from Europe or North America).

From this early euprimate, the earliest anthropoid was probably very similar to the earliest platyrrhine discussed here. The features suggest, if anything, a greater emphasis on horizontal, quadrupedal climbing, perhaps (as suggested by Szalay & Dagosto, 1980, based on the elbow) with mechanical improvements for increased stability. The primary need for increased stability may result from the constraints of increased body size in the early platyrrhines (and possibly the other early anthropoids). *Aotus*, among living platyrrhines, comes closest to exhibiting the mix of postcranial features proposed here for the ancestral platyrrhine, although it clearly has diverged from this in some features (Ford, 1980a, 1986a). *Aotus* is somewhat larger than Dagosto's proposed ancestral euprimate, averaging around 1000 g (780–1250 g range), and is usually described as a generalized arboreal quadruped (Wright, 1981).

Summary and conclusions

Through a cladistic analysis of known primates, especially extinct and extant platyrrhine primates, an hypothesis of the structure of the postcranium of the earliest ancestral platyrrhine was constructed. Through this process, a number of questions are raised about the evolution of primates.

An analysis of the functional implications of the features making up the total morphological pattern within the joints examined (aspects of the shoulder, elbow, hip, knee, upper ankle, subtalar, and midtarsal joints) suggests that the ancestral platyrrhine was an arboreal quadrupedal monkey, emphasizing branch running and walking with a semi-pronated, grasping limb. This primate would not have included a significant component of leaping in its locomotor repertoire, and would have preferred horizontal over vertical supports, although it may have engaged in infrequent leaping and climbing. Among living platyrrhines, *Aotus* is most similar to the hypothesized ancestor in structure and behavior, although it does exhibit some derivations from this ancestor. The ancestral platyrrhine was probably close to the size of *Aotus*, approximately 1000 g.

Consideration of other primates and recently proposed models leads to the conclusion that, in many features, the structure proposed for the ancestral platyrrhine is also appropriate for the earliest euprimate common ancestor. This first euprimate would have been an arboreal quadruped of small size (less than 500 g), emphasizing grasping locomotion on horizontal supports. The few features that indicate a derived nature for platyrrhines suggest an increased emphasis on quadrupedalism on horizontal supports with improvements for increased stability, most likely associated with an increase in body size. This may also be true of the earlier, common anthropoid ancestor; further study of *Qatrania* is needed to address this issue. Strepsirhines and adapids changed from the common euprimate ancestor by emphasizing the use of vertical supports for climbing, probably also in association with increased body size. Later, various members of this group independently developed a large component of leaping in their locomotor repertoire (some with reduced body size), as have a number of other primates. Omomyids retained the ancestral euprimate small body size but developed increased leaping behaviors.

Thus, early in the divergence of euprimates, three alternative adaptive strategies developed: (1) increase body size and switch to vertical supports for climbing (strepsirhines); (2) increase body size but maintain and improve quadrupedalism on horizontal supports (at least platyrrhines, probably anthropoids); and (3) retain small size but switch to leaping (omomyids and tarsiiforms). As Dagosto (1988) notes (although her model differs in some respects from the one presented here), this does not allow a clear statement on the Eocene ancestors of anthropoid primates. It does rule out adapids as too derived, and the model given here would also rule out known omomyids with specializations for leaping as too derived. Thus, anthropoids must have developed either: (1) from an Eocene member of known groups that was far less specialized postcranially than any fossil yet discovered; or (2) from an as yet undiscovered Eocene euprimate stock, perhaps located in Africa or East Asia, that would have a postcranium of a generalized grasping arboreal quadruped using horizontal supports, in most (but not all) respects similar to the list of features given in Table 1, but a small animal (less than 500 g). A partial living behavioral analog might be some of the more quadrupedal species of *Saguinus* (Thorington, 1968; Castro & Soini, 1977; Fleagle & Mittermeier, 1980; Terborgh, 1983; Yoneda, 1984), but with several anatomical differences, including flattened nails. Similarities of many of the Fayum primates to the ancestral platyrrhine morphotype and the presence of a very small primate at that site (*Qatrania*) are, therefore, even more intriguing for the light they and future African finds may eventually shed on euprimate origins.

From the South American fossil record it appears that known specimens are already derived, either in the direction of living members of the Platyrrhini or in their own, distinctive ways (Ford, 1986a). Material from the Fayum leaves open the possibility that platyrrhines arrived in South America subsequent to their development in Africa as a separate anthropoid group. This study does allow a prediction for testing in the fossil record, the prediction that somewhere, probably in South America in the early Oligocene or very late Eocene but possibly in Africa, there existed a form with the postcranium described in Table 1, and that a fossil of this platyrrhine monkey may one day be discovered.

Acknowledgements

I would like to thank all those that have provided me access to specimens in their care during earlier studies, from which this is drawn. I thank Albert Allen and Karen Schmitt for preparing Figures 2–5, Karen Schmitt for Figures 9 and 10, John Vercillo for photography, John Richardson for general assistance in preparing all figures, and Dr J. S. Farris for providing a copy of the Wagner79 program for cladistic analysis, which was used during initial stages of the compilation of the cladogram on which this article is based. I also thank Elizabeth Strasser and Dr Marian Dagosto for inviting me to participate in their lively symposium and for the stimulating work presented by the other authors involved. The idea for this paper was conceived in 1981; its final form is immeasurably improved by work done by others in the intervening years. I thank L. Davis, B. Holt, E. Strasser, Drs M. Dagosto, J. Fleagle, M. D. Rose, and two anonymous reviewers for critical comments and discussion on this paper.

References

Barnett, C. H. (1970). Talocalcaneal movements in mammals. *J. Zool., Lond.* **160**, 1–7.

Barnett, C. H. & Napier, J. R. (1952). The axis of rotation at the ankle joint in man: its influence upon the form of the talus and the mobility of the fibula. *J. Anat.* **86**, 1–9.

Barnett, C. H. & Napier, J. R. (1953). The rotatory mobility of the fibula in Eutherian mammals. *J. Anat.* **87**, 11–21.

Beard, K. C. & Godinot, M. (1988). Carpal anatomy of *Smilodectes gracilis* (Adapiformes, Notharctinae) and its significance for lemuriform phylogeny. *J. hum. Evol.* **17**, 71–92.

Berge, C. H. & Jouffroy, F. K. (1986). Morpho-functional study of *Tarsius*'s foot as compared to the galagines's: what does an "elongate calcaneus" mean? In (D. M. Taub & F. A. Ring, Eds.) *Current Perspectives in Primate Biology*, pp. 146–158. New York: Van Nostrand Reinhold.

Campbell, B. (1939). The shoulder anatomy of the moles: a study in phylogeny and adaptation. *Amer. J. Anat.* **64**, 1–39.

Carleton, A. (1941). A comparative study of the inferior tibio-fibular joint. *J. Anat.* **76**, 45–55.

Cartmill, M. (1978). The orbital mosaic in prosimians and the use of variable traits in systematics. *Folia primatol.* **30**, 89–114.

Castro, R. & Soini, P. (1977). Field studies on *Saguinus mystax* and other callitrichids in Amazonian Peru. In (D. G. Kleiman, Ed.) *The Biology and Conservation of the Callitrichidae*, pp. 73–78. Washington, D.C.: Smithsonian Institution Press.

Conroy, G. C. (1976). Primate postcranial remains from the Oligocene of Egypt. *Contrib. Primatol.* **8**, 1–134.

Dagosto, M. (1983). Postcranium of *Adapis parisiensis* and *Leptadapis magnus* (Adapiformes, Primates): adaptational and phylogenetic significance. *Folia primatol.* **41**, 49–101.

Dagosto, M. (1985). The distal tibia of primates with special reference to the Omomyidae. *Int. J. Primat.* **6**(1), 45–75.

Dagosto, M. (1986). The joints of the tarsus in the strepsirhine primates: functional, adaptive, and evolutionary implications. Ph.D. Dissertation, City University of New York.

Dagosto, M. (1988). Implications of postcranial evidence for the origin of euprimates. *J. hum. Evol.*, **17**, 35–56.

Davis, L. C. (1987). Morphological evidence of positional behavior in the hindlimb of *Cebupithecia sarmientoi* (Primates: Platyrrhini). M.A. Thesis, Arizona State University.

Decker, R. L. & Szalay, F. S. (1974). Origin and function of the pes in the Eocene Adapidae (Lemuriformes, Primates). In (F. A. Jenkins, Jr., Ed.) *Primate Locomotion*, pp. 261–291. New York: Academic Press.

Delson, E. & Andrews, P. (1975). Evolution and interrelationships of the catarrhine primates. In (W. P. Luckett & F. S. Szalay, Eds) *Phylogeny of the Primates*, pp. 405–446. New York: Plenum Press.

Eldredge, N. & Cracraft, J. (1980). *Phylogenetic Patterns and the Evolutionary Process*. New York: Columbia University Press.

Erikson, G. E. (1963). Brachiation in New World monkeys and in anthropoid apes. *Symp. Zool. Soc. London* **10**, 135–164.

Fleagle, J. G. (1977). Locomotor behavior and skeletal anatomy of sympatric Malaysian leaf-monkeys (*Presbytis obscura* and *Presbytis melalophos*). *Yrbk phys. Anthrop.* **20**, 440–453.

Fleagle, J. G. (1983). Locomotor adaptations of Oligocene and Miocene hominoids and their phyletic implications. In (R. L. Ciochon & R. S. Corruccini, Eds) *New Interpretations in Ape and Human Ancestry*, pp. 301–324. New York: Plenum Press.

Fleagle, J. G. & Bown, T. M. (1983). New primate fossils from Late Oligocene (Colhuehuapian) localities of Chubut Province, Argentina. *Folia primatol.* **41**, 240–266.

Fleagle, J. G. & Mittermeier, R. A. (1980). Locomotor behavior, body size, and comparative ecology of seven Surinam monkeys. *Am. J. phys. Anthrop.* **52**, 301–314.

Fleagle, J. G., Mittermeier, R. A., & Skopec, A. L. (1981). Differential habitat use by *Cebus apella* and *Saimiri sciureus* in Central Surinam. *Primates* **22**(3), 361–367.

Fleagle, J. G. & Simons, E. L. (1978). Humeral morphology of the earliest apes. *Nature* **276**, 705–707.

Fleagle, J. G. & Simons, E. L. (1982). The humerus of *Aegyptopithecus zeuxis:* a primitive anthropoid. *Am. J. phys. Anthrop.* **59**(2), 175–193.

Fleagle, J. G. & Simons, E. L. (1983). The tibio-fibular articulation in *Apidium phiomense*, an Oligocene anthropoid. *Nature* **301**, 238–239.

Ford, S. M. (1980*a*). A systematic revision of the Platyrrhini based on selected features of the postcranium. Ph.D. Dissertation, University of Pittsburgh.

Ford, S. M. (1980*b*). Phylogenetic relationships of the Platyrrhini: the evidence of the femur. In (R. L. Ciochon & A. B. Chiarelli, Eds) *Evolutionary Biology of the New World Monkeys and Continental Drift*, pp. 317–330. New York: Plenum Press.

Ford, S. M. (1986*a*). Systematics of the New World monkeys. In (D. Swindler, Ed.) *Comparative Primate Biology, Vol. 1: Systematics, Evolution, and Anatomy*, pp. 73–135. New York: Alan R. Liss.

Ford, S. M. (1986*b*). Subfossil platyrrhine tibia (Primates: Callitrichidae) from Hispaniola: a possible further example of island gigantism. *Am. J. phys. Anthrop.* **70**, 47–62.

Ford, S. M. & Morgan, G. S. (1986). A new ceboid femur from the Late Pleistocene of Jamaica. *J. Vert. Paleo.* **6**(3), 281–289.

Gebo, D. L. (1986*a*). The anatomy of the prosimian foot and its application to the primate fossil record. Ph.D. Dissertation, Duke University.

Gebo, D. L. (1986*b*). Anthropoid origins—the foot evidence. *J. hum. Evol.* **15**, 421–430.

Gebo, D. L. (1986*c*). Miocene lorisids—the foot evidence. *Folia primatol.* **42**, 217–225.

Gebo, D. L. & Dagosto, M. (1988). Foot anatomy, climbing, and the origin of the Indriidae. *J. hum. Evol.* **17**, 135–154.

Gebo, D. L. & Simons, E. L. (1987). Morphology and locomotor adaptations of the foot in early Oligocene anthropoids. *Am. J. phys. Anthrop.* **74**, 83–101.

Gingerich, P. D. (1980). Eocene Adapidae, paleobiogeography, and the origin of South American Platyrrhini. In (R. L. Ciochon & A. B. Chiarelli, Eds) *Evolutionary Biology of the New World Monkeys and Continental Drift*, pp. 123–138. New York: Plenum Press.

Godinot, M. & Jouffroy, F. K. (1984). La main d'*Adapis* (Primates, Adapidae). In (E. Buffetaut, J. M. Mazin, & E. Salmion, Eds) *Actes du symposium paléontologique G. Cuvier*, pp. 221–242. Montbeliard.

Grand, T. I. (1967). The functional anatomy of the ankle and foot of the slow loris. *Am. J. phys. Anthrop.* **26**, 207–218.

Grand, T. I. (1968). The functional anatomy of the lower limb of the howler monkey (*Alouatta caraya*). *Amer. J. phys. Anthrop.* **28**, 163–182.

Grand, T. I. & Lorenz, R. (1968). Functional analysis of the hip joint in *Tarsius bancanus* (Horsfield, 1821) and *Tarsius syrichta* (Linnaeus, 1758). *Folia primat.* **9**, 161–181.

Gregory, W. K. (1920). On the structure and relations of *Notharctus*, an American Eocene primate. *Mem. Amer. Mus. nat. Hist.* n.s. **3**(2), 49–243.

Hall-Craggs, E. C. B. (1965). An analysis of the jump of the lesser galago. *J. Zool. (Lond.)*, **147**, 20–29.

Hill, W. C. O. (1959). The anatomy of *Callimico goeldii* (Thomas), a primitive American primate. *Trans. Amer. Phil. Soc.* **49**(5), 1–116.

Hill, W. C. O. (1960). *Primates. Comparative Anatomy and Taxonomy. Vol. IV, Cebidae. Part A*. Edinburgh: Edinburgh University Press.

Hoffstetter, R. (1977). Origine et principales dichotomies des Primates Simiiformes (= Anthropoidea). *C. R. Acad. Sc. Paris, s.D*, **284,** 2095–2098.

Howell, A. B. (1944). *Speed in Animals*. Chicago: University of Chicago Press.

Jenkins, F. A., Jr. (1973). The functional anatomy and evolution of the mammalian humeroulnar articulation. *Am. J. Anat.* **137,** 281–298.

Jenkins, F. A., Jr. & Camazine, S. M. (1977). Hip structure and locomotion in ambulatory and cursorial carnivores. *J. Zool., Lond.* **181,** 351–370.

Jenkins, F. A., Jr. & Parrington, F. R. (1976). The postcranial skeleton of the Triassic mammals *Eozostrodon, Megazostrodon* and *Erythrotherium. Phil. Trans. Royal Soc. London, B. Biol. Sci.* **273,** 387–431.

Jolly, A. (1966). *Lemur Behavior*. Chicago: University of Chicago Press.

Jolly, C. J. (1967). The evolution of baboons. In (H. Vagtborg, Ed.) *The Baboon in Medical Research*, pp. 427–457. Austin: University of Texas Press.

Jouffroy, F. K. & Gunther, M. M. (1985). Interdependence of morphology and behavior in the locomotion of galagines. In (S. Kondo, Ed.) *Primate Morphophysiology, Locomotor Analyses and Human Bipedalism*, pp. 201–234. Tokyo: University of Tokyo Press.

Jungers, W. L. (1977). Hindlimb and pelvic adaptations to vertical climbing and clinging in *Megaladapis*, a giant subfossil prosimian from Madagascar. *Yrbk phys. Anthrop.* **20,** 508–524.

Kay, R. F. (1980). Platyrrhine origins: a reappraisal of the dental evidence. In (R. L. Ciochon & A. B. Chiarelli, Eds.) *Evolutionary Biology of the New World Monkeys and Continental Drift*, pp. 159–188. New York: Plenum Press.

Kay, R. F. (1984). On the use of anatomical features to infer foraging behavior in extinct primates. In (P. S. Rodman & J. G. H. Cant, Eds) *Adaptations for Foraging in Nonhuman Primates*, pp. 21–53. New York: Columbia University Press.

Kay, R. F. & Cartmill, M. (1977). Cranial morphology and adaptations of *Palaechthon nacimienti* and other Paramomyidae(Plesiadapoidea, Primates) with a description of a new genus and species. *J. hum. Evol.* **6,** 19–53.

Kinzey, W. G. (1986). New World primate field studies: what's in it for Anthropology? *Ann. Rev. Anthropol.,* **15,** 121–148.

Kinzey, W. G., Rosenberger, A. L. & Ramirez, M. (1975). Vertical clinging and leaping in a Neotropical anthropoid. *Nature* **255,** 327–328.

Langdon, J. H. (1986). Functional morphology of the Miocene hominoid foot. *Contrib. Primatol.* **22,** 1–225.

Le Gros Clark, W. E. & Thomas, D. P. (1951). Associated jaws and limb bones of *Limnopithecus macinnesi. Fossil Mammals of Africa, No. 3*. London: British Museum (Natural History).

Lessertisseur, J. & Saban, R. (1967). Squelette appendiculaire. In (P-P Grassé, Ed.) *Traité de Zoologie: Anatomie, Systématique, Biologie. Tome XVI: Mammifères: Anatomie et Reproduction. Fasc. I: Téguments et squelette*, pp. 709–1078. Paris: Masson et Cie.

Lewis, O. J. (1980). The joints of the evolving foot. I. The ankle joint. *J. Anat. (Lond.)* **130,** 527–543.

MacKinnon, J. (1974). The behavior and ecology of wild orangutans (*Pongo pygmaeus*). *Anim. Behav.* **22,** 3–74.

Maple, T. L. & Hoff, M. P. (1982). *Gorilla Behavior*. New York: Van Nostrand Reinhold Co.

Martin, R. D. (1972). Adaptive radiation and behavior of the Malagasy lemurs. *Phil. Trans. Royal Soc. Lond.* **264,** 295–352.

Matthew, W. D. (1937). Paleocene faunas of the San Juan Basin, New Mexico. *Trans. Amer. Phil. Soc.*, n.s., **30.**

Morbeck, M. E. (1972). A re-examination of the forelimb of the Miocene Hominoidea. Ph.D. Dissertation, University of California, Berkeley.

Morbeck, M. E. (1983). Miocene hominoid discoveries from Rudabanya: implications from the postcranial skeleton. In (R. L. Ciochon & R. S. Corruccini, Eds.) *New Interpretations in Ape and Human Ancestry*, pp. 369–404. New York: Plenum Press.

Moynihan, M. (1976). *The New World Primates*. Princeton: Princeton University Press.

Napier, J. R. (1964). The evolution of bipedal walking in the hominids. *Arch. Biol. (Liege)* **75,** 673–708.

Napier, J. R. & Davis, P. R. (1959). The forelimb skeleton and associated remains of *Proconsul africanus. Fossil Mammals of Africa, No. 16*. London: British Museum (Natural History).

Napier, J. R. & Walker, A. C. (1967). Vertical clinging and leaping—a newly recognized category of locomotor behavior in primates. *Folia primatol.* **6,** 204–219.

Novacek, M. J. (1980). Cranioskeletal features in tupaiids and selected Eutheria as phylogenetic evidence. In (W. P. Luckett, Ed.) *Comparative Biology and Evolutionary Relationships of Tree Shrews*, pp. 35–93. New York: Plenum Press.

Novacek, M. J. (1986). The skull of leptictid insectivorans and the higher-level classification of eutherian mammals. *Bull. Amer. Mus. nat. Hist.*, **183**(1), 1–112.

Orlosky, F. (1973). Comparative dental morphology of extant and extinct Cebidae. Ph.D. Dissertation, University of Washington, Seattle.

Prost, J. H. (1976). Postural behavior in chimpanzees. *Yrbk phys. Anthrop.* **20,** 454–465.

Rasmussen, D. T. (1986). Anthropoid origins: a possible solution to the Adapidae-Omomyidae paradox. *J. hum. Evol.* **15**, 1–12.

Reed, C. A. (1951). Locomotion and appendicular anatomy in three soricoid insectivores. *The American Midland Naturalist* **45**(3), 513–671.

Reeser, L. A. (1984). Morphological affinities of new fossil talus of *Dolichocebus gaimanensis*. *Am. J. phys. Anthrop.* **63**, 206–207 (abstract).

Ribeiro, A. de M. (1940). Commentaries on South American primates. *Mem. Inst. Oswaldo Cruz (Rio de Janeiro, Brasil)* **35**(4), 779–851.

Rollinson, J. & Martin, R. D. (1981). Comparative aspects of primate locomotion with special reference to arboreal cercopithecines. *Symp. Zool. Soc. Lond.* **48**, 377–427.

Rose, K. D. (1987). Climbing adaptations in the Early Eocene mammal *Chriacus* and the origin of Artiodactyla. *Science* **236**, 314–316.

Rose, K. D. & Fleagle, J. G. (1981). The fossil history of non-human primates in the Americas. In (A. F. Coimbra-Filho & R. Mittermeier, Eds.) *Ecology and Behavior of Neotropical Primates, Vol. 1.*, pp. 111–168. Rio de Janeiro: Academia Brasileira de Ciencias.

Rose, K. D. & Walker, A. C. (1985). The skeleton of early Eocene *Cantius*, oldest lemuriform primate. *Am. J. phys. Anthrop.* **66**, 73–89.

Rose, M. D. (1983). Miocene hominoid postcranial morphology—monkey-like, ape-like, neither, or both? In (R. L. Ciochon & R. S. Corruccini, Eds.) *New Interpretations in Ape and Human Ancestry*, pp. 405–417. New York: Plenum Press.

Rose, M. D. (1988). Another look at the anthropoid elbow. *J. hum. Evol.* **17**, 193–224.

Rosenberger, A. L. (1979). Phylogeny, evolution and classification of New World monkeys (Platyrrhini, Primates). Ph.D. Dissertation, City University of New York.

Rosenberger, A. L., Strasser, E. & Delson, E. (1985). Anterior dentition of *Notharctus* and the Adapid-Anthropoid hypothesis. *Folia primat.* **44**, 15–39.

Rosenberger, A. L. & Szalay, F. S. (1980). On the tarsiiform origins of the Anthropoidea. In (R. L. Ciochon & A. B. Chiarelli, Eds.) *Evolutionary Biology of the New World Monkeys and Continental Drift*, pp. 139–157. New York: Plenum Press.

Savage, D. E. & Waters, B. T. (1978). A new omomyid primate from the Wasatch Formation of Southern Wyoming. *Folia primatol.* **30**, 1–29.

Schlosser, M. (1907). Beitrag zur Osteologie und systematichen Stellung der Gattung *Necrolemur*, sowie zur Stammesgeschichte der Primaten uberhaupt. *Neuen Jahrb. Min. Geol. Paleont.*, pp. 197–226.

Schmid, P. (1979). Evidence of microchoerine evolution from Dielsdorf (Zurich Region, Switzerland)—a preliminary report. *Folia primatol.* **31**, 301–311.

Schwartz, J. H. (1986). Primate systematics and a classification of the order. In (D. Swindler, Ed.) *Comparative Primate Biology, Vol. 1: Systematics, Evolution, and Anatomy*, pp. 1–41. New York: Alan R. Liss.

Simons, E. L. (1972). *Primate Evolution*. New York: Macmillan.

Simons, E. L. & Kay, R. F. (1983). *Qatrania*, new basal anthropoid primate from the Fayum, Oligocene of Egypt. *Nature* **304**, 624–626.

Strasser, E. (1988). Pedal evidence for the origin and diversification of cercopithecid clades. *J. hum. Evol.* **17**, 225–245.

Susman, R., Badrian, N. L., & Badrian, A. J. (1980). Locomotor behavior of *Pan paniscus* in Zaire. *Am. J. phys. Anthrop.* **53**, 69–80.

Szalay, F. S. (1966). The tarsus of the Paleocene leptictid *Prodiacodon* (Insectivora, Mammalia). *Amer. Mus. Novit.*, No. 2267, 13 pp.

Szalay, F. S. (1975). Phylogeny, adaptations and dispersal of the tarsiiform primates. In (W. P. Luckett & F. S. Szalay, Eds) *Phylogeny of the Primates*, pp. 357–404. New York: Plenum Press.

Szalay, F. S. (1977). Ancestors, descendants, sister groups, and testing phylogenetic hypotheses. *Syst. Zool.* **26**, 12–18.

Szalay, F. S. & Dagosto, M. (1980). Locomotor adaptations as reflected on the humerus of Paleogene primates. *Folia primatol.* **34**, 1–45.

Szalay, F. S. & Decker, R. L. (1974). Origins, evolution, and function of the tarsus in Late Cretaceous eutherians and Paleocene primates. In (F. A. Jenkins, Jr., Ed.) *Primate Locomotion*, pp. 223–259. New York: Academic Press.

Szalay, F. S. & Drawhorn, G. (1980). Evolution and diversification of the Archonta in an arboreal milieu. In (W. P. Luckett, Ed.) *Comparative Biology and Evolutionary Relationships of Tree Shrews*, pp. 133–169. New York: Plenum Press.

Szalay, F. S., Tattersall, I. & Decker, R. L. (1975). Phylogenetic relations of *Plesiadapis*—the postcranial evidence. *Contr. Primatol.* **5**, 136–166.

Tardieu, C. (1981). Morpho-functional analysis of the articular surfaces of the knee-joint in primates. In (A. B. Chiarelli & R. S. Corruccini, Eds) *Primate Evolutionary Biology*, pp. 68–80. Berlin: Springer-Verlag.

Tattersall, I. (1982). *The Primates of Madagascar*. New York: Columbia University Press.

Terborgh, J. (1983). *Five New World Primates: A Study in Comparative Ecology.* Princeton: Princeton University Press.

Thorington, R. W., Jr. (1968). Observations of the tamarin *Saguinus midas. Folia primatol.,* **9,** 95–98.

Walker, A. C. (1974). Locomotor adaptations in past and present prosimian primates. In (F. A. Jenkins, Jr., Ed.) *Primate Locomotion,* pp. 349–381. New York: Academic Press.

Wiley, E. O. (1981). *Phylogenetics: The Theory and Practice of Phylogenetic Systematics.* New York: Wiley.

Wright, P. C. (1981). The night monkeys, genus *Aotus.* In (A. F. Coimbra-Filho & R. Mittermeier, Eds) *Ecology and Behavior of Neotropical Primates, Vol. 1.* pp. 211–240. Rio de Janeiro: Academia Brasileira de Ciencias.

Yoneda, M. (1984). Comparative studies on vertical separation, foraging behavior, and traveling mode of saddle-backed tamarins (*Saguinus fuscicollis*) and red-chested moustached tamarins (*Saguinus labiatus*) in Northern Bolivia. *Primates* **25,** 414–422.

M. D. Rose

Department of Anatomy, New Jersey

Another look at the anthropoid elbow

Department of Anatomy, New Jersey
Medical School, 100 Bergen St,
Newark, NJ 07103, U.S.A.

Received 29 May 1987
Revision received 30 October
1987 and accepted
25 November 1987

Publication date June 1988

Keywords: elbow joint,
humeroulnar joint, humeroradial
joint, proximal radioulnar joint,
kinematics, Aegyptopithecus,
Miocene hominoids,
anthropoids, cebids, catarrhines,
cercopithecoids, hominoids.

Functional features of the articular surfaces of the joints of the elbow region
are described for a number of extant and fossil anthropoids. At the
humeroradial and proximal radioulnar joints of extant hominoids a
complex of features occurs that guarantees stability throughout extensive
pronation–supination movements. By contrast, in extant non-hominoids
the morphology of these joints is such as to permit more limited movement
and to provide a close-packed position in full pronation. Ramapithecines
and *Oreopithecus* resemble extant hominoids in these features, while *Proconsul*
exhibits some features of both groups of extant anthropoids. The other
Oligocene and Miocene anthropoids investigated all resemble extant
non-hominoids. Major differences in the humeroulnar joint relate mostly to
different means of stabilizing the joint. In extant hominoids, *Oreopithecus*,
and ramapithecines the joint is truly trochleiform and is stabilized against
forces acting in any direction. In other Oligo-Miocene anthropoids except
Proconsul, and in extant non-hominoids, the main joint surface is more
cylindrical, and is stabilized by an anterodistally projecting flange medially
and a posteriorly projecting flange laterally. These features are best
developed in cercopithecoids. It is suggested that in cebids the humeroulnar
joint is translatory, but that it is non-translatory in all the extant and fossil
catarrhines considered.

A number of autapomorphies in the elbow regions of hominoids and of
cercopithecoids are tabulated and grouped in terms of their functional
significance. On the basis of this distribution of features, and using *Cebus* as
an out-group, it is suggested that *Oreopithecus*, ramapithecines, and *Proconsul*
are hominoids. On the basis of the features examined, it is not possible to
determine whether it is cercopithecoids or the taxon to which *Dendropithecus*
and *Pliopithecus* belong that is the sister group of hominoids. *Aegyptopithecus*
is the sister group of all other catarrhines.

Journal of Human Evolution (1988) **17**, 193–224

Introduction

The anatomy of the elbow region in primates has received considerable attention in the
literature. This is due to the fact that there are a number of morphological patterns in this
region among, particularly, extant anthropoids. In addition, the region is relatively well
represented in the anthropoid (especially catarrhine) fossil record. It is therefore an
attractive region for attempts to link structure to function, to link function to positional
behavior, and to elucidate relationships among extant and extinct anthropoid taxa.

Although there is only a single joint cavity present at the elbow, it is anatomically and
functionally divided into the humeroulnar, humeroradial, and proximal radioulnar
articulations. Within each of these there are further functional divisions. Load is
transferred between the arm and the forearm at the elbow. This transfer presumably takes
place predominantly medially or laterally to the extent that either the humeroulnar or
humeroradial joint has the greater contact surface. Stabilizing features are emphasized on
whichever articulation is predominantly involved in load transfer. In general, stabilizing
features will minimize the tendency of the forearm, in response to unbalanced compressive
or tensile forces, to adduct or abduct to the elbow, or of the ulna to rotate around its long
axis. Additional features relate to stability during movement, or in particular joint
positions (close-packing). Flexion–extension movement and its range are principally
determined by the morphology of the humeroulnar joint. Pronation–supination movement

0047–2484/88/01/20193 + 32 $03.00/0

and its range are determined, at the elbow, by the morphology of the proximal radioulnar joint and to a lesser extent, of the humeroradial joint.

Materials and methods

This study concentrates on osteological features of the articular surfaces of the joints of the elbow region. Variation in morphology is explored by using data from small samples of a variety of representative extant anthropoids ($n = 10$, five males and five females) and from a number of fossil species (listed in Tables 1A, 1B, and 2). Both qualitative and quantitative features of this variation are examined. Measurements taken are illustrated in Figures 1 and 2. Comparative osteometrics are expressed in the form of simple indices. While reference is made to the effects of absolute size on a number of quantitative features, these effects are not examined in any detail. Although functional aspects of morphology are emphasized, some reference is made to the significance of the data for phyletic relationships among the taxa considered. A standard suite of taxa has generally been used in constructing the figures. However, although *Cercopithecus* is used as a representative cercopithecid genus, examples from other genera have been used in some figures, especially

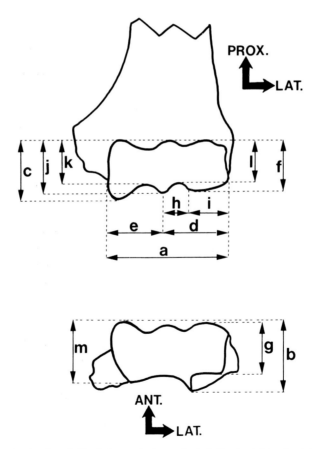

Figure 1. Anterior (top) and distal (bottom) views of a left *Pan troglodytes* distal humerus, showing measurements taken (see also Tables 1A and 1B).

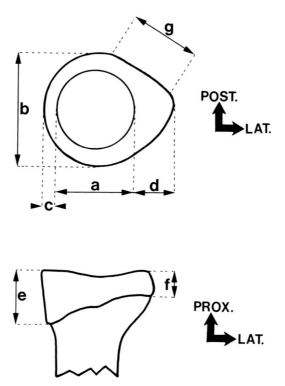

Figure 2. Proximal (top) and anterior (bottom) views of a left *Presbytis entellus* proximal radius, showing measurements taken (see also Table 2).

where individual specimens include particularly good examples of the morphological features being illustrated.

The brachioantebrachial joint as a whole

The distal humerus is expanded mediolaterally to accommodate the articular surface. Compared to the midshaft mediolateral diameter, the articular surface itself is approximately equally wide in all anthropoids except cercopithecines, where the surface is relatively narrow (Table 1A, Index 1). In terms of biepicondylar width, all cercopithecids have a relatively un-expanded distal humerus (Harrison, 1987). There is considerable variation in the overall proportions of the distal articular surface. While there is a trend for the anteroposterior dimension especially to become relatively longer as absolute size increases, cercopithecines are again outstanding in having a distal articular surface that is relatively deep anteroposteriorly (Table 1A, Index 2; Figure 3) and proximodistally long (Table 1A, Index 3; Figures 4, 5). By contrast, in all the fossil species these dimensions are relatively short. As expressed here these proportions are largely determined by trochlear features that will be discussed below. The relative mediolateral expansion of the area for articulation with the radius compared to that of the trochlea is another feature that is generally related to absolute size (Harrison, 1987). The trochlea is relatively wider in larger-bodied species (Table 1A, Index 4; Figures 4, 5). However, it is nevertheless the case that loading across the elbow will be maximal on the ulnar side in species with a relatively

Table 1A　　　　　　　　　　　　**Humeral indices**

Index number Index name Index definition*	Index 1 Shaft/articular width nX100/a	Index 2 Articular depth bX100/a	Index 3 Articular height cX100/a	Index 4 Humeroradial width dX100/e	Index 5 Capitular height fX100/g	Index 6 Zona width hX100/i
Pan troglodytes†	45·2‡ 42–50 2·7	64·5 61–70 3·4	53·5 49–58 2·7	108·2 91–120 9·5	108·2 88–119 9·8	24·5 19–29 3·4
Hylobates	48·6 44–56 3·7	64·3 59–71 3·6	52·4 50–55 1·8	123·5 110–141 10·0	97·2 87–105 6·9	25·4 20–31 4·2
Cercopithecus	62·5 56–69 4·5	75·3 71–84 3·7	64·8 59–79 6·8	136·4 119–157 13·1	128·3 114–143 11·3	37·7 31–43 3·9
Colobus	47·2 40·52 3·8	63·9 56–74 5·6	50·4 47–55 3·0	164·2 149–184 11·2	115·3 100–131 9·0	33·4 28–38 3·0
Ateles	43·1 37·47 3·4	55·1 49–60 3·8	40·0 35–45 3·0	149·2 130–186 17·4	106·9 98–117 7·7	29·7 27–32 2·2
Cebus	45·5 41–49 2·5	57·6 53–63 2·6	46·7 42–52 3·4	148·8 129–187 19·2	112·5 100–125 8·3	29·9 24–35 4·0
Aegyptopithecus 　*zeuxis* DPC-1275	42·8	47·9	48·4	136·4	115·0	31·1
Pliopithecus 　*vindobonensis* Individual 1	47·8	52·0	35·8	156·0	111·5	31·9
Dendropithecus 　*macinnesi* KNM-RU1675 KNM-RU2097 KNM-CA405	— 48·5 —	— 54·2 —	— 46·6 —	106·9 132·6 105·6	94·3 103·4 —	— 35·1 —
Proconsul africanus KNM-RU2036AH KNM-RU2036AK	40·2 —	50·5 —	37·1 —	100·8 —	94·5 93·2	25·7 —
Proconsul nyanzae KNM-RU7696	—	—	—	—	89·8	20·1
?Rudapithecus RUD53	—	47·0	44·2	88·9	97·4	25·2
Sivapithecus sp GSP6663	—	—	—	—	88·8	22·1
?Gigantopithecus GSP12271	—	—	—	—	91·4	17·2
Indet. KNM-FT2751	47·7	65·0	46·8	93·4	111·8	30·0

* These are the measurements illustrated in Figure 1.

† $n = 10$ (five 5 of each sex) for samples of extant anthropoids. Except for *Pan troglodytes* the samples are made up of specimens from several species of each genus.

‡ The numbers listed in each cell are mean (above), range (middle), and s.D. (below).

Table 1B   **Humeral indices (*continued*)**

Index number Index name	Index 7 Lateral keel development kX100	Index 8 *Zona* height (j − 1)X100	Index 9 Trochlear depth	Index 10 Trochlear height	Index 11 Trochlear waisting
Index definition*	0.5X(c + j)	j	mX100/e	cX100/e	jX100/k
Pan troglodytes†	59·8‡	21·2	112·6	110·9	148·2
	51–69	15–28	100–128	102–118	129–178
	5·2	4·4	8·1	5·6	14·8
Hylobates	72·6	13·0	124·6	116·6	125·1
	66–79	10–19	112–133	103–130	111–144
	4·4	3·0	6·2	8·9	12·0
Cercopithecus	89·9	2·0	158·6	152·1	98·6
	83–98	0–4	144–173	141–165	91–107
	5·1	1·2	8·4	8·5	5·5
Colobus	91·1	5·9	141·6	137·9	96·6
	84–96	1–9	127–163	129–149	92–101
	4·6	2·8	11·4.	7·0	3·2
Ateles	93·1	4·9	112·1	99·1	101·8
	88–99	1–10	101–128	86–110	93–109
	3·9	2·8	9·9	7·2	4·4
Cebus	91·2	5·0	120·9	124·5	95·7
	83–97	2–8	102–144	110–141	91–102
	5·6	2·3	17·0	11·2	3·8
Aegyptopithecus zeuxis DPC-1275	86·3	4·1	97·3	87·2	103·2
Pliopithecus vindobonensis Individual 1	83·7	8·7	124·3	97·1	107·8
Dendropithecus macinnesi KNM-RU1675	90·6	10·6	106·9	86·1	102·6
KNM-RU2097	82·3	8·2	131·1	109·5	105·1
KNM-CA405	—	9·2	—	—	102·8
Proconsul africanus KNM-RU2036AH	78·4	15·7	89·6	78·5	119·4
KNM-RU2036AK	—	17·7	—	—	—
Proconsul nyanzae KNM-RU7696	—	23·8	—	—	—
?Rudapithecus RUD53	72·7	22·0	86·8	98·6	119·8
?Gigantopithecus GSP12271	—	24·0	—	—	—
Indet. KNM-FT2751	84·0	13·0	92·2	85·0	109·2

* These are the measurements illustrated in Figure 1.

† *n* = 10 (five of each sex) for samples of extant anthropoids. Except for *Pan troglodytes* the samples are made up of specimens from several species of each genus.

‡ The numbers listed in each cell are mean (above), range (middle), and s.D. (below).

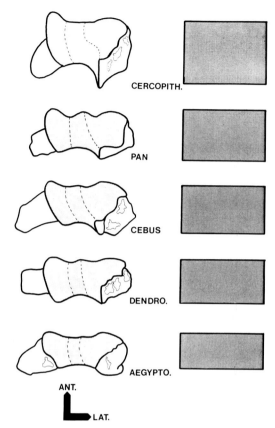

Figure 3. *Left column:* distal views of left humeri drawn to the same mediolateral width of the articular surface in each case. The drawing of the *Dendropithecus* specimen is reversed. The articular surface is stippled. *Right column:* rectangles that enclose the articular surface, arranged in order of anteroposterior depth.

broad trochlea. To the extent that, in some positions of the joint, loading will tend to cause displacement of the elements of the joint, features related to stability might be emphasized in humeroulnar joints of this type.

The humeroradial and proximal radioulnar joints

The major flexors and extensors of the elbow insert into the ulna and, as will be discussed below, the curvatures of the humeroulnar joint surfaces result in uniaxial movement. Thus, during flexion–extension the movement of the radial head on the humeral capitulum is largely dictated by movement of the ulna on the humeral trochlea. However, pronation–supination excursion is largely determined by rotation of the radial head within the confines of the radial notch of the ulna and the annular ligament. During these movements the proximal surface of the radial head also spins on the humeral capitulum. Although it is not considered here, the distal radioulnar joint is also important in determining pronation–supination excursion (O'Connor, 1975, 1976; O'Connor & Rarey, 1979; Sarmiento, 1985). As far as distal forelimb function in general is concerned, forearm rotation is also linked to movements taking place at the wrist and midcarpal joints (Jenkins

& Fleagle, 1975; Jenkins, 1981; Rose, 1983; Sarmiento, 1983). There are two major functional patterns evident in the humeroradial region of the anthropoid elbow. Individual morphological features will be considered first; their functional implications will then be discussed.

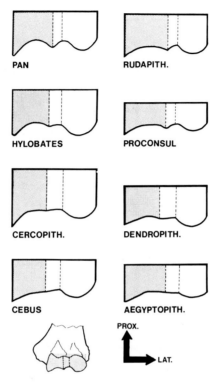

Figure 4. Diagrammatic anterior views of left distal humeral articular surface, drawn to the same mediolateral width in each case. The diagram for *Dendropithecus* is reversed. The trochlea, for articulation with the ulna, is stippled. On the area for articulation with the radius, the *zona conoidea* and the capitulum are separated by a dashed line.

Structural features. The distal humeral articular surface for contact with the radius is divided into two morphologically distinct regions (Figure 4). The capitulum proper lies laterally and articulates with a central fossa on the proximal surface of the radial head. Lying medial to this, between the capitulum and the trochlea, is the *zona conoidea*, which articulates with the peripheral part of the proximal radial surface and, variably, with the articular surface on the side of the radial head.

In extant hominoids the capitulum is globular and relatively equally curved in all directions. Figure 6 illustrates this in lateral view, where it is evident that the capitulum is approximately equal in proximodistal and anteroposterior extent (see also Table 1A, Index 5). Although this feature is somewhat variable, there is a tendency among other extant anthropoids for the capitulum to be longer proximodistally than anteroposteriorly. This is particularly evident among cercopithecines, where the contour of the capitulum includes a less curved, anterodistally facing region (Figure 6, *Cercopithecus*). This may result

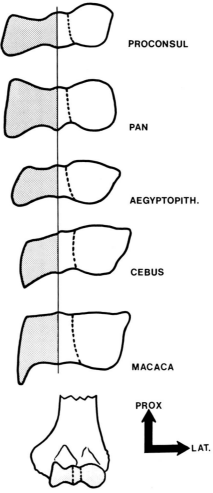

Figure 5. Anterior views of the articular surface of left distal humeri, drawn to the same mediolateral width in each case. The trochlea is stippled. The drawings are aligned along the lateral margin of the trochlea, in order to demonstrate differences in relative trochlea width in the species illustrated.

in an increase in contact between the radius and the humerus when the elbow is in a position of partial flexion, as occurs during the maximal load bearing phase of quadrupedal progression. In cebids this anterodistal flattening of the capitulum is minimal. Among the fossil specimens ramapithecines (*Sivapithecus*, *Gigantopithecus* and *Rudapithecus*—from the Middle and Late Miocene) and *Proconsul* (from the Early and Middle Miocene) have capitula that resemble extant hominoids in this feature (Figure 6, *Pan* and *Proconsul*). *Aegyptopithecus* (from the Oligocene), and *Dendropithecus*, *Pliopithecus* and the indeterminate KNM-FT2751 specimen (from the Early and Middle Miocene) all have capitula that are most similar to cebids in this feature (Figure 6, *Cebus* and *Aegyptopithecus*).

In extant non-hominoid anthropoids the anterior aspect of the capitulum is extended proximolaterally to form a variably developed tail (Figure 5, *Aegyptopithecus*, *Cebus*, and *Macaca*). Among the fossil specimens this feature is present on the capitula of *Aegyptopithecus*, *Pliopithecus* and *Dendropithecus*.

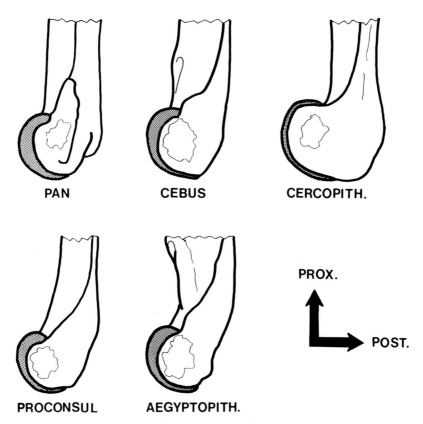

Figure 6. Lateral views of left distal humeri, drawn to the same anteroposterior depth of the capitulum in each case. The capitular articular surface is stippled.

Among anthropoids there is considerable variation in the mediolateral width of the *zona conoidea* compared to that of the capitulum (Table 1A, Index 6; Figures 7, 8). Compared to the cebid condition, the *zona conoidea* is usually wider in cercopithecids and narrower in hominoids. Among the fossil species *Proconsul* and ramapithecine humeri resemble extant hominoids in this feature, while *Aegyptopithecus*, *Pliopithecus*, and *Dendropithecus* resemble non-hominoids. The *zona conoidea* also varies in the degree to which it forms a proximally recessed gutter between the capitulum and the trochlea (Table 1B, Index 8; Figures 4, 7, 8, 9). Among extant anthropoids this gutter is relatively deeper in the hominoid samples than in other anthropoids. Among the fossil specimens, ramapithecines and *Proconsul* again conform to the hominoid pattern. Thus in extant non-hominoids, *Aegyptopithecus*, *Pliopithecus*, and *Dendropithecus*, the *zona conoidea* forms a mediolaterally wide, proximodistally shallow, and mostly proximally facing surface, while in extant hominoids, ramapithecines, and *Proconsul* it forms a narrow, deep, and proximolaterally facing gutter. *Oreopithecus* (from the Late Miocene) resembles extant hominoids in all of the above mentioned distal humeral features (Harrison, 1987; Sarmiento, 1987).

All of these humeral features are matched by corresponding features of the radial head. Thus, in extant hominoids the circular fossa on the proximal surface of the radial head is relatively wide, corresponding to the relatively wide humeral capitulum (Table 2, Index 1;

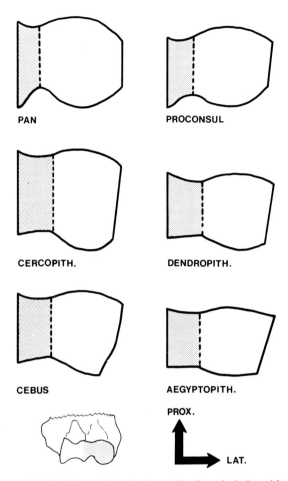

Figure 7. Anterior views of left distal humeri showing the surface for articulation with the radius, drawn to the same mediolateral width in each case. The diagram for *Dendropithecus* is reversed. The surface of the *zona conoidea* is stippled.

Figure 10). By contrast, this area is relatively smaller in non-hominoids, although there is considerable variation in the expression of this feature. Among the fossil specimens the relative size of this area, as expressed in radial Index 1, is within the range of non-hominoids or in the region of overlap between hominoids and non-hominoids. The peripheral part of the proximal surface of the radial head, which articulates with the *zona conoidea*, is broader laterally than it is elsewhere (Table 2, Index 2; Figure 10). This asymmetry is least in hominoids and *Ateles*, and differs from the much more asymmetrical situation in other anthropoids, where there is, in effect, a lateral lip to the proximal surface. This marked asymmetry results in the radial head having an oval outline in proximal view. All of the fossil radii (which do not include any ramapithecine representatives) are of the more asymmetrical type.

When the radial head is viewed from the side the surface articulating with the *zona conoidea* can be seen to extend from the proximal surface onto the side of the head. There is variation in the distal limit to this extension, and in the degree to which the extension takes

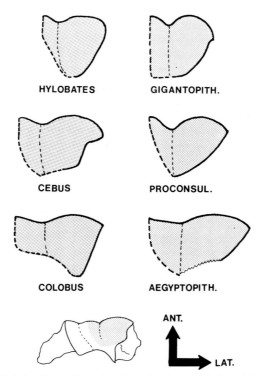

Figure 8. Distal views of left distal humeri showing the surface for articulation with the radius, drawn to the same anteroposterior depth in each case. The drawing of the *Gigantopithecus* specimen is reversed. The division between the *zona conoidea* and the capitulum is indicated with a dotted line.

place around the circumference of the head (Figure 11). The ultimate effect of these features is to produce a bevel around the periphery of the radial head. While the variation in these features among anthropoid species has not been investigated quantitatively, except for the individual specimens illustrated in Figure 11, the differences are striking to visual inspection. In hominoids the bevel reaches relatively far distally, and is present all around the circumference of the head. The angulation of this bevelling is such as to result in good contact between the surface and that of the proximolaterally oriented major surface of the *zona conoidea* (Figure 12, *Pan:* see also Figure 10, *Pan* and Figure 11, *Hylobates*). In other anthropoids the bevelling is minimal and only involves the lateral lip (Figure 11, *Colobus*). Except for *Proconsul* the fossil specimens conform to the non-hominoid pattern (Figure 11, *Dendropithecus*). In *Proconsul*, the degree of bevelling and the degree of extension around the circumference of the head is intermediate between that of the two groups of extant anthropoids (Figure 11, *Proconsul*).

The remainder of the articular surface on the side of the radial head lies distal to the bevelled surface. It articulates with the radial notch of the ulna, and the annular ligament. Among extant anthropoids this surface extends relatively far distally in non-hominoids, as it does in the available fossil specimens (Table 2, Index 3; Figure 11). The region of maximal distal extension is on the medial and anterior aspects of the radial head (Figures 2 and 11). It is this area that articulates with the radial notch of the ulna during pronation. Posterolaterally the surface is proximodistally shorter. This area only contacts the annular ligament. This variation is again most emphasized in non-hominoid extant species, and in the fossil sample, with the exception of *Proconsul* (Table 2, Index 4; Figures 11, 12, 13). In proximal view the outline of the radial head is flattened in this posterolateral region (Table 2, Index 5). This region is indicated as a straight edge in Figure 10. The length of this flattened border is apparently determined by the amount of the circumference of the side of the head that is occupied by the proximodistally abbreviated articular surface, but also by the extent to which there is a pronounced lateral lip to the proximal articular surface. Thus

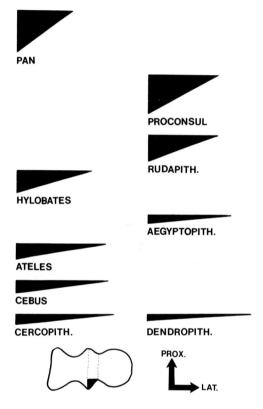

Figure 9. Diagrammatic anterior views showing the proportions of the left *zona conoidea* relative to the mediolateral width of the distal humeral articular surface in each case. The diagram for *Dendropithecus* is reversed.

in *Hylobates*, where the lateral lip is minimally developed and the region of maximal proximodistal extent of the surface on the side of the head extends most of the way around the circumference of the head, the outline of the head is basically circular, with virtually no flattening. The region of flattening is relatively short in *Ateles*, which has an abbreviated lateral lip, and in *Proconsul*, which has a minimal amount of proximodistally short surface on the side of the head. In other extant non-hominoids, and in the rest of the fossil specimens, where the two determining factors are maximally developed, the flattened border is relatively long. In cercopithecids, an uneven curvature to the radial notch of the ulna corresponds to the marked asymmetry of the outline of the radial head.

Functional features. Among extant anthropoids the major functional differences in the lateral side of the elbow occur between hominoids and non-hominoids. A number of studies have established that there is a greater range of pronation–supination in hominoids than in non-hominoids (Sullivan, 1961; Tuttle, 1967, 1969*a,b*; Yalden, 1972; Jenkins & Fleagle, 1975; O'Connor, 1975; Ziemer, 1978; O'Connor & Rarey, 1979). In hominoids the greater amplitude of movement is associated with three morphological features of the radial head:

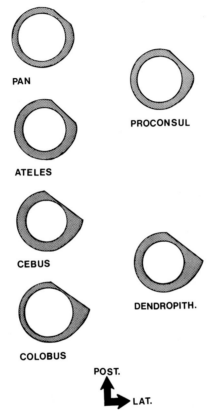

Figure 10. Diagrammatic proximal views of left radial heads, drawn to the same anteroposterior width. The diagrams for *Proconsul* and *Dendropithecus* are reversed. The surface for articulation with the humeral *zona conoidea* is stippled. More realistic versions of this view, superimposed on the distal humerus, are included in Figure 13.

there is a general symmetry to the radial head; the strongly developed bevel for articulation with the *zona conoidea* is present all round the circumference of the head; the surface for articulation with the radial notch of the ulna extends most of the way around the side of the head. In non-hominoids and most of the fossil specimens two morphological features of the radial head result in more limited movement. The bevel for articulation with the *zona conoidea* is restricted to the posterolateral side of the head, and the surface for articulation with the radial notch of the ulna is restricted to the anteromedial side of the head. The expression of these features on the radial head of *Proconsul* is intermediate between that of extant hominoids and that of the other group. This, together with other features, suggests

Table 2   **Radial indices**

Index number Index name Index definition*	Index 1 Capitular area size aX100/b	Index 2 Lateral lip width cX100/d	Index 3 Height of side of head eX100/b	Index 4 Asymmetry of side of head fX100/e	Index 5 Straight border length gX100/b
Pan troglodytes†	79·7‡ 76–86 2·7	78·6 65–98 9·9	39·9 36–42 2·1	64·8 56–74 6·8	52·6 41–62 6·3
Hylobates	73·8 70–83 4·6	80·5 65–92 9·7	44·0 38–47 2·7	78·0 66–90 7·7	0·0§ — 0·0
Cercopithecus	66·1 60–71 3·9	53·6 41–60 5·7	48·7 46–51 1·3	54·5 43–65 7·8	78·7 71–85 4·0
Colobus	71·3 65–79 5·0	43·1 33–49 4·7	48·3 44–55 3·3	57·7 51–67 5·7	82·6 74–93 6·5
Ateles	69·4 62–78 4·9	77·7 67–84 5·3	49·5 45–56 3·4	52·2 35–68 11·2	50·5 43–64 6·7
Cebus	65·9 57–71 4·5	55·7 41–67 8·3	52·4 48–56 3·4	50·0 35–64 8·8	91·5 84–100 5·4
Pliopithecus *vindobonensis* Individual 1	65·2	43·0	56·5	44·6	60·0
Dendropithecus *macinnesi* KNM RU2098	60·2	50·0	59·0	32·7	63·2
Proconsul africanus KNM-RU2036CE	74·9	54·6	50·3	65·0	40·5
?Limnopithecus KNM-CA412	58·8	58·1	59·6	35·9	—
KNM-SO1010	58·7	—	56·4	27·3	52·3
Indet. KNM-MO63	57·8	57·9	52·7	29·3	74·2
KNM-SO1009	78·3	40·0	46·8	—	—
KNM-RU2081	58·6	54·1	—	—	64·7

   * These are the measurements illustrated in Figure 2.

   † $n = 10$ (five of each sex) for samples of extant anthropoids. Except for *Pan troglodytes* the samples are made up of specimens from several species of each genus.

   ‡ The numbers listed in each cell are mean (above), range (middle), and s.d. (below).

   § Measurement g is zero on *Hylobates* radii.

that the amplitude of forearm pronation–supination may have been similarly intermediate (see also Rose, 1983).

    Most of the features analyzed above are also related to stability of the humeroradial and proximal radioulnar joints during pronation–supination movements. In the extant hominoids considered here there is a high degree of stability at these joints *in all positions of pronation–supination of the forearm* (see also Jenkins, 1973). Thus some part of the bevel of the radial head is firmly lodged in the groove of the *zona conoidea* throughout the movement (Figure 12, *Pan*). This configuration provides good contact during compressive load transfer across the elbow, but also resists dislocation of the radial head from the humerus when the elbow is subjected to compression, tension, or torque. It additionally stabilizes the radial head as it rotates *during* pronation or supination. There is similarly good contact between the side of the radial head and the radial notch of the ulna at all stages of movement, due to the same features that provide for extensive pronation–supination movements (Figure 13, *Pan*). Some aspects of this functional complex in hominoids have been discussed in detail by Sarmiento (1985). He notes that the complex is particularly

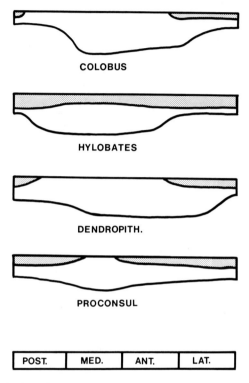

Figure 11. Diagrammatic view of the articular surface on the side of the left radial head, as it would appear if removed from the circumference of the head and laid out flat, and drawn to the same circumferential length. The diagrams for *Dendropithecus* and *Proconsul* are reversed. The extension onto the side of the head of the surface for articulation with the humeral *zona conoidea* is stippled.

well developed in *Hylobates*, where the prevention of radial head dislocation is at a premium, because of the rapid and powerful forearm rotation that occurs during brachiation. Sarmiento also points out that in *Pongo pygmaeus*, forearm rotation occurs during more deliberate and controlled activities. In this species the part of the complex involving the *zona conoidea* has a different morphology from that described above. Sarmiento suggests that soft tissue features provide adequate stability to the region in this species. Features related to the articulation of the *zona conoidea* are also present in muted form in humans. However, the possession of the complex described here is clearly favorable for elbow function during a range of suspensory and climbing activities, as well as during quadrupedal progression. Both Sarmiento (1987) and Harrison (1987) point out that *Oreopithecus* shares all of the features described here for extant hominoids.

The situation in non-hominoids is strikingly different from that in hominoids. In hominoids there is a universal stability; in non-hominoids there is a *particular* stable position, resulting from close-packing in full pronation. In pronation the broader area of the lateral lip provides extensive contact with the *zona conoidea*. Because this is basically a transversely aligned articulation the close-packing provides stability during

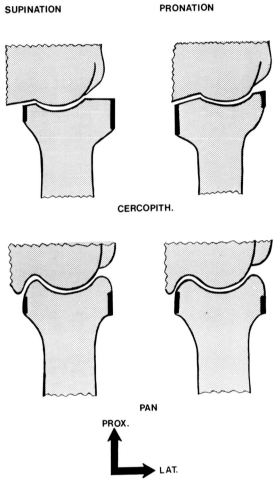

Figure 12. Diagrammatic anterior views of frontal sections through left humeroradial joints. Thick lines on the sides of the radial heads represent sections through the surface articulating with the radial notch of the ulna and the annular ligament. For ease of representation, the radius is shown as having rotated through 180° between the positions of supination and pronation for each species illustrated.

compressive-load transfer across the elbow. It is possible that the tilt to the radial head, present in most non-hominoid anthropoids, is also involved in this mechanism (Le Gros Clark & Thomas, 1951; Conroy, 1976; Harrison, 1982; Rose, 1983). By contrast, in supination only the mediolaterally narrow medial lip of the radial head is in contact with the *zona conoidea* (Figure 12, *Cercopithecus*). Some additional stability may nevertheless be provided in this position by contact between the lateral lip of the radial head and the tail-like proximolateral extension on the anterior surface of the capitulum (Figure 13, *Presbytis*).

There is a similar position of particular stability at the cercopithecid proximal radioulnar joint. In supination there is only limited contact between the periphery of the radial head and the more mediolaterally aligned part of the radial notch of the ulna (Figure 13). By contrast, in pronation the more oval-outlined part of the radial head articulates

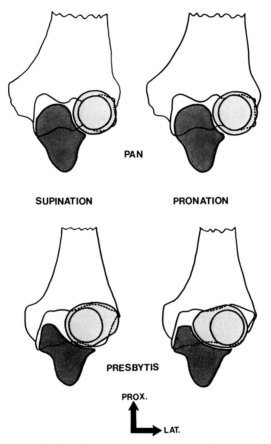

SUPINATION   PRONATION

PRESBYTIS

Figure 13. Position of the radial head in supination and pronation. The left elbow flexed to 90° between the arm and forearm. In each drawing the distal humerus is seen in anterior view. Proximal is towards the top of the page. A section through the ulna in the region of the radial notch (heavy stippling) is superimposed on the outline of the humeral trochlea. Its posterior border is towards the bottom of the page. An outline of the radial head is superimposed on the outline of the humeral *zona conoidea* and capitulum.

with the more anteroposteriorly aligned part of the radial notch as well as its more mediolaterally aligned part. In this position it is the most proximodistally extended part of the articular surface on the periphery of the radial head that contacts the radial notch (Figure 13). The less proximodistally extended area, where the outline of the head is flattened, contact is the annular ligament only, throughout the range of pronation–supination movement. In hominoids there is good contact between the side of the radial head and the radial notch throughout the extensive range of pronation–supination.

Although the range of movement is more restricted in cebids there is, nevertheless, good contact between the articular surfaces throughout the movement range.

In cercopithecids, additional stability results from the fact that, due to the orientation of the radial notch, the radial head lies more anteriorly than laterally to the ulna, especially when it is in the pronated position. In hominoids it is positioned laterally (Figure 13); in cebids it is situated in an intermediate position.

Thus, in extant hominoids, a complex of features at the humeroradial and proximal radioulnar joints result in stability of the elbow region in all positions, throughout all movements and, probably, under conditions of net tension or compression. In cercopithecids a different complex of features produces a particularly stable configuration when the elbow is loaded in a partially flexed position and the forearm is fully pronated, such as occurs during the load bearing phase of quadrupedal progression. Cebids in general share features of the humeroradial joint with cercopithecids, but exhibit a less specialized condition at the proximal radioulnar joint. However, while *Ateles* resembles other cebids in features related to close packing in pronation, it shows some features related to movement that are convergent on the hominoid condition. Thus there is a relatively circular radial head on which the lateral lip is abbreviated (Table 2, Index 2; Figure 10) and on which the area for articulation with the radial notch for the ulna extends relatively far around the circumference of the head, so that the flattened part of the outline of the head is short (Table 2, Index 5; Figure 10). This implies that there is extensive pronation–supination, as in hominoids.

There are interesting variations in the lateral elbow region of the fossils considered here. In *Aegyptopithecus*, *Pliopithecus*, *Dendropithecus*, and *Limnopithecus*, the resemblances are clearly with the non-hominoid group of extant species, particularly cebids. This is especially interesting with respect to *Dendropithecus*, where similarities to the elbow region of *Ateles* have been pointed out (Le Gros Clark & Thomas, 1951; Fleagle, 1983). However, *Dendropithecus* does not share with *Ateles* those features of the radial head interpreted here as relating to extensive pronation–supination movements. Thus, in terms of functional features of the lateral elbow region, *Dendropithecus* is clearly similar to more quadrupedal anthropoids rather than to more suspensory and/or climbing anthropoids.

Proconsul exhibits an interesting combination of features. Both the humerus and radius resemble extant hominoids in features related to general stability and possible range of pronation–supination movements (Table 1, Indices, 4, 5, 6, 7; Table 2, Index 5; Figures 4, 5, 6, 7, 8, 9). However, the degree of development of the lateral lip of the proximal surface of the radial head, and the development of the area for maximal contact with the radial notch of the ulna suggest that there was still a position of particular stability of the elbow in full pronation (Table 2, Indices 2 and 3; Figures 10, 11). These features suggest a behavioral pattern in which quadrupedal activities were important, but the forelimb was nevertheless used in a more eclectic way than in most extant non-hominoid anthropoids (see also Rose, 1983).

The elbow evidence for ramapithecines is limited to the distal humerus. Nevertheless, the morphological pattern of these humeri suggests the presence of a functional complex of the extant hominoid type considered here (Table 1, Indices 4, 5, 6, 7; Figures 4, 8, 9: See also Pilbeam *et al.* 1980; McHenry & Corrucini, 1983; Morbeck, 1983; Rose, 1983; Senut, 1986*a,b*). While the degree of development of these humeral features makes it reasonable to predict that the radial head will prove to be similarly extant hominoid-like, the example of *Proconsul* serves to make this prediction a tentative one.

The humeroulnar joint

Structure and function. The anthropoid humeroulnar joint is close to being purely uniaxial. It allows flexion–extension movement to take place, but not abduction–adduction, or medial and lateral rotation, either as accompaniments to flexion–extension (conjunct movements), or as independent (adjunct) movements. In kinematic terms, all of these movements are rotations. While the term *trochlea* is used to name this region of the distal humerus, it is only truly trochleiform (i.e. pulley-shaped) in hominoids. In other anthropoids it is more cylindrical (Table 1B, Indices 7 and 11; Figures 3, 4, 5). The uniaxial nature of the joint results from the fact that both of the above types of surface have regular curvature.

A conical surface is also uniaxial. Szalay & Dagosto (1980) note that a conical trochlea is characteristic of the humeri of Plesiadapiformes.

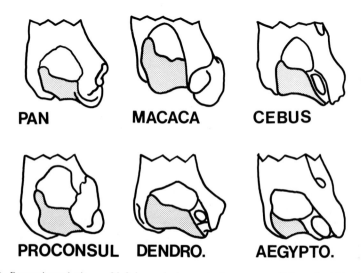

Figure 14. Posterolateral views of left humeri, drawn to the same mediolateral width of the distal humerus. Proximal is towards the top of the page, and the lateral epicondyle is towards the bottom right of each drawing. The trochlea articular surface is stippled.

Harrison (1982) gives mean values for excursion at the humeroulnar joint of between 190° and 215° in various hominoids and between 160° and 170° in non-hominoid anthropoids, although there is considerable overlap in the ranges. The principal differences are in the range of extension. Harrison notes that this is limited by the degree of anterior projection of the olecranon beak at the proximal end of the trochlear notch of the ulna. Other features that show different expression in different anthropoid taxa are therefore probably related to other functions of the joint, such as load bearing, stability, and translatory movements.

Among extant anthropoids the humeral trochlea is somewhat anteroposteriorly deeper than it is mediolaterally wide except in cercopithecids, where the trochlea is considerably deeper for its width (Table 1B, Index 9; Figure 3). By contrast, all of the fossil specimens have relatively shallow trochleas. There is a variable development of keels bounding the trochlea medially and laterally. In hominoids the medial keel represents the maximum flaring of the trochleiform shape of the trochlea, and is thus equally well developed all around the margin of the trochlea. In other anthropoids the medial keel is more in the form of a flange attached to the periphery of the cylindrical trochlea and is more pronounced anteriorly and distally. In cercopithecids the keel is particularly protuberant distally, especially when compared to the mediolateral width of the trochlea (Table 1B, Index 10; Figures 4 and 5). The articular surface of the flange faces laterodistally. In cebids the flange is less pronounced than in cercopithecids, and its articular surface faces more distally than laterally (Figures 4 and 5). Among the fossil specimens those of *Aegyptopithecus*, *Pliopithecus*, and *Dendropithecus* most resemble cebids in the morphology of the medial keel, while *Proconsul*, *Rudapithecus* and KNM-FT2751 are more hominoid-like (see also Senut, 1986*a*).

As with the medial keel, the hominoid lateral keel extends all round the lateral trochlea, although it usually presents less of a flare on the trochlea than does the medial keel (Figures 4 and 5). Distally the keel separates the trochlea surface medially from the *zona conoidea* surface laterally. Posteriorly, the lateral keel is continuous with a vertical, medially facing articular area extending proximally onto the lateral wall of the olecranon fossa (Figures 3 and 14). In non-hominoids the lateral keel is reduced or absent, and only the articular surface extending into the olecranon fossa is present. In cebids the extension into the olecranon fossa is minimally developed and the whole lateral keel complex is only represented by a posteromedially facing extension on the posterolateral corner of the trochlea. *Aegyptopithecus*, *Pliopithecus*, and *Dendropithecus* most resemble cebids in these features, while *Proconsul africanus* and *Rudapithecus* have modestly developed keels of the hominoid type. As Senut (1986*b*) points out, in the KNM-RU7696 *Proconsul nyanzae* specimen, which only preserves the most lateral part of the trochlea, the lateral keel does not seem to protrude beyond the main trochlear surface, although there is a hominoid-type *zona conoidea* on the lateral side of the keel. There is a well developed extension of the keel onto the lateral wall of the olecranon fossa in this specimen. The incomplete *Gigantopithecus* and *Sivapithecus* specimens are undoubtedly hominoid-like in this region.

There are morphological features of the trochlear notch of the ulna that correspond to the humeral trochlear features described above. In hominoids the articular surface of the trochlear notch slopes away from a median ridge which is roughly semicircular in outline in medial or lateral view (Figure 15, *Pan*). This ridge articulates with the depth of the trough of the trochleiform humeral trochlea. In non-hominoids the central part of the trochlear notch is flat, or only gently convex, corresponding to the more cylindrical humeral trochlea.

In hominoids the lateral part of the trochlear notch contacts the lateral keel of the humeral trochlea and the medially facing humeral surface that extends into the olecranon fossa (Figures 14 and 15, *Pan*). The proximal part of this surface of the notch faces almost directly laterally. It is distally elongated and is continuous with the articular surface of the notch that lies directly medial of the radial notch. This distal portion articulates with the well-developed lateral keel surrounding the humeral trochlea distally and anteriorly. In non-hominoids the morphology of this part of the trochlear notch mirrors the fact that the humeral lateral keel is minimally developed. The ulnar surface is thus situated at the proximolateral corner of the trochlear notch. In cercopithecids this surface is set at a sharp angle to the rest of the surface, and faces almost directly laterally so as to articulate with the medially facing humeral surface that extends into the olecranon fossa (Figure 15, *Macaca*, bottom diagram). The distal border of this ulnar surface forms a sharp edge, and there is a distinct, non-articular notch separating it from the more distally placed confluence of the trochlear notch and radial notch (Figure 15, *Macaca*, top and bottom diagrams). In cebids

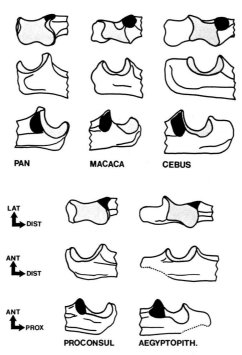

Figure 15. Views of left ulnae. In both the three-column and two-column sections the top rows are anterior views, the middle rows are medial views, and the bottom rows are lateral views. The articular surface of the trochlea notch is lightly stippled, the articular surface of the radial notch is heavily stippled.

this part of the trochlear notch is more continuous with the rest of the notch and faces distolaterally. It thus mirrors the corresponding humeral features.

Among extant anthropoids there is a similar pattern of variation in those parts of the trochlear notch that articulate with the medial trochlear keel (Figure 15, top two rows of diagrams). In non-hominoids, the distal part of the trochlear notch is relatively narrow mediolaterally, corresponding to the relatively narrow trochlea. The medial part of this portion of the notch, which articulates with the medial keel of the humeral trochlea, is angulated to face medioproximally. In hominoids this region of the trochlear notch conforms to the trochleiform shape of the humeral trochlea and is correspondingly wide mediolaterally. Its most medial part articulates with the medial keel of the humeral trochlea and is continuous proximally with an area that articulates with the posterior part of the keel. Among the fossil specimens the ulnar trochlear notches of *Aegyptopithecus*, *Pliopithecus*, *Dendropithecus*, and also *Proconsul* are most similar to those of cebids (Figure 15). Although the degree of convexity of the portion of the notch that articulates with the body of the humeral trochlea is greater than is general in extant non-hominoid taxa, the distinctive median ridge seen on extant hominoid trochlear notches is lacking in these specimens.

Functionally, the cercopithecid humeroulnar joint is in the form of a relatively mediolaterally constricted uniaxial working surface. It is stabilized by an anterodistal flange on the medial side, and a posterior flange on the lateral side. The ulnar and humeral components of both the medial and lateral flanges are in relatively extensive contact at about 90° of flexion–extension: the medial components are in maximum contact towards full flexion, and the lateral components are in maximum contact towards full extension. A detailed functional interpretation of these features is difficult without good data on the direction in which loading forces act at the elbow at different phases of the gait cycle. However, it would seem that the architecture is such that, at peak loading during the support phase of quadrupedal progression, both medially or laterally acting shearing or axially rotating forces are adequately resisted.

It is generally accepted that there are numerous specialized features within the cercopithecid forelimb that are derived with respect to the primitive catarrhine condition (see Fleagle, 1983 for a discussion of the evolution of this viewpoint). The overall pattern, which is partially convergent on that of more cursorially adapted mammals (e.g., Jenkins, 1973), is that of a series of stabilized links, the movements of which result in predominantly parasagittal excursion of the limb as a whole. At the humeroulnar joint (and to an extent at the humeroradial and proximal radioulnar joints also) these specializations represent exaggerations of features found in cebids. In functional terms the more muted cebid pattern may be part of an overall pattern of forelimb usage in which movements are not so constrained to a parasagittal plane, and movements at the shoulder, wrist, and hand may help to reduce forces tending to displace the elements of the elbow joints during peak loading. Stabilization via soft tissue structures might also be important in cebids. Despite these variations among anthropoids, Szalay and Dagosto (1980) point out that in *all* anthropoids the proportions of the trochlea and the degree of development of the medial keel result in a more stable configuration that occurs in other euprimates. They relate this to a general emphasis on quadrupedalism in anthropoid positional repertoires. Jenkins (1973) notes that the pattern of a trochlea bearing posterior and anteriodistal flanges occurs (presumably convergently) in a number of, especially, terrestrial therian mammalian groups. He categorizes it as the *advanced therian* pattern.

In hominoids the trochleiform humeroulnar joint is a morphology that provides great stability: it is able to resist forces acting about any axis and under conditions of net compression or net tension. Jenkins (1973) also makes this point. Sarmiento (1985, 1987) stresses the fact that a morphology of this type is particularly well suited to prevent humeroulnar displacement when the forelimb rotates about its long axis, especially during suspensory activities. The humerus and ulna thus act as a unit when rotation takes place at the shoulder. The ulna is also prevented from rotating about its own long axis while the radius pronates and supinates. Thus, in hominoids, there is an emphasis on universal stability at both the humeroulnar joint and the humeroradial joint.

Among the fossil species the humeroulnar joints of *Aegyptopithecus. Pliopithecus*, and *Dendropithecus* are most similar to those of cebids. *Proconsul* is more equivocal, showing some extant hominoid-like features on the humerus, but a more cebid-like pattern on the ulna (see also e.g., Napier & Davis, 1959; Morbeck, 1972, 1975; Preuschoft, 1973; Fleagle, 1983; Rose, 1983). The ramapithecine evidence is sparse and fragmentary, but the indications, especially from the complete *Rudapithecus* distal humerus, are of a much more extant-hominoid pattern than in *Proconsul*. Both Harrison (1987) and Sarmiento (1987) describe an extant hominoid-like morphology of the humeroulnar joint of *Oreopithecus*.

There are further features of the anthropoid humeroulnar joint which, although not pursued quantitatively here, seem to be significant on qualitative examination. These relate to the possibility of translatory movement accompanying flexion–extension at the elbow (Figures 16 and 17). In all anthropoids the medial and lateral borders of the humeral

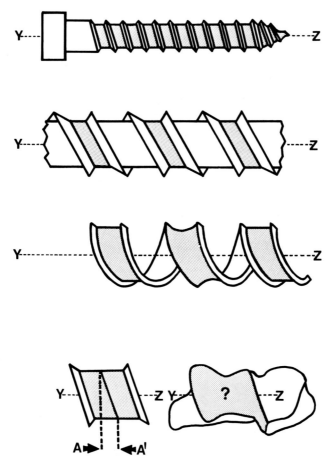

Figure 16. Principles of screw translation. Y–Z represents the axis for movements associated with the surfaces illustrated. *Top diagram:* side view of a screw surface. A second surface, moving on the stippled surface of the screw, would translate along the length of the screw as it rotated around successive turns of the screw. *Second diagram:* side view of a screw surface wound in an open spiral around a cylinder. The stippled part of the surface follows the contour of the cylinder, and has projecting flanges at its margins. The angle between each turn of the surface and the Y–Z axis is the pitch angle. It determines how much translation accompanies a given amount of rotation of a second surface moving on the screw surface. *Third diagram:* oblique view of the same screw surface as illustrated in the second diagram, but with the cylinder removed, illustrating the spiral nature of the surface. On the second turn of the spiral the flanges have been blended with the rest of the surface and the surface itself has been made concave. Compare with the bottom right diagram. *Bottom left diagram:* Side view of a half turn of a screw surface. A solid line bisects the surface: the angle between it and the Y–Z axis is the pitch angle for screw translation. A second surface, rotating 180° around the half turn, translates through the distance A–A' parallel to the Y–Z axis. *Bottom right diagram:* distal view of a *Pan troglodytes* left humerus. Anterior is towards the top of the page. The trochlea surface is stippled. Y–Z is the axis for flexion–extension at the humeroulnar joint. There are apparent similarities with the screw surfaces illustrated in the other diagrams, but the presence of a pitch angle has not been established. This analysis is continued in Figure 17.

trochlea migrate laterally as they pass distally and, especially, as they pass posteriorly on the distal humerus (Figure 3). This gives the impression that the articular surface as a whole spirals around the mediolaterally directed axis of the humeroulnar joint. If this is so, then the ulna would translate medially during flexion and translate laterally during

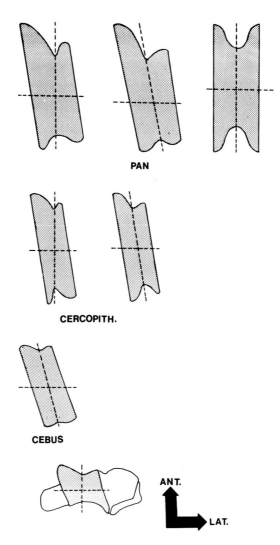

Figure 17. Diagrammatic distal views of left humeral trochlea surfaces as they would appear if removed from the distal humerus and laid out flat. Horizontal dotted lines represent the axis for flexion–extension at the humeroulnar joint. Vertical and slanting dotted lines follow the central, or most concave part of the trochlear surface. *Left column:* the pattern actually found in the species illustrated. In *Pan* and *Cercopithecus* the two dotted lines cross at 90°: there is thus no pitch angle and no translatory movement accompanies flexion or extension, despite the fact that the medial and lateral margins of the trochlea are oriented obliquely with respect to the axis of the joint. In *Cebus* a pitch angle is present and the trochlea is thus a screw surface. Medial translation of the ulna will accompany flexion and lateral translation of the ulna will accompany extension. *Middle column: Pan* and *Cercopithecus* trochleas as they would appear if they were screw surfaces. *Right diagram (Pan):* the trochlea as it would appear if it was perfectly symmetrical. The whole surface is aligned anteroposteriorly and there is equal development of both the medial and lateral flanges. In fact, in all three of the species illustrated, the medial flange is more extensive anteriorly and the lateral flange is more extensive posteriorly. This feature is least developed in *Cebus*, due to the pitch angle of the central part of the trochlea. This feature is disscussed further in the text.

extension (Figure 16). However, in extant catarrhines, the depth of the trochlear groove is directed anteroposteriorly, and does not spiral as it passes around the distal humerus. It is thus situated at right angles to the axis of the humeroulnar joint. There is no pitch angle, so screw translation cannot occur (Figure 17). Among anthropoids it is only in cebids that the main articular surface itself forms a portion of a screw surface, and some translatory movement is possible. The functional consequences of this postulated action in cebids are not clear. However, the movements may serve to redistribute load, or change the configuration of the humeroulnar joint as the direction of forces acting on it changes during the stance phase of progression. The features underlying screw translation may thus be linked to the features mentioned above that enable cebids to have less well-developed marginal flanges than in other anthropoids.

There is a further consequence of the extant catarrhine configuration. Because of the difference in angulation of the margins and the central part of the articular surface, the portions of the joint providing stability are more emphasized medially at the anterior end of the surface, and more emphasized laterally at the posterior end (Figure 17). This even holds to an extent in cebids, where the pitch angle of the central part of the trochlea is somewhat less than the angle of the trochlear margins. This may imply that forces tending to displace the humerus laterally on the ulna are maximal in flexion, and that in extension forces tending to medially displace the humerus are maximal. In cercopithecids this feature adds further morphological emphasis to the asymmetrical development of the stabilizing flanges discussed above. All of the fossil humeri, including that of *Aegyptopithecus*, are of the non-translatory type, as in extant catarrhines.

Implications for elucidating relationships among anthropoid taxa

This study is limited in a number of ways. It is based on the examination of a limited number of anthropoid species and only articular features of a single region have been considered. Small sample sizes have been used for osteometrics and there is overlap between species ranges for many quantitative features. Nevertheless, many of the points discussed above have relevance for a consideration of the relationships among catarrhine taxa. These points generally support the pattern of relationships elucidated from features of the anthropoid elbow region by Fleagle (1983). However, characters not investigated by Fleagle have been included in the analysis presented above. It has also been shown that, although a relatively large number of individual morphological features are present in the elbow region, subsets of these features form complexes associated with specific functions. Finally, it has been shown that particular clusters of functions are unique to various anthropoid taxa (see Tables 3, 4, 5 and 6 for summaries).

The elbow region of extant hominoids exhibits many unique features which, as Table 3 clearly indicates, are autapomorphies (features that are uniquely derived for the taxon). These features form clusters that are associated with all of the aspects of elbow and forearm function considered (Table 6). Of the features listed in Table 3, only one (H7a, a non-translatory humeroulnar joint) is a hominoid/cercopithecoid synapomorphy (shared derived feature). This feature is therefore important, within the elbow region, for defining the catarrhine condition.

Table 3   List of functional features of the elbow region and their distribution among extant anthropoids

Feature		Present in *Cebus*	Present in Cercopithecids	Present in hominoids
Humeral features				
H1	Trochlea cylindrical	+	+	−
H1a	Trochlea trochleiform	−	−	+
H2	Trochlea relatively narrow mediolaterally	+	+	−
H2a	Trochlea relatively broad mediolaterally	−	−	+
H3	Trochlea not relatively deep anteroposteriorly	+	−	+
H3a	Trochlea relatively deep anteroposteriorly	−	+	−
H4	Medial keel most developed distally and faces distolaterally	+	−	−
H4a	Medial keel most developed distally and faces laterally	−	+	−
H4b	Medial keel equally developed all around trochlea	−	−	+
H5	Lateral keel not well developed	+	+	−
H5a	Lateral keel well developed	−	−	+
H6	Limited extension of articular surface into olecranon fossa	+	+	−
H6a	Considerable extension of articular surface into olecranon fossa	−	−	+
H7	Humeroulnar joint translatory	+	−	−
H7a	Humeroulnar joint non-translatory	−	+	+
H8	Capitulum asymmetrical but not flattened anterodistally	+	−	−
H8a	Capitulum asymmetrical and flattened anterodistally	−	+	−
H8b	Capitulum globular or cylindrical	−	−	+
H9	Capitulum has proximolateral tail	+	+	−
H9a	Capitulum has no proximolateral tail	−	−	+
H10	*Zona conoidea* relatively wide mediolaterally	+	+	−
H10a	*Zona conoidea* relatively narrow mediolaterally	−	−	+
H11	*Zona conoidea* relatively shallow proximodistally	+	+	−
H11a	*Zona conoidea* relatively deep proximodistally	−	−	+
H12	Humeral distal articular surface relatively wide	+	−	+
H12a	Humeral distal articular surface relatively narrow	−	+	−
Radial features				
R1	Radial head fossa for articulation with humeral capitulum relatively small	+	+	−
R1a	Radial head fossa for articulation with humeral capitulum relatively large	−	−	+
R2	Radial head oval	+	+	−
R2a	Radial head circular	−	−	+
R3	Slightly bevelled surface for articulation with humeral *zona conoidea*	+	+	−
R3a	Markedly bevelled radial surface for articulation with humeral *zona conoidea*	−	−	+
R4	Bevelling does not extend all around margin of radial head	+	+	−
R4a	Bevelling extends all around margin of radial head	−	−	+
R5	Outline of radial head has a straight border posterolaterally	+	+	−
R5a	Outline of radial head has short or absent straight border	−	−	+
R6	Surface for articulation with radial notch of the ulna does not extend all the way round the radial head	+	+	−
R6a	Surface for articulation with the radial notch of the ulna extends all the way round the radial head	−	−	+

Table 3—*continued*

Feature		Present in *Cebus*	Present in Cercopithecids	Present in hominoids
Ulnar features				
U1	Trochlear notch is mediolaterally narrow distally	+	+	−
U1a	Trochlear notch is mediolaterally wide distally	−	−	+
U2	Olecranon beak is relatively long	+	+	−
U2a	Olecranon beak is relatively short	−	−	+
U3	Surface for articulation with medial humeral trochlea keel restricted to distal end of trochlea notch	+	+	−
U3a	Surface for articulation with medial humeral trochlea keel occupies the whole length of the trochlear notch	−	−	+
U4	Trochlear notch does not have a marked median ridge	+	+	−
U4a	Trochlear notch has a marked median ridge	−	−	+
U5	Limited but continuous, distolaterally facing area for articulation with lateral part of humeral trochlea	+	−	−
U5a	Limited, laterally facing area for articulation with lateral side of humeral trochlea	−	+	−
U5b	Extensive area for articulation with humeral lateral keel and surface in humeral olecranon fossa	−	−	+
U6	Radial notch faces anterolaterally	+	−	−
U6a	Radial notch faces anteriorly	−	+	−
U6b	Radial notch faces laterally	−	−	+
U7	Radial notch evenly curved	+	−	+
U7a	Radial notch unevenly curved	−	+	−

Cercopithecids exhibit a relatively small number of features that are clearly autapomorphies (Table 5). These features are mostly associated with humeroulnar and proximal radioulnar joint stability when the elbow is in a partially flexed position and the forearm is fully pronated. These particular configurations are especially important during the stance phase of quadrupedal locomotion. The numerous symplesiomorph (shared primitive) characters of cercopithecids and *Cebus* (the cebid example used here) relate to the restriction of movement in the elbow and forearm, stability of the humeroradial joint in full pronation, and to the fairly equal load-bearing function of the radial and ulnar sides of the elbow region (Tables 3, 4 and 5).

The evidence from the ramapithecines *Rudapithecus*, *Sivapithecus*, and *Gigantopithecus* is limited to the humerus. However, together with *Proconsul*, they clearly share a number of synapomorphies with extant hominoids. These include the following features listed in Table 3: H1a, H2a, H4b, H5a, H6a, H8b, H9a, H10a, and H11a. Most of these features are also present in the oreopithecid *Oreopithecus* (Harrison, 1987; Sarmiento, 1987). These derived features all imply that most aspects of elbow function in these fossil species were specialized in the same way as in extant hominoids (see also Pilbeam *et al.*, 1980; Morbeck, 1983; Rose, 1983; Senut, 1986*a,b*; Harrison, 1987; Sarmiento, 1987). As discussed by Harrison (1987) and Sarmiento (1987), the radius of *Oreopithecus* also shares synapomorphies with the radii of extant hominoids (Table 3, features R2a, R3a, R4a, R5a). The situation in *Proconsul* is different. While the radius exhibits one feature in common with extant hominoids (Table 3, R5a) it shares a number of symplesiomorphic features with cercopithecids and cebids (Table 3, R2, R3, R6). These features are associated with the close packing of the joints of the radial head in pronation. On the basis of the humeral features, all of the above mentioned taxa are considered to be hominoids (see also Feldesman, 1982; Fleagle, 1983). This view contrasts to that of Harrison (1982,

Table 4   **Summary of functional features of the elbow region in *Cebus***

Positional activities	Forelimb use	General elbow/ forelimb function	Specific elbow/ forelimb function	Structural features*
			Limited pronation/ supination	Humeroradial joint: R1 R2 R4
		Limited movement of elbow and forearm		Proximal radioulnar joint: R5 R6
			Limited extension at humeroulnar joint	U2 H6
Quadrupedal and climbing activities used	Forelimb used mostly in compression. Movement not restricted to a particular plane. Forearm frequently held in full pronation		Structural features affecting humeroulnar joint stability minimally developed	H3 H5/U5† H6 U3
			Translatory humeroulnar joint compensates for lack of other structural features?	H7
		Elbow region maximally stable in full pronation	Humeroradial joint maximally stable in full pronation	H10 H11/R1/R2 R3
			Additional stable position in full supination?	H9/R2/R3 R1
			Stability not related to flexion/extension position	H8
			Radioulnar joint slightly more stable in full pronation	R2 U6
		Load shared equally by humeroulnar and humeroradial joints		Humeroulnar joint: H1/U4 H2 ?H7 Humeroradial joint: H10/R3 H11/R2

* These are the features defined in Table 3.
† Features separated by a / are reciprocal features of two bones associated with the same function.

Table 5 **Summary of functional features of the Cercopithecid elbow region**

Positional activities	Forelimb use	General elbow/ forelimb function	Specific elbow/ forelimb function	Structural features*
			Limited pronation/ supination	Humeroradial joint: R1 R2 R4 Proximal radioulnar joint: R5 R6 *U7a*†
		Limited movement of elbow and forearm	Pronation/supination maximal when elbow partly flexed	*H8a*
			Limited extension at humeroulnar joint	U2
Quadrupedalism predominates	Forelimb used mostly in compression. Movement mostly parasagittal, with the forearm held in full protonation		Humeroulnar joint maximally stable when partly flexed	*H3a* *H4b*/U3‡ H6/*U5a*
		Elbow region maximally stable in partial flexion and full pronation	Humeroradial joint maximally stable in full protonation	H10 H11/R1/R2 R3
			Additional stable position in full supination?	H9/R2/R3 R1
			Humeroradial joint maximally stable when elbow partly flexed	*H8a*
			Radioulnar joint maximally stable in full protonation	R2/*U7a* *U6a*
		Loading along the long axis of the limb. Load shared equally by humeroulnar and humeroradial joints		Humeroulnar joint: H1/U4 H2 *H12a* Humeroradial joint: H10/R3 H11/R2

* These are the features defined in Table 3.
† Italicized features are uniquely derived (autopomorphous) for cercopithecids.
‡ Features separated by a / are reciprocal features of two bones associated with the same function.

Table 6   Summary of functional features of the hominoid elbow region

Positional activities	Forelimb use	General elbow/ forelimb function	Specific elbow/ forelimb function	Structural features
			Extensive pronation/ supination	Humeroradial joint: *R2a*† *R4a* Proximal radioulnar joint: *R5a* *R6a* *?U6b* U7
		Extensive movement of elbow and forearm	Full range of pronation/ supination in all positions of flexion/extension	*H8b* *R1a*
			Extensive extension at humeroulnar joint	*H6a* *H8b* *U2a*
Climbing, suspensory, and quadrupedal activities all used	Forelimb used in a a variety of ways, in both compression and tension		Humeroulnar joint stable in all positions of flexion/extension	*H1a/U1a/* *U4a*‡ *H4b/U3a* *H5a/U5b*
		Universal stability of elbow region	Humeroradial joint stable in all positions of flexion/extension	*H8b* *H9a* *H10a/R1a* *H11a/R3a*
			Humeroradial joint stable in all positions of pronation/ supination	*R2a* *R4a* *R5a*
		Humeroulnar joint bears most load		*H2a* *?H7a*

* These are the features defined in Table 3.
† Italicized features are uniquely derived (autapomorphous) for hominoids.
‡ Features separated by a / are reciprocal features of two bones associated with the same function.

1987), who considers *Proconsul* to be part of the sister group of all extant catarrhines. Harrison reaches this conclusion on the basis of a consideration of features not considered here, and on a different interpretation of some of the features that have been considered here. It also contrasts with the now largely discredited view of *Proconsul* as the sister group of all, or particular extant great apes (as stated, or implied by, e.g., Lewis, 1971; Conroy & Fleagle, 1972; Zwell & Conroy, 1973).

As in all other catarrhines, the humeroulnar joint in *Aegyptopithecus*, *Pliopithecus* and *Dendropithecus* is non-translatory. These taxa share none of the autapomorphies of hominoids or of cercopithecoids. Apparent close similarities of these taxa (and *Proconsul*) to cercopithecoids are only evident if the cebid evidence is not considered and if a limited range of morphometrics is utilized (e.g., McHenry & Corruccini, 1975). Due to the presence of a number of features (e.g., the entepicondylar foramen), *Aegyptopithecus* is, within catarrhines, clearly less derived than any of the other taxa considered (see also Fleagle, 1983) and therefore forms the sister group of all the other catarrhines considered here. Neither cercopithecids, nor the taxon to which *Pliopithecus* and *Dendropithecus* belong, share synapomorphies with hominoids to the exclusion of the other taxon. Thus it is not possible to determine, on the basis of the features considered here, which of them is the sister group of the hominoids. The above mentioned features apart, *Aegyptopithecus*, *Pliopithecus*, and *Dendropithecus* share a large number of symplesiomorphies with generalized cebids such as *Cebus*. These indicate a similarity in the functioning of the elbow region.

Acknowledgements

I would like to thank E. Strasser and M. Dagosto for inviting me to prepare this paper. I am especially grateful for the extremely helpful comments and suggestions of J. Fleagle, T. Harrison, and E. Sarmiento, who reviewed the submitted version of this paper. G. Musser, Chairman of the Department of Mammalogy at the American Museum of Natural History, kindly allowed me access to the collections in his care. The research reported here was supported by NSF grant BNS-8418909 to the author.

References

Clark, W. E., Le Gros & Thomas, D. P. (1951). Associated jaws and limb bones of *Limnopithecus macinnesi*. *Fossil Mammals of Africa*, No. 3, pp. 1–27. London: British Museum (Natural History).
Conroy, G. C. (1976). Primate postcranial remains from the Oligocene of Egypt. *Contrib. Primatol.* **8**, 1–134.
Conroy, G. C. & Fleagle, J. G. (1972). Locomotor behavior of living and fossil pongids. *Nature* **237**, 103–104.
Feldesman, M. R. (1982). Morphometric analysis of the distal humerus of some Cenozoic catarrhines: the late divergence hypothesis revisited. *Am. J. phys. Anthrop.* **59**, 73–95.
Fleagle, J. G. (1983). Locomotor adaptations of Oligocene and Miocene hominoids and their phyletic implications. In (R. L. Ciochon & R. S. Corruccini, Eds) *New Interpretations of Ape and Human Ancestry*, pp. 301–324. New York: Plenum.
Fleagle, J. G. & Kay, R. F. (1983). The phyletic position of Oligocene hominoids. In (R. L. Ciochon & R. S. Corruccini, Eds) *New Interpretations of Ape and Human Ancestry*, pp. 181–210. New York: Plenum.
Fleagle, J. G. & Simons, E. L. (1982). The humerus of *Aegyptopithecus zeuxis*: a primitive anthropoid. *Am. J. phys. Anthrop.* **59**, 175–193.
Harrison, T. (1982). Small-bodied apes from the Miocene of East Africa. Ph.D. Dissertation, University of London.
Harrison, T. (1987). A reassessment of the phylogenetic relationships of *Oreopithecus bambolii* Gervais. *J. hum. Evol.* **15**, 541–583.
Jenkins, F. A., Jr. (1973). The functional anatomy and evolution of the mammalian humeroulnar articulation. *Am. J. Anat.* **137**, 281–298.
Jenkins, F. A., Jr. (1981). Wrist rotation in primates: A critical adaptation for brachiators. *J. zool. Soc. Lond.* **48**, 429–451.
Jenkins, F. A., Jr. & Fleagle, J. G. (1975). Knuckle walking and the functional anatomy of the wrist in living apes. In (R. H. Tuttle, Ed.) *Primate Functional Morphology and Evolution*, pp. 213–227. The Hague: Mouton.
Lewis, O. J. (1971). Brachiation and the early evolution of the Hominoidea. *Nature* **230**, 577–578.
McHenry, H. M. & Corruccini, R. S. (1975). Distal humerus in hominoid evolution. *Folia primatol.* **23**, 227–244.
McHenry, H. M. & Corruccini, R. S. (1983). The wrist of *Proconsul africanus* and the origin of hominoid postcranial adaptations. In (R. L. Ciochon & R. S. Corruccini, Eds) *New Interpretations of Ape and Human Ancestry*, pp. 252–367. New York: Plenum.
Morbeck, M. E. (1972). A re-examination of the forelimb of the Miocene Hominoidea. Ph.D. Dissertation, University of California.
Morbeck, M. E. (1975). *Dryopithecus africanus* forelimb. *J. hum. Evol.* **4**, 39–46.

Morbeck, M. E. (1983). Miocene hominoid discoveries from Rudabánya: implications from the postcranial skeleton. In (R. L. Ciochon & R. S. Corruccini, Eds) *New Interpretations of Ape and Human Ancestry*, pp. 369–404. New York: Plenum.

Napier, J. R. & Davis, P. R. (1959). The Forelimb Skeleton and associated remains of *Proconsul africanus. Fossil Mammals of Africa*, No. 16, pp. 1–69. London: British Museum (Natural History).

O'Connor, B. L. (1975). Functional morphology of the cercopithecoid wrist and inferior radioulnar joints, and their bearing on some problems in evolution of the Hominoidea. *Am. J. phys. Anthrop.* **43**, 113–121.

O'Connor, B. L. (1976). *Dryopithecus (Proconsul) africanus:* quadruped or non-quadruped? *J. hum. Evol.* **5**, 279–284.

O'Connor, B. L. & Rarey, K. E. (1979). Normal amplitudes of radioulnar pronation and supination in several genera of anthropoid primates. *Am. J. phy. Anthrop.* **51**, 39–44.

Pilbeam, D., Rose, M. D., Badgley, C. & Lipschutz, B. (1980). Miocene hominoids from Pakistan. *Postilla* **181**, 1–94.

Preuschoft, H. (1973). Body posture and locomotion in some East African Miocene Dryopithecinae. In (M. H. Day, Ed.) *Human Evolution: Symp. Soc. Study hum. Biol.* **11**, 13–46.

Rose, M. D. (1983). Miocene hominoid postcranial morphology: monkey-like, ape-like, neither, or both? In (R. L. Ciochon & R. S. Corruccini, Eds) *New Interpretations of Ape and Human Ancestry*, pp. 405–417. New York: Plenum.

Sarmiento, E. E. (1983). Some functional specializations in the skeleton of *Oreopithecus bambolii. Am. J. phys. Anthrop.* **60**, 248–249.

Sarmiento, E. E. (1985). Functional Differences in the Skeleton of Wild and Captive Orang-Utans and Their Adaptive Significance. Ph.D. Dissertation, New York University.

Sarmiento, E. E. (1987). The phylogenetic position of *Oreopithecus* and its significance in the origin of Hominoidea. *Am. Mus. Novitates* **2881**, 1–44.

Senut, B. (1986a). Distal humeral osseous anatomy and its implications for hominoid phylogeny. *Anthropos* **23**, 3–14.

Senut, B. (1986b). New data on Miocene hominoid humeri from Pakistan and Kenya. In (J. G. Else and P. C. Lee, Eds) *Proc. Tenth Congr. Int. Primatol. Soc.* Vol. 1, pp. 151–161. Cambridge: Cambridge University Press.

Sullivan, W. E. (1961). Skeleton and joints. In (C. G. Hartman & W. L. Straus, Eds) *The Anatomy of the Rhesus Monkey*, pp. 43–84. New York: Hafner.

Szalay, F. S. & Dagosto, M. (1980). Locomotor adaptations as reflected on the humerus of Paleogene primates. *Folia primatol.* **34**, 1–45.

Tuttle, R. H. (1967). Knuckle-walking and the evolution of hominoid hands. *Am. J. phys. Anthrop.* **26**, 171–206.

Tuttle, R. H. (1969a). Quantitative and functional studies on the hands of the Anthropoidea. I. The Hominoidea. *J. Morph.* **128**, 309–364.

Tuttle, R. H. (1969b). Knuckle-walking and the problem of human origins. *Science* **166**, 953–961.

Yalden, D. W. (1972). The form and function of the carpus in some arboreally adapted mammals. *Acta. Anat.* **82**, 383–406.

Ziemer, L. K. (1978). Function and morphology of forelimb joints in the woolly monkey *Lagothrix lagotricha. Contrib. Primatol.* **14**, 1–130.

Zwell, M. & Conroy, G.C. (1973). Multivariate analysis of the *Dryopithecus africanus* forelimb. *Nature* **244**, 373–375.

Elizabeth Strasser

Department of Anthropology, Graduate School, City University of New York, 33 West 42 Street, New York, NY 10036, U.S.A.

Received 19 January 1988
Revision received 16 February 1988
and accepted 18 February 1988

Publication date June 1988

Keywords: Cercopithecidae, Cercopithecinae, Colobinae, foot morphology, tarsus, metatarsus

Pedal evidence for the origin and diversification of cercopithecid clades

Cercopithecids share 12 characters which are derived relative to the catarrhine morphotype. These features are related to remodelling the foot for cursorial locomotion on the ground. The origin of cercopithecids is thus linked with the exploitation of a new niche by the cercopithecid ancestor. The two extant cercopithecid subfamilies can be distinguished from each other by additional derived modifications of the foot. Cercopithecines are characterized by three features related to foot stabilization during dorsiflexion and eversion and to reduced axial rotation of the hallux. These suggest a continued commitment to terrestrial life after the ancestral colobine re-invaded the rain-forest habitat. A complex of nine derived character-states are shared by colobines. This complex can be broken into three components: those that involve modifications which enhance inversion and supination at the lower ankle and transverse tarsal joints; those involved with shortening of the tarsus; and, those related to realigning the morphological, and probably functional, axis of the digit rays. It is hypothesized that these are related to the acquisition of a novel pedal position by the colobine ancestor. This may have allowed colobines to secondarily re-invade the rainforest. In this setting they could avoid competition with hominoids by occupying and leaping between the peripheries of tree crowns.

Journal of Human Evolution (1988) **17**, 225–245

Introduction

It is increasingly evident that higher-level primate taxa are distinguished by suites of postcranial characters as well as those traditionally recognized in the cranium and dentition (e.g., Szalay & Dagosto, 1980). This suggests that the acquisition of novel locomotor strategies is coupled with the exploitation of new food resources. While the dentition of cercopithecids has long been recognized as quite derived for catarrhines (Szalay & Delson, 1979) their postcranium has been considered primitive and homogeneous, especially since the numerous studies of Schultz (1930, 1963a,b, 1970). Jolly (1965) and Fleagle (1976, 1977) were among the first to analyze the cercopithecid skeletal and muscular anatomical diferences which accompany different locomotor behaviors in the wild. Since those seminal works, more students have focused their research on the postcranium. These newer studies (e.g., Fleagle, 1983; Langdon, 1986; Rose, 1988) are demonstrating that the cercopithecid morphotype may be a derived one for anthropoids and shows a diversity comparable to that found in most other primate families.

This paper: (1) documents pedal features which are typical of the family Cercopithecidae and its two extant subfamilies; (2) discusses pedal features in terms of their phyletic value by reference to the inferred catarrhine morphotype; and (3) discusses the functional implications of cercopithecid pedal morphotypes. The combination of functional analysis with phylogenetic analysis provides valuable insights into the origin of cercopithecid higher taxa.

Materials and methods

A descriptive and comparative anatomical approach is coupled with morphometric tests to delineate differences in relative size, shape and orientation of articular surfaces and bony

0047–2484/88/01/20225 + 21 $03.00

Table 1   Skeletal sample and body weights of the anthropoid species included in this study[a]

Species[b]	Number of adult skeletons	Body weight (g)[c]	Species	Number of adult skeletons	Body weight (g)
Nasalis larvatus	15	15220	*Cercocebus torquatus*	19	11176
Pygathrix nemaeus	9	10896	*Cercocebus galeritus*	12	—
Rhinopithecus roxellanae	7	—	*Cercocebus albigena*	24	7091
Semnopithecus entellus	9	11085	*Macaca cyclopis*	2	—
Trachypithecus obscura	15	6617	*Macaca thibetana*	4	—
Trachypithecus cristata	25	6264	*Macaca sylvana*	5	—
Presbytis frontata	4	—	*Macaca nemestrina*	13	7733
Presbytis melalophus	8	6748	*Macaca mulatta*	21	6942
Presbytis aygula	6	6680	*Macaca fuscata*	3	—
Presbytis rubicunda	22	5935	*Macaca nigra*	11	5954
Presbytis hosei	4	5562	*Macaca fascicularis*	25	4301
Colobus guereza	34	8824	*Hylobates syndactylus*	12	11050
Colobus angolensis	6	8541	*Hylobates concolor*	6	—
Colobus polykomos	6	7777	*Hylobates hoolock*	3	6800
Colobus satanas	3	—	*Hylobates klossi*	2	5830
Procolobus badius	29	8059	*Hylobates lar*	41	5753
Procolobus verus	10	4145	*Hylobates agilis*	6	5740
Erythrocebus patas	9	8500	*Pongo pygmaeus*	7	59750
Cercopithecus albogularis	20	7378	*Pan troglodytes*	20	45900
Cercopithecus nictitans	19	6500	*Pan paniscus*	7	39100
Cercopithecus mitis	20	5791	*Alouatta palliata*	9	8940
Cercopithecus aethiops	31	4203	*Alouatta seniculus*	6	6690
Cercopithecus cephus	23	3286	*Alouatta villosa*	2	–
Cercopithecus ascanius	20	3913	*Ateles geoffroyi*	7	7730
Cercopithecus petaurista	5	–	*Ateles paniscus*	4	–
Cercopithecus mona	16	2733	*Ateles fusciceps*	4	–
Cercopithecus campbelli	5	–	*Lagothrix lagotricha*	10	7255
Cercopithecus pogonias	20	–	*Pithecia pithecia*	4	1800
Cercopithecus diana	4	–	*Aotus trivirgatus*	2	746
Miopithecus talapoin	17	1250	*Callicebus moloch*	2	681
Mandrillus sphinx	8	26900	*Cebus capucinus*	7	2235
Mandrillus leucophaeus	9	–	*Cebus apella*	3	2635
Papio hamadryas ursinus	8	26716	*Saimiri oerstedii*	4	700
Papio h. anubis	23	22022	*Callimico goeldi*	1	472
Papio h. hamadryas	12	18633	*Leontopithecus rosalia*	2	517
Papio h. cynocephalus	20	18415	*Callithrix jacchus*	2	329
Theropithecus gelada	11	17575	*Saguinus midas*	2	588

[a] Species for which no body weight data are available are excluded from the bivariate plot but included in χ^2 and t-test analyses.

[b] Taxonomy for cercopithecid species follows Napier (1981, 1985) except for *Papio*. *Papio* subspecies are treated as species because of differences in body weight.

[c] Body weight estimates are pooled for sexes and are taken from the following sources: Napier, 1981, 1985; Jungers, 1985; Rosenberger, 1979 and from specimen labels of skeletons examined.

segments. Measurements were made on more than 850 wild-lived specimens housed in museum collections (see Table 1 for species examined) and are defined in table and figure legends. Body-weight data were taken from the literature (Rosenberger, 1979; Napier, 1981, 1985; Jungers, 1985) and pooled for sexes. When species sample sizes were equivalent for different sources the mean weight was taken. If there was a large difference in sample size, then the weight reported for the larger sample was used. Body weight data are used in a bivariate plot of calcaneal traits. For cercopithecid subclades χ^2 tests were performed on the distribution of two proportional traits: the shape of the cuboid and sizes of the ectocuneiform-cuboid facets. An index of the length of the fourth digit relative to the third in cercopithecids was examined using *t*-tests. Statistical analyses were performed using SAS procedures on a personal computer.

The classification of primates used here follows Strasser & Delson (1987) for Cercopithecidae, Andrews (1985) for other catarrhines and Rosenberger (1979) for Platyrrhini. The characters discussed below are a subset of a larger group currently under study (Strasser, in prep.). For the sake of convenience they are described as 23 separate characters although most of them are simply aspects of complex functional units (Table 2). Pedal morphotypes are constructed for Colobinae, Cercopithecinae, Cercopithecidae and Catarrhini by determining the most common morphology for all the extant genera of each taxon for which feet are available (Table 3). The characters which comprise each pedal morphotype are evaluated using the outgroup criterion of the cladistic method to determine their relative phyletic significance. Thus, characters which distinguish the feet of cercopithecids from those of hominoids are analyzed with regard to their distribution in platyrrhines, and where necessary, strepsirhines. The polarity of features which distinguish cercopithecid subfamilies is considered with reference to hominoids and platyrrhines. The functional implications of these morphotypes are determined where possible using the principles discussed by MacConaill (1973) and Kapandji (1970).

Familial morphotype

A character complex of twelve features serves to distinguish cercopithecids from other catarrhines (see Tables 2–3). These are related to: (1) stabilization of the foot at the upper and lower ankle joints; (2) increasing the power of the lever arm for the triceps surae muscles; (3) substantial pronation of the forefoot; and (4) reduced axial rotation of the hallux. Such a constellation of features is most likely related to a positional repertoire that emphasizes cursoriality rather than pedal grasp-climbing or suspension.

Inferior tibiofibular and upper ankle joints
In cercopithecids, the inferior tibiofibular joint is basically syndesmotic. Some specimens of many species, except for the highly terrestrial forms, may have a tiny, anteriorly positioned tibial articular facet (Ford, 1980, 1988). The inferior tibiofibular joint is generally synovial in other extant catarrhines (Carleton, 1941). In a survey of 24 hylobatid individuals, while three had no facet, two had a tiny anteriorly positioned facet, 14 had an anteriorly deep facet which tapered posteriorly and five exhibited a large semilunar facet extending the complete length of the epiphysis. The great apes also have a semilunar facet which is deepest anteriorly or centrally and tilted at approximately 45° to the long axis of the tibia, undoubtably due to the valgus orientation of the tibia. The morphotypic catarrhine condition is inferred to have been synovial since all New World monkeys present a well developed synovial inferior tibiofibular joint which ancestrally extends the anteroposterior length of the distal tibia (Ford, 1980, 1988). Given this distribution, cercopithecids are clearly derived at this joint.

At the upper ankle joint, cercopithecids have a slightly wedge-shaped, highly asymmetrical astragalar trochlea, with the lateral border rising substantially above the medial (Figure 1A,B). This is most marked in the large-bodied terrestrial forms but characterizes all cercopithecids, albeit some less dramatically. Most hominoids have a more symmetrical trochlea with hylobatids and *Pan* showing very little asymmetry. *Gorilla*

Table 2

Definition of character-states analysed

Characters	Ancestral (for anthropoids)	Derived
Inferior Tibiofibular Joint	1. Synovial	1'. Syndesmotic
Trochlear Asymmetry	2. Symmetrical	2'. Asymmetrical
Trochlear Wedging	3. Absent	3'. Moderate
Curvature Proximal Calcaneal Facet	4. Gentle	4'. Tight
Proximal Shape Proximal Astragalocalcaneal Facet	5. Lateral Side > Medial, narrow	5'. Lateral Side > Medial, wide 5". Lateral Side = Medial, wide
Tilt Proximal Calcaneal Facet	6. Acute	6'. Oblique
Pressure Facet for Fibulocalcaneal Ligament	7. Absent	7'. Present
Distal Astragalocalcaneal and Sustentacular Facets	8. Confluent	8'. Separate
Relative Length Proximal Astragalocalcaneal Facet[a]	9. Intermediate	9'. Short 9". Secondarily Intermediate or Long
Length Proximal Calcaneal Body	10. Short	10'. Long
Facet for Peroneus Longus Sesamoid	11. Absent	11'. Present
Shape Entocuneiform	12. Dorsal Head, Helical Groove	12'. Kidney-shaped, small groove 12". Kidney-shaped, flat
Astragalotibial Malleolar Facet	13. Extends 3/4 Down Astragalus	13'. Extends to Plantar Border
Distal Astragalocalcaneal Facet	14. Weakly Curved	14'. Strongly Curved
Curvature Proximal Astragalar Facet	15. Weak	15'. Strong 15". Very Strong
Astragalar Head	16. Little Lateral Rotation	16'. Strong Lateral Rotation
Facet for Plantar Calcaneonavicular Ligament	17. Medial < Lateral	17'. Medial > Lateral
Length Plantar Edge Entocuneiform Navicular Facet	18. Longer than Dorsal	18'. Much Longer than Dorsal
Relative Length Cuboid	19. Length > 110% Width	19'. Length 100–110% > Width 19". Length < 100% Width
Size Proximal Ectocuneiform-Cuboid Facet	20. Larger than Distal	20'. Smaller than Distal 20". Absent
Proportion Digit 4	21. Digit 4 < Digit 3	21'. Digit 4 = > Digit 3
Insertion Flexor Fibularis on Hallux	22. Present	22'. Weak-Absent
Groove for Flexor Tibialis on Astragalus	23. Absent	23'. Present

[a] These character-states do not form a morphocline

Table 3   Distribution of character-states defined in Table 2. For each group the primitive state is given. Autapomorphies are discussed in the text

	1	2	3	4	5	6	7	8	9	10	11	12	13	14	15	16	17	18	19	20	21	22	23
Atelids	1	2	3	4	5	6	7	8	9	10	11	12	13	14	15	16	17	18	19	20	21	22	23
Cebids	1	2	3	4	5	6	7	8	9	10	11	12	13	14	15	16	17	18	19	20	21	22	23
Catarrhine morphotype	1	2	3	4	5'	6	7	8	9	10	11	12	13	14	15	16	17	18	19	20	21	22	23
Hylobatids	1	2	3	4	5'	6	7	8	9"	10	11	12	13	14	15	16	17	18	19'	20	21	22	23
Great apes	1	2	3	4	5'	6	7	8	9"	10	11	12	13	14	15"	16	17	18	19'	20	21	22	23
Cercopithecid morphotype	1'	2'	3'	4'	5'	6'	7'	8'	9'	10'	11'	12'	13'	14'	15'	16'	17'	18'	19'	20'	21'	22'	23
Cercopithecine morphotype	1'	2'	3'	4'	5"	6'	7'	8'	9'	10'	11'	12"	13'	14'	15'	16'	17'	18'	19'	20'	21'	22'	23
Cercopithecini	1'	2'	3'	4'	5"	6'	7'	8'	9'	10'	11'	12"	13'	14'	15'	16'	17'	18'	19'	20'	21'	22'	23
Papionini	1'	2'	3'	4'	5"	6'	7'	8'	9'	10'	11'	12"	13'	14'	15'	16'	17'	18'	19'	20'	21'	22'	23
Colobine morphotype	1'	2'	3'	4'	5"	6'	7'	8'	9'	10'	11'	12'	13'	14	15"	16'	17'	18'	19'	20'	21'	22'	23'
Presbytina	1'	2'	3'	4'	5"	6'	7'	8'	9'	10'	11'	12'	13'	14	15"	16'	17'	18'	19'	20'	21'	22'	23'
Colobina	1'	2'	3'	4'	5"	6'	7'	8'	9'	10'	11'	12'	13'	14	15"	16'	17'	18'	19'	20'	21'	22'	23'

and *Pongo* do have a more asymmetrical trochlea but not to the degree seen in cercopithecids (Langdon, 1986). Hylobatids generally have an extremely parallel-sided astragalar trochlea while the remaining hominoids have a moderately wedged one, except for *Gorilla* which has an extremely wedged trochlea. Among the New World monkeys, only atelines have a wedge-shaped trochlea and present trochlear asymmetry but in their case it is the medial side which rises above the lateral. All other platyrrhines have a symmetrical, parallel-sided trochlea. Based on the distribution of these features the more typical platyrrhine and hylobatid condition is considered the primitive one for catarrhines with wedging and asymmetry evolving independently in the catarrhine taxa which exhibit it.

Functionally, the wedge-shaped and asymmetrical astragalar trochlea of cercopithecids should increase the amount of abduction which accompanies dorsiflexion of the foot. The widening of the crural mortise which must accompany this movement (Barnett & Napier, 1953) is probably accomplished by some separation of the tibia and fibula at the inferior tibiofibular joint. This movement would be allowed by the stretching of the interosseous ligaments. This may provide more stability than would tibiofibular rotation at a synovial joint.

Lower ankle joint

Cercopithecids uniformly present a lower ankle joint complex which is clearly unique among anthropoids. The proximal calcaneal facet for the astragalar body is tightly curved (i.e., the radius of curvature is small and the angle of the arc subtended large). It is wide and squared off proximally so that the medial and lateral ends are equally long (Figure 1D,E). It is also tilted obliquely to the dorsoplantar axis of the calcaneus so that the facet faces more dorsally than medially (Langdon, 1986). The proximal facet on the astragalus is correspondingly tightly curved and squared off (Figure 1C). Cercopithecids also consistently have a pressure facet lateral to the proximal calcaneal facet (Figure 1D; Olivier & Fontaine, 1957) which is produced by the posterior fibulocalcaneal ligament when it becomes taut during eversion of the foot. In the distal compartment of the lower ankle joint the facets for the head and neck are discontinuous and oriented acutely to each other (Figure 1D,E; Langdon, 1986).

The proportions of the calcaneus are distinctive in cercopithecids. The proximal facet is short relative to calcaneal length (Szalay, 1975; Langdon, 1986; Szalay & Langdon, 1986) as a bivariate plot of the index RELPAF (length of proximal facet relative to calcaneal length) against body weight clearly demonstrates (Figure 2). Additionally, the proximal calcaneal body is relatively very long, thus increasing the power arm of the triceps surae muscles (Figures 1C–D, 3).

The apes have a more gently curved proximal calcaneal facet (i.e., the radius of curvature is large and the angle of the arc small). It is relatively long and wide (Figures 1D–E, 2, 3; Langdon, 1986), and tapers proximally due to the lateral side being longer than the medial (Figure 1E). The corresponding astragalar facet is shallow and C-shaped because of its tapered proximal end (Figure 1C). The African apes are similar to cercopithecids in the tilt of the proximal facet while in *Pongo* and hylobatids it faces more medially [Langdon, 1986; Rose, 1986). An accessory pressure facet is never seen later to the proximal calcaneal facet. The distal astragalocalcaneal articulation is a single, long and gently curving concavity which supports the astragalar head and neck. The proximal and distal portions of the calcaneus are relatively short (Figure 3) with the exception of *Pan* in which the proximal calcaneus is relatively long.

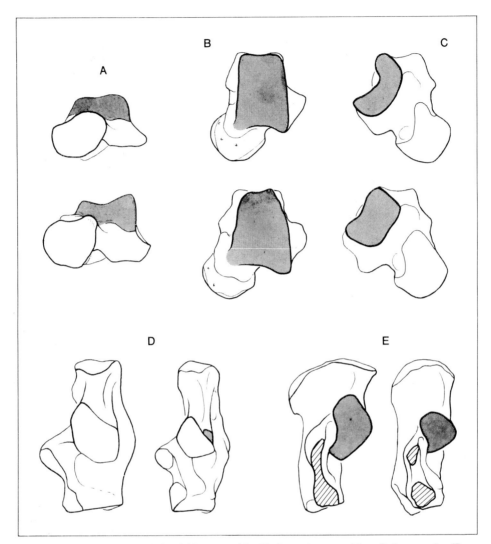

Figure 1. Selected features distinguishing cercopithecids from other catarrhines. Left astragalus. Top, *Hylobates*. Bottom, *Cercocebus*. A: distal view, lateral right. Note trochlear asymmetry in *Cercocebus*. B: dorsal view, distal bottom. Note wedged trochlea in *Cercocebus*. C: plantar view, distal bottom. Note squared-off proximal facet in *Cercocebus*. Left calcaneus. Left, *Hylobates*. Right, *Cercocebus*. Proximal top. D: Dorsal view, medial left. Note pressure facet (stippled) for fibulocalcaneal ligament. E: Medial view. Note short proximal (stippled) and separate (parallel lines) sustentacular (top) and distal (bottom) astragalocalcaneal facets in *Cercocebus*.

All platyrrhines also have a gently curved but narrow proximal calcaneal facet. They are variable in the distribution of its relative length: most atelids are like apes in having a proportionately long facet while *Aotus*, *Callicebus* and cebids have a relatively short facet (Figures 2–3). The proximal calcaneal facet is steeply tilted to face medially (Langdon, 1986). The reciprocal astragalar facet is shallow and C-shaped as in hominoids. An accessory pressure facet is rarely seen in these animals. In the distal compartment the sustentacular and distal astragalocalcaneal facets are usually confluent and gently curved.

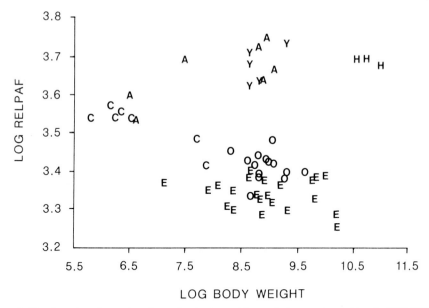

Figure 2. Bivariate plot of mean length proximal calcaneal facet relative to calcaneal length (RELPAF) to mean body weight for taxa listed in Table 1. Values on both axes are transformed to log$_e$. Note that most atelids (A), hylobatids (Y) and great apes (H) are similar in having a relatively long proximal calcaneal facet. Cercopithecids [Colobines (O), Cercopithecines (E)] are at the bottom of the plot with a very short proximal calcaneal facet. The majority of colobines have a longer facet than do cercopithecines of the same body weight. Most cebids (C), *Aotus* (A) and *Callicebus* (A) lie between the other clades. Note that *Cebus* (C) is similar to colobines.

However, many individual cebines and pitheciines can have separate facets but they are never oriented acutely to each other as in Old World monkeys. The atelids (except for *Aotus* and *Callicebus*) are like hominoids in having a relatively short proximal and distal calcaneus (Figure 3). The remaining platyrrhines combine the short proximal calcaneal facet with a short proximal but long distal calcaneus.

The distribution of the accessory pressure facet, the morphology of the proximal calcaneal and astragalar facets, and the length of the proximal calcaneus unequivocally points to the cercopithecid condition being derived for catarrhines. The situation is not so immediately apparent for the relative length of the proximal calcaneal facet or the separation of the distal astragalocalcaneal facets. In the case of the former, reference to the next sister-taxon, Strepsirhini, provides some resolution to this problem. Calcaneal proportions in lemurids, and especially indriids, suggest that the condition found in cebids (and *Aotus* and *Callicebus*), is primitive for anthropoids (see also Szalay & Langdon, 1986). This includes a proximal calcaneal body and facet which are relatively short and a distal calcaneus which is relatively long (Figure 3). From a cebid-like calcaneal ancestory, the suspensory atelids and hominoids have converged in their relative elongation of the proximal facet coupled with shortening of the distal calcaneal body. Such a relatively long and gently curved facet would allow considerable mobility at the lower ankle joint as well as offer a large surface area for loading during climbing activities (Langdon, 1986). Cercopithecids have diverged from cebids in the opposite direction. The relative length of the proximal facet for any given body weight is shorter (Figure 2) and the proximal calcaneal body is relatively longer (Figure 3).

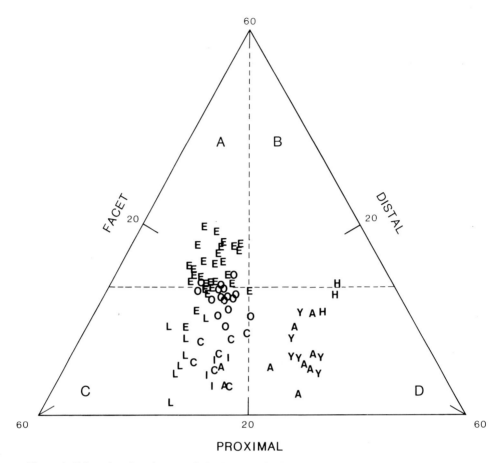

Figure 3. Triangular plot of mean relative lengths of calcaneal segments for taxa listed in Table 1. Conventions as in Figure 2 with addition of lemurs (L) and indriids (I). Data for strepsirhines taken from Gebo & Simons (1987, Table 5; *Lepilemur* and *Lemur coronatus* excluded). *Facet* relative length of proximal calcaneal facet; *distal* relative length of calcaneus distal to proximal facet; *proximal* relative length of calcaneus proximal to proximal facet. All axes extend from 20% of calcaneus length (20) to 60% (60). Plot divided into four quadrants: A, long proximal, short facet, long distal; B, long proximal; long facet, short distal; C, short proximal, short facet, long distal; D, short proximal, long facet, short distal. Note that cercopithecines cluster in quadrant A; colobines cluster along the border between A and C; cebids, lemurids, indriids, *Aotus* (A) and *Callicebus* (A) cluster in quadrant C; atelids and most apes cluster in quadrant D; and, *Pan* (H) falls on border between B and D.

The most homoplastic feature of the lower ankle joint complex appears to be the condition of the sustentacular and distal astragalocalcaneal facets. Separate facets are often found in *Saimiri, Cebus*, some pithecines, *Apidium* (Gebo & Simons, 1987), *Pliopithecus* (Zapfe, 1960) and *Homo*. As a result, it is difficult to weigh them strongly if considered as an isolated feature, but as argued by Gebo & Simons (1987), when considered with the proximal calcaneal facet, they form an important part of an integrated functional complex (see also Langdon, 1986 and Rose, 1986). In this context, all aspects of the cercopithecid lower ankle joint described above are considered derived relative to the primitive catarrhine condition which is likely to have been like that of cebids.

The lower ankle joint of cercopithecids appears to be designed to reduce the screw-like

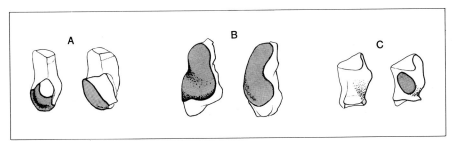

Figure 4. Selected features distinguishing cercopithecids from other anthropoids. Left entocuneiform. Left, *Aotus*. Right, *Cercocebus*. A: Dorsal view, distal bottom. Note hallucal facet (stippled) faces primarily medially in *Cercocebus*. B: Distal view, medial left. Note dorsoplantar elongation of dorsal convexity and reduction of helical groove in *Cercocebus*. Left cuboid. Left, *Hylobates*. Right, *Cercocebus*. C: Lateral view, dorsal left, proximal top. Note facet (stippled) for peroneus longus sesamoid in *Cercocebus*.

movement typical in primates, which is important for inversion of the foot (Szalay, 1975; Lewis, 1981; Rose, 1986). The morphology of both the proximal and distal astragalocalcaneal facets provides stability at the expense of mobility during inversion and eversion (Langdon, 1986). This constrained movement at the lower ankle joint is clearly compensated for in cercopithecids by increasing mediolateral deviation of the foot at the upper ankle joint. These stabilizing morphologies, combined with a long proximal calcaneal body, suggest powerful loading of the foot during cursorial activities.

Cuboid

In cercopithecids the lateral surface of the cuboid consistently presents a strongly-developed facet for the sesamoid of the peroneus longus muscle (Figure 4C; Olivier & Fontaine, 1957), the primary everter of the foot. Hylobatids rarely show a facet for this sesamoid and when they do it is weakly developed. Among the larger hominoids, the great apes never present a facet while humans occasionally do. A similar distribution has recently been reported for the presence of the sesamoid itself within the tendon of *m. peroneus longus* (Le Minor, 1987) except that it was found that the sesamoid was consistently present in both feet of the six *Hylobates* examined. It was reported to be absent in both feet of five great ape specimens, unilaterally and bilaterally present in 64 out of 520 humans (a slightly higher frequency than reported by Pfitzner, cited in Manners-Smith, 1908) and missing in four of the 127 cercopithecids examined. Manners-Smith (1908) reports that the *os peroneum* is represented by a cartilaginous thickening in platyrrhines and that there is a facet on the cuboid in Atelidae and *Cebus* but not in callitrichines while Le Minor (1987) states that the *os peroneum* is not found in any other primate family. Such a facet was not observed in any New World monkey, all the genera of which are indistinguishable in this regard. Therefore, the results of the histological study of Le Minor (1987) appear to be stronger evidence than the report of Manners-Smith. Given the distribution of the peroneal facet, an *os peroneum* is considered here to be a derived feature for catarrhines and the *consistent* development of a cuboid facet for this sesamoid is a cercopithecid synapomorphy.

Entocuneiform

The distal entocuneiform in cercopithecids is quite compressed dorsally so that the dorsal two-thirds of the hallucal facet faces primarily medial while the plantar one-third wraps around the bone to face distally (Figure 4A,B). The medial edge of the hallucal facet is convex and semilunar in outline and the bulk of the articular surface is evenly convex in all directions (Figure 4B; Strasser & Delson, 1987; Szalay & Dagosto, 1988). In colobines, there is a wide shallow furrow located peripherally on the medial and plantar edge of the facet whereas in cercopithecines this is strongly reduced if not eliminated. Extant hominoid genera are distinctly autapomorphic in their entocuneiform morphology (Szalay & Langdon, 1986; Szalay & Dagosto, 1988; Strasser, in prep.) and without the addition of fossils, which broaden the range, it would be difficult to determine the primitive catarrhine condition. *Proconsul* (Langdon, 1986), *Oreopithecus* (Szalay & Langdon, 1986) and *Pliopithecus* (Zapfe, 1960) are similar to each other in having a shallow furrow which spirals around the dorsal half of the hallucal facet medially and plantarly (Szalay & Langdon, 1986). While there is certainly a great deal of variation among platyrrhines, all those examined here share a common groundplan of a distal entocuneiform in which there is a dorsal globular head which faces both distally and medially and is surrounded medially and distally by a wide, spiraling furrow [Figures 4A,B; see Szalay & Dagosto (1988) for a more specific discussion of morphologies within New World monkeys]. Thus, as concluded by Lewis (1972), a dorsal convexity surrounded by a spiraling groove is common in Anthropoidea and is distinctly derived relative to other haplorhines and strepsirhines (Szalay & Dagosto, 1988; Strasser, in prep.).

Lewis (1972) observed that the helical groove is significant in its consequences for hallucal movement: during opposition of the hallux to the lateral digits, the hallux conjunctly rotates around its long axis by sliding around the dorsal convexity while travelling along the groove. The medial and plantar walls of the furrow serve as a buttress against displacement in those directions. Since the catarrhine morphotype is reconstructed to have retained the primitive anthropoid condition in the construction of its distal entocuneiform it is likely that some conjunct hallucal rotation was important in their grasp [see Szalay & Dagosto (1988) for a discussion of the evolution of the anthropoid hallucal joint from the more constrained, euprimate condition]. The ancestral cercopithecid compressed the entocuneiform dorsally so that the dorsal articular surface faced primarily medially, expanded the dorsal convexity plantarwards to encompass most of the articular surface but, like colobines, retained a peripheralized helical furrow. Thus, adjunct rotation would have been emphasized over conjunct rotation in cercopithecids as a result of their ovoid, kidney-shaped entocuneiform.

Subfamilial morphotypes

There are also distinctive differences between cercopithecid subfamilies which, taken together with familial characteristics, complete our picture of the ancestral cercopithecid as well as provide clues to the nature of the origin of the subfamilies.

Cercopithecine morphotype
Cercopithecines present several features which indicate that the foot is typically stressed during positions which involve extreme plantarflexion and inversion as well as dorsiflexion and eversion. The feature which indicates support during the extremes of flexion is found at the upper ankle joint. Typically in anthropoids the astragalar facet (malleolar cup) for the medial malleolar pestle flares slightly medially and extends 3/4 the way down the astragalar body, terminating before the plantar border is reached (Figure 5A; Gebo, 1986). This is characteristic of New World monkeys and most hominoids. Cercopithecids (and

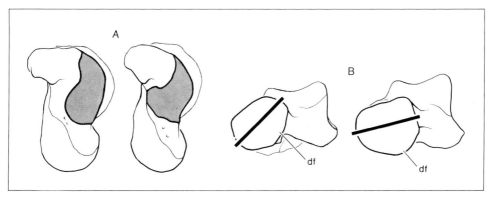

Figure 5. Selected features distinguishing cercopithecid subfamilies. Left astragalus. Left, *Colobus*. Right, *Cercocebus*. A: medial view, distal bottom. Note tibial malleolar facet (stippled) contact with sustentacular facet in *Cercocebus*. B: distal view, medial left. Note deep distal calcaneal facet (df) in *Cercocebus* and more strongly rotated head (bold line) in *Colobus*.

Pan) have a better developed malleolar cup, more deeply concave and medially flaring (Harrison, 1986). In colobines the malleolar cup extends as far plantad as in other anthropoids. In cercopithecines the malleolar cup approximates or defines the plantar border of the bone (Figure 5A). As a result, the astragalus in cercopithecines is medially more completely surrounded by the crural mortise and the tibial pivot may maintain a large degree of contact even during plantarflexion of the foot. This can also be seen on the distal tibia, where the malleolar pestle dominates the medial articular surface.

At the lower ankle joint in cercopithecines, the distal astragalocalcaneal facet (df) on the astragalus is strongly convex and acutely angled to the plantar calcaneonavicular ligament facet. As a result, the astragalar head appears deeper, in distal view, on its lateral side (Figure 5B). No other extant anthropoid presents such an angulated facet. Functionally, this angulation may effectively secure the astragalus against the calcaneus during inversion of the foot (Langdon, 1986) as well as during eversion and pronation. It prevents the astragalus from being displaced proximally by joint reactant forces while being loaded. Both the well developed tibial malleolar facet and highly curved distal astragalocalcaneal facet may stabilize the proximal tarsus, for example, during the pushoff phase of running.

Cercopithecines also tend to have a flatter entocuneiform than do colobines which would limit the degree of axial rotation that could occur. There is little or no plantar and medial buttressing present on the entocuneiform suggesting that the hallux is not stressed by grasping postures to the extent seen in colobines.

Colobine morphotype

In colobines, the foot appears to be designed for a position which involves supination of the forefoot and an increased span to its grasp.

Lower ankle joint. Colobines are characterized by a strongly curved proximal astragalocalcaneal facet, a feature first documented by Olivier & Fontaine (1957). This curvature is obvious in lateral view (Figure 6A) and appears to be the result of proximally elongating the facet. Perhaps correlated with this is the relatively longer proximal calcaneal facet seen in most colobines (Figures 2, 3; Szalay, 1975). A proximally elongated and strongly curved facet would increase the area of contact at the lower ankle joint and restrict the amount of proximal sliding that would result from joint reactant forces during inversion.

Proximal elongation of the proximal astragalar facet also has a morphological consequence on the relations of the *m. flexor fibularis* sulcus to the proximal edge of the trochlea (Figure 6F). In cercopithecines these two regions are continuous while in colobines there is a break between them and the sulcus appears dorsoplantarly elongated. Other catarrhines are similar to cercopithecids in having little discontinuity between the *m. flexor fibularis* sulcus and the trochlea, while the proximal astragalar facet is not strongly curved.

Transverse tarsal joint. Colobines also have a complex of traits which indicate that the forefoot (distal tarsus and digits) is habitually supinated. One of the most obvious of these is the strong lateral rotation of the astragalar head (Figure 5B; Olivier & Fontaine, 1957). As Elftman and Manter (1935) discussed for humans, astragalar head torsion contributes to a habitually supinated set to the foot. This is because lateral rotation of the head and neck effects medial elevation of the tarsus (especially of the ecto- and mesocuneiforms) and digits distal to it.

A morphological feature which is a consequence of astragalar head and neck torsion in colobines is seen in the astragalar facet for the plantar calcaneonavicular ligament (Figure 6B). In colobines it is more limited laterally than in cercopithecines extending instead proximally, medially and dorsally where the head contacts the ligament. This gives the astragalar head a more rounded appearance medially. Because there is less contact laterally between the astragalar neck and the plantar calcaneonavicular ligament the development of articular cartilage on the astragalar neck can be suppressed. This takes various expressions, ranging from a depressed area, thinly covered by cartilage, to a pitted surface to a large nonarticular invagination between the sustentacular and more distal facets (Figure 6B).

Distal tarsals. With greater torsion of the astragalar head in colobines, the navicular is oriented more vertically than in cercopithecines. This brings the facet for the entocuneiform into a more proximal position. As a result, the plantar portion of the cup-like facet on the entocuneiform neck is elongated proximally to maintain contact with the retracted navicular facet (Figure 6C). The more typical condition, found in most cercopithecines, apes and New World monkeys, is to have a navicular facet with a slightly longer plantar than dorsal edge. This is retained even in the larger papionins which have extremely robust entocuneiforms (Jolly, 1965).

Gabis (1960) noted that colobines have a relatively shorter distal tarsus than do cercopithecines while Schultz (1963b) found that the tarsus as a whole is short relative to total foot length. This shortening is reflected in the shape of the cuboid as well as in the proximal contact between the cuboid and ectocuneiform (Strasser & Delson, 1987). In African colobines the majority of individuals have a cuboid which is shorter than its proximal width (Figure 7). A cuboid which is between 1–10% longer than it is wide is likely the primitive one for cercopithecids since it is found in a wide variety of anthropoids (hylobatids, cebids, most Miocene catarrhines). Within cercopithecines it is of interest to note that the majority of Cercopithecini have a relatively long cuboid whereas most papionins are like presbytinans (Strasser, in prep., will discuss generic distinctions).

The shorter tarsus in colobines is associated with a reduction in the proximal contact between the cuboid and ectocuneiform (Figures 6D, 8). The common anthropoid condition, and the one retained in cercopithecines, is to have a large proximal and smaller

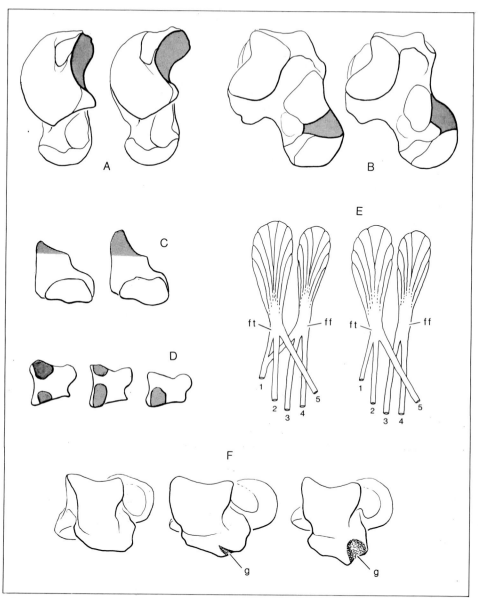

Figure 6. Selected features distinguishing cercopithecid subfamilies. Left astragalus, proximal top. Left, *Cercocebus*. Right, *Presbytis*. A: lateral view, dorsal left. Note more strongly curved proximal astragalocalcaneal facet (stippled) in *Presbytis*. B: plantar view, medial right. Note proximally elongate facet for plantar calcaneonavicular ligament (stippled) and lateral suppression of facet in *Presbytis*. C: Left entocuneiform, medial view. Left, *Cercocebus*. Right, *Colobus*. Proximal top, plantar left. Note plantar elongation of facet for navicular in *Colobus*. D: Left ectocuneiform, lateral view. Left, *Cercocebus*. Middle, *Presbytis*. Right, *Colobus*. Proximal top, dorsal left. Note large proximal, small distal facets (stippled) in *Cercocebus*, reversed relationship in *Presbytis* and loss of proximal contact in *Colobus*. E: Left extrinsic digital flexor muscles, dorsal view. Left, *Cercopithecus*. Right, *Colobus*. Proximal top, medial left. Note that flexor fibularis (ff) loses its contribution to the first digit (1) in *Colobus*. Flexor tibialis (ft) is solely responsible for managing the first digit. F: Left astragalus, proximal view. Left, *Cercocebus*. Middle, *Colobus*. Right, *Presbytis*. Plantar bottom, medial right. Specimens rotated medially to show flexor tibialis groove (g) at best advantage. Note no groove in *Cercocebus*, swelling distal to groove in *Colobus* and large size of groove in *Presbytis*.

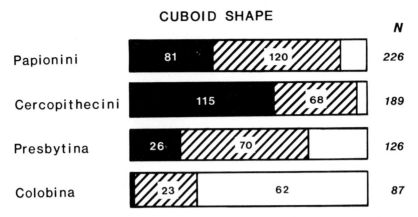

Figure 7. Bar chart illustrating frequencies for cuboid shape. Black, long [length (maximum medial proximodistal diameter) > 110% width (maximum proximal mediolateral diameter)]. Parallel lines, intermediate (length 100%–109% > width). White, short (length < 100% width). Values within bars are number of specimens for each category. N, total sample size per taxon. For all taxa χ^2 value is 239·5 with 6 degrees of freedom (df), significant (P) at 0·0001. Cercopithecine tribes significantly differ from each other ($\chi^2 = 28\cdot857$, df 2, $P = 0\cdot0001$) as do colobine subtribes (χ^2 50·0, 2 df, $P = 0\cdot0001$).

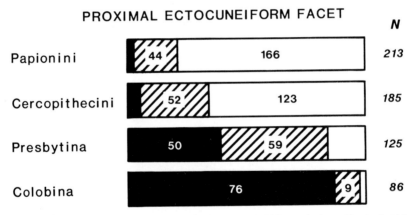

Figure 8. Bar chart illustrating frequencies for size proximal cuboid facet relative to distal cuboid facet on ectocuneiform. Black, no proximal facet. Parallel lines, area (proximodistal diameter × dorsoplantar diameter) proximal facet < area distal facet. White, area proximal facet > area distal facet. Same conventions as in Figure 7. For all taxa χ^2 value is 394·7 with 6 df, $P = 0\cdot0001$. Cercopithecine tribes not significantly different from each other at 0·01 level (χ^2 6·3, 2 df) while colobine subtribes are significantly different (χ^2 48·6, 2 df, $P = 0\cdot0001$).

distal contact between the cuboid and ectocuneiform. Presbytinans reverse this pattern with most individuals having a smaller proximal facet or none at all (Olivier & Fontaine, 1957). In African colobines, the majority of individuals have no proximal contact between these two bones. This complete loss of a facet has no obvious functional significance because there is little movement between the distal tarsal bones in any case. It is more likely a consequence of the shortening of the cuboid.

Digits. In the skeletal and muscular morphology of the digits there are further differences between colobines and cercopithecines. In anthropoids, the primitive morphological axis

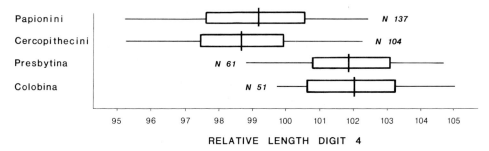

Figure 9. Box and whisker plot relative length digit 4 [(lengths digit 4 ÷ by digit 3) × 100]. Digit lengths calculated by adding length metatarsal to length proximal phalanx. Vertical bar, mean. Box, 1 standard deviation to each side of mean. Horizontal line, range. *N*, total sample size per taxon. Cercopithecine tribes significantly different from each other (Student's *t* value −2·46, 239 df, *P* = 0·01). Colobine subtribes not significantly different from each other (Student's *t* value 0·54, 110 df, *P* = 0·60).

of the foot passes along the third digit, as inferred by its greatest length, robusticity and arrangements of muscles around it (Morton, 1924; Lessertisseur & Jouffroy, 1973). This condition is retained in cercopithecines but colobines have shifted this axis more laterally, between the third and fourth digits. As a result, the fourth digit is equal in length or longer than the third (Figure 9; Morton, 1924; see Strasser & Delson, 1987 for illustration). This may reflect a different sort of pedal grasp in colobines that may be part of their leaping locomotor repertoire, also suggested by the reduction in their hallux (see below) and supination at the transverse tarsal joint.

Accompanying the shift in the morphological axis of the skeletal digit rays is a modification in the extrinsic flexor muscles of the digits. Primitively, *m. flexor fibularis* (ff) contributes fibers to the hallux, as well as primarily operating the third and fourth digits (Figure 6E; Straus, 1930; Lewis, 1962). In African colobines *m. flexor fibularis* has eliminated its contribution to the hallux (Polak, 1908; Lewis, 1962; Strasser, in prep.). The same is true for *Presbytis* while *Trachypithecus* has a relatively weak connection with the hallux, judging from the relative sizes of the tendons of insertion. Thus, in at least the colobines surveyed here, there is a tendency for *m. flexor fibularis* to devote itself completely to managing the central rays while *m. flexor tibialis* (ft) is left to operate the hallux alone, albeit with assistance from *m. flexor accesorius* (Straus, 1930; Lewis, 1962). The loss of *m. flexor fibularis* insertion is probably due to the reduction of the hallucal phalanges in colobine monkeys (Midlo, 1934; Schultz, 1963*b*). This may be an epigenetic effect or a developmental consequence of selection for a reduced thumb in colobines (see Jouffroy & Lessertisseur, 1976).

The increased stress put upon *m. flexor tibialis* is reflected more proximally on the astragalus. In African and most Asian colobines, the retinacular tunnel for this muscle is reinforced by bony walls thus more securely restricting and channeling the muscle during its activity. In African colobines this reinforcement takes the form of a large swelling distal to the groove (g) for *m. flexor fibularis*. In *Presbytis*, *Trachypithecus* and *Semnopithecus* the edges of the groove are discrete ridges (Figure 6F; Olivier & Fontaine, 1957). *Nasalis, Pygathrix* and *Rhinopithecus* have the poorest expression of the groove with a distal tubercle present for the retinaculum but not swollen as in African colobines. Among non-cercopithecids, only *Pan* shows a similar, but certainly convergent, arrangement to colobines.

All of the colobine tarsal features discussed above suggest an habitually supinated forefoot. Along with the short tarsus, the elongation of the fourth digit suggests that the foot is positioned in such a manner that the lateral digits are bearing more stress than the central or medial ray. This may be related to increasing the span of the pedal grasp between the hallux and lateral digits. That the hallux is involved in this position is suggested by the buttressing of the distal entocuneiform. Observations of captive or free-ranging animals, specifically addressing foot postures, are required to determine the exact nature of this pedal position.

Discussion

The inferred cercopithecid pedal morphotype is characterized by a set of features which are derived relative to the catarrhine morphotype. These include: (1) an essentially syndesmotic inferior tibiofibular joint; (2) an asymmetrical astragalar trochlea; (3) a wedge-shaped astragalar trochlea; (4) a squared-off proximal calcaneal facet; (5) a tightly-curved proximal calcaneal facet; (6) a relatively short proximal calcaneal facet; (7) an obliquely tilted proximal astragalar facet; (8) a pressure facet for the fibulocalcaneal ligament; (9) separate and acutely oriented sustentacular and distal astragalocalcaneal facets; (10) an elongated proximal calcaneal body; (11) a consistently well-developed facet for the tendon of *m. peroneus longus*; and, (12) a kidney-shaped hallucal facet on the entocuneiform.

For several reasons it is of interest to determine whether these features evolved sequentially or as a unit in cercopithecids. If a model of sequential evolution were to fit the data, then we would not expect to find the whole suite of characters in the cercopithecid ancestors or a sister-taxon. In this case it would be difficult to falsify a phylogenetic hypothesis positing such an ancestor based on isolated features. On the other hand, if all of these features form an integrated unit, inseparable one from the other, and a model of simultaneous reorganization of the foot were supported, then a candidate for cercopithecid ancestry must present the whole suite to qualify as such.

Given the data presented above, it appears that the salient features of the cercopithecid foot form an integrated unit which evolved simultaneously, since they are found in all cercopithecids. Isolated aspects of this complex appear in other anthropoids, but in no case are more than three of the twelve features found together. In cercopithecids, the extreme asymmetry of the upper ankle joint allows the lateral displacement of the foot required during locomotion. Indeed, this appears to be the primary site of mediolateral displacement of the foot. This asymmetry compensates for the constraints imposed on the range of movement possible at the lower ankle joint. This latter region is usually the site of the most pronounced mediolateral movements in primates, but in cercopithecids it is remodelled to function primarily as a hinge (Szalay, 1975; Langdon, 1986). Additionally, the peroneal facet on the cuboid suggests that pronating the foot is a significant part of positional behaviors, as during the push off phase of locomotion. The reduction in axial rotation of the hallux, which results from the kidney-shaped entocuneiform, also suggests limited hallux-dependent grasping abilities, such as would not be required for locomotion on the ground.

The data presented here support the hypothesis that the origin of cercopithecids involved a shift to semi-terrestriality (Delson, 1973, 1975; Szalay, 1975; contra Napier, 1970 and Temerin & Cant, 1983). Benefit (1987) has recently suggested that this shift was the result of early cercopithecids exploiting a frugivorous niche in a riverine mixed forest environment which first appeared in Africa during the Miocene.

On the basis of pedal material two fossil taxa have been suggested to be relatives of cercopithecids. The first, *Oreopithecus*, may well share dental similarities with cercopithecids (Szalay & Delson, 1979) but its foot shows none of the synapomorphies characteristic of that family (Harrison, 1986; but see Szalay, 1975). Indeed, Szalay & Langdon (1986) have described it as phenetically most similar to *Pan* while Sarmiento (1987) argues that these similarities are synapomorphies uniting *Oreopithecus* with extant hominoids. There is no reason to consider *Oreopithecus* a collateral relative of cercopithecids until dental and/or cranial homologies are demonstrated between cercopithecids and *Oreopithecus* and its postulated postcranial synapomorphies with hominoids are proven to be non-homologous.

Gebo & Simons (1987) have resurrected the hypothesis that the second fossil taxon, *Apidium*, is ancestral to cercopithecids, implying that the shift to terrestriality occurred prior to the Miocene (see Fleagel & Kay, 1988 for a discussion of the problems associated with this interpretation). *Apidium* does share with cercopithecids the same relative calcaneal proportions and separate distal astragalocalcaneal facets (Gebo & Simons, 1987). But the tilt of the proximal calcaneal facet is much steeper than in cercopithecids and the proximal astragalar facet is more C-shaped, like that of non-cercopithecid anthropoids. These morphologies yield a greater range of inversion and eversion at the lower ankle joint than is seen in cercopithecids. Additionally, in *Apidium*, the upper ankle joint is distinctly different: the trochlea is absolutely symmetrical in height and its sides are parallel (Conroy, 1976; Gebo & Simons, 1987). This trochlear morphology is completely in line with the function of its derived inferior tibiofibular joint, suggesting a positional repertoire which emphasizes leaping (Fleagle & Simons, 1983). Finally, the navicular attributed to *Apidium* (Gebo & Simons, 1987; Fleagle & Kay, 1988; Fleagle & Simons, in prep.) is morphologically completely unlike that of cercopithecids (Strasser, in prep.) and more similar to the long navicular of *Saimiri* or *Aotus* (but see Gebo & Simons, 1987). Since only a few aspects of the morphotypic cercopithecid foot are found in *Apidium* it appears more parsimonious to interpret *Apidium* as having converged on the cercopithecid condition at the lower ankle joint.

At a lower taxonomic level, there are several distinctive differences between colobines and cercopithecines. Cercopithecines have two features interpreted here as associated with stabilizing the foot during dorsiflexion and eversion of the foot: a more secure medial tibio-astragalar articulation and an angulated distal astragalocalcaneal articulation which prevents proximal sliding of the astragalus during eversion. They are also characterized by decreased adjunct rotation of the hallux due to flattening of the distal entocuneiform surface. These pedal features may be related to increasing stability of the foot during running on the ground.

Colobines share nine characters associated with increased supination of the foot and modification of the functional axis of the digit rays, interpreted here as features associated with a novel grasp. Pedal synapomorphies (shifting the axis within the forefoot laterally, shortening the tarsus and reinforcing the astragalar channel for *m. flexor tibialis*) are as ubiquitous within the subfamily as are digestive tract specializations for folivory. This suggests that the origin of the group, which involved the exploitation of this food resource, also required a novel positional repertoire. Benefit (1987) suggested that the divergence of colobines from cercopithecines can be traced back to their invasion of a stable rainforest environment in which apes were well established. Since adaptations for climbing are evident in Miocene apes (Langdon, 1986) perhaps colobines were forced to the periphery of crowns where leaping between trees was a reasonable alternative to descent along the trunk.

Szalay (1975) first suggested that colobines might be secondarily derived from the ancestral catarrhine condition in having a relatively longer proximal astragalocalcaneal facet. An examination of the distribution of this feature (Figures 2, 3) might initially lead one to conclude that colobines have retained an intermediate condition between cebids (the primitive anthropoid condition) and cercopithecines. However, the increased curvature of the astragalar component of the proximal compartment lends support to Szalay's hypothesis since such a long and strongly curved facet is not found in any other anthropoid, even *Cebus* which otherwise is similar to colobines in this proportion (Figure 2). Given the derived nature of the astragalar half of this joint it seems reasonable, at this time, to conclude that the long proximal astragalocalcaneal facets in colobines are secondarily derived. Heritage aspects of the colobine foot, such as those at the upper ankle joint, may have constrained colobines to remodel their foot at the lower and transverse tarsal joints and digit rays for an arboreally committed lifestyle.

Summary

This paper documents derived pedal features which characterize the family Cercopithecidae and its two extant subfamilies. The salient points developed above are that cercopithecids share 12 pedal synapomorphies: a syndesmotic inferior tibiofibular joint; an asymmetrical astragalar trochlea; a moderately wedged astragalar trochlea; a proximally squared-off proximal calcaneal facet; a tightly-curved proximal astragalocalcaneal facet; a relatively short proximal calcaneal facet; an obliquely tilted proximal calcaneal facet; a pressure facet for the proximal fibulocalcaneal ligament; separate sustentacular and distal astragalocalcaneal facets; an elongated proximal calcaneal body; a consistently developed facet for the sesamoid of peroneus longus; and a kidney-shaped distal entocuneiform. All of these features are related to securing the foot and decreased reliance on the hallux during terrestrial, cursorial locomotion.

Cercopithecines are further derived from the cercopithecid morphotype by extending the contact between the tibial malleolar pestle and astragalus, securing the astragalar head during eversion by increasing the curvature of the distal astragalocalcaneal facet, and decreasing axial rotation of the hallux. These are related to a longer history on the ground than is the case for colobines.

Colobines have remodelled the foot, presumably for a novel grasp during leaping, by acquiring a secondarily elongated proximal astragalocalcaneal facet; laterally rotating the astragalar head; increasing the medial extent of the facet for the plantar calcaneonavicular ligament; plantarly elongating the entocuneiform facet for the navicular; shortening the cuboid; reducing the proximal contact between the cuboid and ectocuneiform; shifting the morphological axis of the digit to between digits three and four; reducing the insertion of flexor fibularis on the hallux, and strengthening the channel on the astragalus for flexor tibialis. These are interpreted as being acquired upon secondarily invading the rainforest biome.

Many of these characters have been known for a long time but most have not been considered in phylogenetic reconstruction. This analysis reveals that even such a purportedly homogeneous and plastic system as the cercopithecid postcranium can reveal characters useful for phylogenetic analysis. Indeed, these are used to examine the hypotheses that *Oreopithecus* and *Apidium* are especially related to cercopithecids.

Acknowledgements

I thank Drs Eric Delson, Fred Szalay, and especially Esteban Sarmiento for their stimulating discussions with me on aspects of this work; Drs Marian Dagosto, Dan Gebo, Mike Rose and an anonymous reviewer for their careful review of this paper and constructive comments; Lorraine Meeker and Chester Tarka of the Department of Vertebrate Paleontology, American Museum of Natural History, for generously providing provocative advice on preparation of the illustrations; and the curators and staffs of the Departments of Mammalogy, Anthropology and/or Paleontology of the following institutions for access to collections in their care: American Museum of Natural History, Museum of Comparative Zoology, National Museum of Natural History, Field Museum of Natural History, The University of Wisconsin-Milwaukee Tappen collection, British Museum (Natural History), Powell-Cotton Museum, Musée Royal de l'Afrique Centrale, Rijksmuseum van Natuurlijke Historie, Zoologisk Museum (Copenhagen), Zentrum Anatomie der Georg-August-Universität (Göttingen), Natur-Museum Senckenberg, Zentrum der Morphologie der Klinikum der Johann Wolfgang Goethe-Universität, Naturhistorisches Museum Wien, Naturhistorisches Museum Basel, Anthropologisches Institut und Museum der Universität Zürich-Irchel, Muséum National d'Histoire Naturelle (Paris) and the National Museums of Kenya.

Research upon which this work is based was supported by an NSF Dissertation Improvement grant (BNS-8407911), a C.U.N.Y. Mina Rees Award, a Smithsonian Predoctoral Fellowship and a Wenner-Gren grant.

References

Andrews, P. (1985). Family group systematics and evolution among catarrhine primates. In (E. Delson, Ed.) *Ancestors: The Hard Evidence*, pp. 14–22. New York: Alan R. Liss.

Barnett, C. H. & Napier, J. R. (1953). The rotatory mobility of the fibula in eutherian mammals. *J. Anat.* **87**, 11–21.

Benefit, B. (1987). The molar mrophology, natural history and phylogenetic position of the middle Miocene monkey *Victoriapithecus* and their implications for understanding the evolution of Old World monkeys. Ph.D. Dissertation, New York University.

Carleton, A. (1941). A comparative study of the inferior tibio-fibular joint. *J. Anat.* **76**, 45–55.

Conroy, G. C. (1976). Primate postcranial remains from the Oligocene of Egypt. *Contrib. Primatol.* **8**, 1–134.

Dagosto, M. (1986). The joints of the tarsus in strepsirhine primates: functional, adaptive, and evolutionary implications. Ph.D. Dissertation, City University of New York.

Delson, E. (1973). Fossil colobine monkeys of the circum-Mediterranean region and the evolutionary history of the Cercopithecidae (Primates, Mammalia). Ph.D. Dissertation, Columbia University.

Delson, E. (1975). Evolutionary history of the Cercopithecidae. *Contrib. Primatol.* **5**, 167–217.

Elftman, H. & Manter, J. (1935). The evolution of the human foot, with especial reference to the joints. *J. Anat.* **70**, 56–67.

Fleagle, J. (1976). Locomotor behavior and skeletal anatomy of sympatric Malaysian leaf monkeys *Presbytis obscura* and *Presbytis melalophos*. *Yrbk phys. Anthrop.* **20**, 440–453.

Fleagle, J. G. (1977). Locomotor behavior and muscular anatomy of sympatric Malaysian leaf monkeys *Presbytis obscura* and *Presbytis melalophos*. *Am. J. phys. Anthrop.* **46**, 297–308.

Fleagle, J. G. (1983). Locomotor adaptations of Oligocene and Miocene hominoids and their phyletic implications. In (R. L. Ciochon & R. S. Corruccini, Eds) *New Interpretations of Ape and Human Ancestry*, pp. 301–324. New York: Plenum.

Fleagle, J. G. & Kay, R. F. (1988). The phyletic position of the Parapithecidae. *J. hum. Evol.* **16**, 483–531.

Fleagle, J. G. & Simons, E. L. (1983). The tibio-fibular articulation in *Apidium phiomense*, an Oligocene anthropoid. *Nature* **301**, 238–239.

Fleagle, J. G. & Simons, E. L. (in prep.). The hindlimb skeleton of *Apidium phiomense*.

Ford, S. M. (1980). A systematic revision of the Platyrrhini based on features of the postcranium. Ph.D. Dissertation, University of Pittsburgh.

Ford, S. M. (1988). Postcranial adaptations of the earliest platyrrhine. *J. hum. Evol.* **17**, 155–192.

Gabis, R. (1960). Les os des membres des singes cynomorphes. *Mammalia* **24**, 577–602.

Gebo, D. L. (1986). The anatomy of the prosimian foot and its application to the primate fossil record. Ph.D. Dissertation, Duke University.

Gebo, D. L. & Dagosto, M. (1988). Foot anatomy, climbing, and the origin of the Indriidae. *J. hum. Evol.* **17**, 139–158.

Gebo, D. L. & Simons, E. L. (1987). Morphology and locomotor adaptations of the foot in early Oligocene anthropoids. *Am. J. phys. Anthrop.* **74**, 83–101.

Harrison, T. (1986). A reassessment of the phylogenetic relationships of *Oreopithecus bambolii* Gervais. *J. hum. Evol.* **15**, 541–583.

Jolly, C. J. (1965). The origins and specializations of the long-faced Cercopithecoidea. Ph.D. Dissertation, University of London.

Jouffroy, F. K. & Lessertisseur, J. (1976). Processus de réduction des doigts (main et pied) chez les primates: modalités, implications génétiques. *Coll. internation. C.N.R.S.* **266**, 381–391.

Jungers, W. L. (1985). Body size and scaling of limb proportions in primates. In (W. L. Jungers, Ed.) *Size and Scaling in Primate Biology*, pp. 345–381. New York: Plenum Press.

Kapandji, I. A. (1970). *The Physiology of Joints. Vol. 2.* New York: Livingston.

Langdon, J. H. (1986). Functional morphology of the Miocene hominoid foot. *Contrib. Primatol.* **22**, 1–226.

Le Minor, J. M. (1987). Comparative anatomy and significance of the sesamoid bone of the peroneus longus muscle (*os peroneum*). *J. Anat.* **151**, 85–99.

Lessertisseur, J. & Jouffroy, F. K. (1973). Tendances locomotrices des primates traduites par les proportions du pied. *Folia primatol.* **20**, 125–160.

Lewis, O. J. (1962). The comparative morphology of *M. flexor accessorius* and the associated long flexor tendons. *J. Anat.* **96**, 321–333.

Lewis, O. J. (1972). The evolution of the hallucial tarsometatarsal joint in the Anthropoidea. *Am. J. phys. Anthrop.* **37**, 13–34.

Lewis, O. J. (1981). Functional morphology of the joints of the evolving foot. *Symp. zool. Soc. Lond.* **46**, 169–188.

MacConaill, M. A. (1973). A structuro-functional classification of synovial articular units. *Ir. J. Med. Sci.* **142**, 19–26.

Manners-Smith, T. (1908). A study of the cuboid and *os peroneum* in the primate foot. *J. Anat.* **42**, 397–414.

Midlo, C. (1934). Form of hand and foot in primates. *Am. J. phys. Anthrop.* O. S. **19**, 337–389.

Morton, D. J. (1924). Evolution of the human foot. II. *Am. J. phys. Anthrop.* O.S. **7**, 1–52.

Napier, J. R. (1970). Paleoecology and catarrhine evolution. In (J. R. Napier & P. H. Napier, Eds) *Old World Monkeys: Evolution, Systematics, and Behavior*, pp. 53–95. New York: Academic Press.

Napier, P. H. (1981). Catalogue of Primates in the British Museum (Natural History) and elsewhere in the British Isles. Part II: Family Cercopithecidae, Subfamily Cercopithecinae. London: British Museum (Natural History).

Napier, P. H. (1985). Catalogue of Primates in the British Museum (Natural History) and elsewhere in the British Isles. Part III: Family Cercopithecidae, Subfamily Colobinae. London: British Museum (Natural History).

Olivier, G. & Fontaine, M. (1957). Les os du pied du Semnopithèque. *Mammalia.* **21**, 142–189.

Polak, C. (1908). *Die Anatomie des Genus Colobus.* Amsterdam: Johannes Muller.

Rose, M. D. (1986). Further hominoid postcranial specimens from the Late Miocene Nagri Formation of Pakistan. *J. hum. Evol.* **15**, 333–367.

Rose, M. D. (1988). Another look at the anthropoid elbow. *J. hum. Evol.* **17**, 193–224.

Rosenberger, A. L. (1979). Phylogeny, evolution and classification of New World monkeys (Platyrrhini, Primates). Ph.D. Dissertation, City University of New York.

Sarmiento, E. E. (1987). The phylogenetic position of *Oreopithecus* and its significance in the origin of Hominoidea. *Am. Mus. Novitates.* No. 2881, 1–44.

Schultz, A. H. (1930). The skeleton of the trunk and limbs of higher Primates. *Hum. Biol.* **2**, 303–438.

Schultz, A. H. (1963a). The relative lengths of the foot skeleton and its main parts in primates. *Symp. Zool. Soc. Lond.* **10**, 199–206.

Schultz, A. H. (1963b). Relations between the lengths of the main parts of the foot skeleton in primates. *Folia primatol.* **1**, 150–171.

Schultz, A. H. (1970). The comparative uniformity of the Cercopithecoidea. In (J. R. Napier & P. H. Napier, Eds) *Old World Monkeys. Evolution, Systematics, and Behavior*, pp. 39–51. New York: Academic Press.

Strasser. E. (in prep.). Pedal morphology of Cercopithecidae and the evolution of the foot in Catarrhini. Ph.D. Dissertation, City University of New York.

Strasser, E. & Delson, E. (1987). Cladistic analysis of cercopithecid relationships. *J. hum. Evol..* **16,** 81–99.

Straus, W. L., Jr (1930). The foot musculature of the highland gorilla (*Gorilla beringei*). *Quar. Rev. Biol.* **5,** 261–317.

Szalay, F. S. (1975). Haplorhine relationships and the status of the Anthropoidea. In (R. Tuttle, Ed.) *Primate Functional Morphology and Evolution*, pp. 3–22. The Hague: Mouton.

Szalay, F. S. & Dagosto, M. (1980). Locomotor adaptations as reflected on the humerus of Paleogene primates. *Folia primatol.* **34,** 1–45.

Szalay, F. S. & Dagosto, M. (1988). Evolution of hallucial grasping in the Primates. *J. hum. Evol.* **17,** 1–33.

Szalay, F. S. & Delson, E. (1979). *Evolutionary History of the Primates*. New York: Academic Press.

Szalay, F. S. & Langdon, J. H. (1986). The foot of *Oreopithecus:* an evolutionary assessment. *J. hum. Evol.* **15,** 585–621.

Temerin, L. A. & Cant, J. G. H. (1983). The evolutionary divergence of Old World monkeys and apes. *Am. Nat.* **122,** 335–351.

Zapfe, H. (1960). Die Primatenfunde aus der miozänen spaltenfüllung von Neudorf an der March (Devinska Nova Ves), Tschechoslowakei. *Mem. Suisses Paleontol.* **78,** 1–293.

William L. Jungers

Department of Anatomical Sciences, School of Medicine, State University of New York at Stony Brook, Stony Brook, NY, 11794-8081, U.S.A. U.S.A.

Received 27 July 1987
Revision received 19 October 1987 and accepted 25 November 1987

Publication date June 1988

Keywords: joint size, body size, scaling, locomotion, hominoids, human evolution.

Relative joint size and hominoid locomotor adaptations with implications for the evolution of hominid bipedalism

The relationship between relative joint size and locomotor adaptations in living hominoids is examined using a variety of analytical strategies: narrow allometry, *a priori* geometrical adjustments, and empirical regression (allometric) adjustments. Regardless of method, the observation that emerges conspicuously is that modern humans possess exceptionally large hindlimb and lumbo-sacral joints for their body size. Full-time terrestrial bipedality precludes the sharing of weight support and propulsion with the forelimbs, and this fundamental difference from the other hominoids is reflected in the relative size of human hindlimb joints. Similar analyses including "Lucy" (A.L. 288-1, *Australopithecus afarensis*) suggest that a modest degree of hindlimb joint enlargement had already taken place at this point in hominid evolution, but that the highly-derived relative joint size characteristic of modern humans had not yet been achieved. This implies that the adaptation to terrestrial bipedalism in early hominids was far from complete and not functionally equivalent to the modern human condition. It is speculated that later enlargement of the hindlimb joints and elongation of the lower extremity represent a major adaptive shift linked to the advent of longer distance travel in human evolution.

Journal of Human Evolution (1988) **17**, 247–265

Introduction

Living hominoid primates vary enormously in body size and in many important aspects of their positional repertoire. They range from just over 5 kg in species of *Hylobates* up to nearly 200 kg in exceptional male gorillas (Jungers, 1984*a*; Jungers & Susman, 1984). Fossil representatives would expand both ends of this range (Fleagle, 1978). Among the extant species are highly specialized, suspensory forms such as the lesser apes and the orang-utan, knuckle-walking quadrupeds and climbers as in the African apes, and one very large-brained species devoted to virtually full-time terrestrial bipedality. Extinct hominoids may have been characterized by other positional elements or by frequency differences in those behaviors practiced today (Rose, 1983; Susman *et al.*, 1984).

Overall body proportions are correlated clearly to the aforementioned differences in locomotor and postural adaptations, especially when body-size differences are taken into account (e.g., Jungers, 1984*b*, 1985). It also seems reasonable to suspect that the size and shape of articular surfaces in the postcranial joints of hominoids should be linked biomechanically to habitual postures and movements. Joint form and size are constrained by (1) the specific types of movements required (MacConaill & Basmajian, 1969; Kapandji, 1970; Norkin & Levangie, 1983); and (2) the necessity of sustaining repetitive loads and the resultant stresses. It has been argued persuasively that a primary goal of joint construction is to diminish the load/area relationship in order to limit damage to the subchondral bone and overlying cartilage, loss of joint congruence and, ultimately, the development of osteoarthritis (Radin, 1982; Radin *et al.*, 1982). In other words, then, the area over which contact takes place must be large enough to sustain recurring loads without creating unduly high stress concentrations because these can lead eventually to joint damage (Currey, 1980; Hodge, *et al.*, 1986).

0047–2484/88/01/20247 + 19 $03.00/0

To date, most inquiries into the relationship between joint size and body size have focused on the "primary signal" (*sensu* Schmidt-Nielsen, 1984) of scaling: the slope in log-log space of joint size as a function of body size (e.g., Jungers & Susman, 1984; Jungers, in press). Alexander (1980, 1981) has proposed a biomechanical model which predicts that maximum articular stresses are probably of the same order of magnitude in animals regardless of body size. These model expectations are based on his findings that maximum joint forces appear to be proportional to (body mass)$^{2/3}$, and skeletal dimensions often scale geometrically. It follows, therefore, that a "null hypothesis" of geometric similarity or isometry for articular measures has a plausible biomechanical foundation as well as a simple dimensional basis. In a preliminary test of this model on primates, Jungers & Susman (1984) found that allometric scaling of linear articular dimensions was the best description of the relationship in most joints of African apes, but Alexander's model predictions could not be rejected conclusively when confidence limits for slopes were taken into account. If all living species of hominoids are examined, however, significant departures from geometric scaling can be demonstrated in several joints of both the fore- and hindlimb (Jungers, in press). Most dimensions examined tended to show some departure from geometric scaling, but confidence intervals usually did include isometry. Although exceptions do exist, Alexander's model based on maximum joint forces still remains viable in general, and its predictions more or less obtain for hominoids specifically. As such, it merits further testing in additional taxonomic assemblages of primates and other vertebrates.

Although the primary signals of scaling relationships permit important insights and inferences, much additional information resides in the "secondary signals" of these same relationships (Schmidt-Nielsen, 1984). Secondary signals are the residuals either from empirically determined scaling functions or from theoretical, *a priori* expectations about size and scaling (Smith, 1984*a*). In the context of joint size scaling, secondary signals represent one measure of *relative joint size*—i.e., joint size residual to body size. It is the goal of the present study to determine whether or not any differences in relative joint size can be linked to adaptive differences among living hominoids in locomotor and postural behavior. Narrow allometric comparisons (*sensu* Smith, 1980, 1984*b*) are also included, and relative joint size and its positional correlates are examined in one early hominid, A.L. 288-1 (*Australopithecus afarensis*).

Materials and methods

Seven living hominoid species are included in the analyses to follow: gorillas, common chimpanzees, bonobos or pygmy chimpanzees, orang-utans, siamang, lar gibbons, and modern humans of European descent. The non-human sample consists exclusively of wild-collected adult individuals with recorded body mass at death. The human skeletal sample is derived from a large cadaver population and also contains only adult individuals of known body mass (Jungers, 1982). Body mass and linear dimensions of the long bones in this human sample are very close to the mean values published for normal young adults (Reed & Stuart, 1959; Anderson *et al.*, 1964). The total sample size is 87 specimens. Sex-specific means were computed and used throughout the study. Average body mass ranged from 5·5 kg in the female gibbons to 164·3 kg in the male gorillas. Further details of the sample can be found in Jungers (in press).

Sixteen linear dimensions of postcranial joints are included in the analyses:

1. Glenoid cavity width (abbreviated in tables as GLW).
2. Average humeral head diameter (HHD), the mean of two chords perpendicular to greater and lesser tubercles.
3. Anterior width of humeral trochlea (TRW).
4. Anterior width of humeral capitulum (CPW).
5. Maximum radial head diameter (RHD).
6. Maximum breadth of distal radial articulation (DRA).
7. Ulnar head diameter (UHD).
8. Acetabulum height (ACT).
9. Average femoral head diameter (FHD), the mean of anteroposterior and craniocaudal diameters.
10. Posterior width of medial femoral condyle (MCW).
11. Posterior width of lateral femoral condyle (LCW).
12. Width of patellar articular surface (PAT).
13. Anteroposterior length of medial tibial condyle (MTB).
14. Anteroposterior length of lateral tibial condyle (LTB).
15. Anteroposterior diameter of distal tibial articulation (DTB).
16. Width of lumbo-sacral articulation (LSW).

Seven of these variables measure size of upper limb joints, eight measure size of lower limb joints, and one measures size of the first sacral vertebral body.

The problem of quantifying *relative* joint size has been approached here from several different angles. First, the common sense approach of "narrow allometry") Smith, 1980, 1984*b*) has been enlisted to provide direct comparisons of joint size in individuals of different angles. First, the common sense approach of "narrow allometry" (Smith, 1980, 1984*b*) has been enlisted to provide direct comparisons of joint size in individuals of different species that have the same body mass. One may compute a percentage difference treated in this fashion because body size ranges do not overlap in many of the taxa, and sampling error is a very real risk when dealing with single specimens of a given species.

The second method—and the one ultimately favored over all others in this study—involves the calculation of joint size residual to isometry. In other words, the null hypothesis of geometric similarity is used as a criterion of subtraction. As was noted above, in addition to dimensional criteria there exist excellent biomechanical reasons for this type of adjustment based on Alexander's model. As will be demonstrated, this approach eliminates most of the variance due simply to body size differences but preserves those size-related differences in joint size due to allometry. In log-log space, a line with a slope of $0 \cdot 333$ is forced through the grand means of body mass and a given joint dimension. Relative joint size is computed as the positive or negative departure from this *a priori* non-empirical equation (Smith, 1984*a*). These secondary signals are then standardized to Z-scores (i.e., mean of zero and standard deviation of $1 \cdot 0$) so that smaller joints may be compared more directly to larger joints (e.g., radial head diameter versus acetabulum height); this simple transformation indicates whether a particular deviation is uncommonly large or small relative to all deviations for that dimension (Smith, 1984*a*).

Measures of relative joint size were also calculated as joint size residual to body size using the best-fitting, empirical relationships between articular dimensions and body mass in logarithmic space. The slope of this line describes the nature of the relative proportional changes in joint size and body size (Smith, 1984b; Jungers, in press). Joint dimensions residual to such a scaling relationship can be described as proportionally or "allometrically" adjusted variables. Given the predictive goal of this technique, least squares regression has been employed to calculate the empirical relationship (Sokal & Rohlf, 1981). For reasons similar to those outlined above with respect to Z-scores, standardized residuals from such regressions are taken as the measure of relative joint size. Initially, means of all taxa and both sexes were used in this type of adjustment (as in the case of the geometric adjustment). However, in least squares fits those points removed most from the grand means have the most "leverage" in determining the slope. Therefore, in order to eliminate the enormous impact of male gorilla values on relative joint size in females, sex-specific allometric adjustments to joint size were also calculated. Comparison of the total sample versus sex-specific analyses revealed only subtle differences, but on balance the sex-specific variables were judged to be an improvement over the total sample residuals. Accordingly, only the former will be considered in detail subsequently.

The geometric and allometric adjustments to relative joint size result each in a 14×16 matrix (14 species/sexes, 16 joint dimensions). Careful inspection of the entire matrix is necessary to detect unusually large or small deviations and to disclose species-specific trends, but this onerous task can be facilitated greatly by multivariate summaries of overall similarity or dissimilarity. Employing what Corruccini (1975) has called non-parametric multivariate methods, overall phenetic resemblance among hominoid taxa in relative joint size can be assessed as follows. A 14×14 symmetric matrix of average taxonomic distances (d) was computed from each type of adjusted variable (Sneath & Sokal, 1973). A clustering strategy was then used to find groups of species/sexes that are more similar within groups than among groups (Neff & Marcus, 1980). The UPGMA (unweighted pair-group method using arithmetic averages) clustering algorithm of the mainframe version of NT-SYS (Numeric Taxonomic System of Multivariate Statistical Programs; Rohlf et al., 1986) was employed for this purpose. A cophenetic correlation can be calculated between the original distance matrix and the distances implied by the resulting phenogram (Sneath & Sokal, 1973). Principal coordinate ordinations of the distance matrices were also performed to reduce and summarize the information about overall resemblances into three easily visualized dimensions (Gower, 1966; Rohlf, 1972; Thorpe, 1980; Reyment et al., 1984). Again, it is possible to compute a matrix correlation between the original distance matrix and the distances implied by the three-dimensional ordination. One may then try to "relate the multivariate results back to the original anatomy" (Corruccini, 1978, p. 141) and draw functional inferences by computing correlations between the adjusted variables and each of the principal coordinate axes. In other words, it is possible to dissect out those relative joint dimensions primarily responsible for ordering groups along each axis. This part of the analysis utilized both NT-SYS and SAS statistical software packages. It should also be noted that non-metric multidimensional scaling ordinations were also performed, but the differences from the principal coordinate analysis were found to be trivial and do not merit separate discussion.

Results

A sample of the many possible narrow allometric comparisons (i.e., contrasts among individuals of equivalent body size) in joint size among living hominoids is presented in

Table 1. One representative upper limb dimension, the articular width (AW) of the distal humerus (sum of TRW and CPW) is contrasted with one lower limb dimension, femoral head diameter (FHD). Humans are used as the reference species to calculate percentage differences. Compared to common chimpanzees and orang-utans of the same size (mass), humans possess a small AW (i.e., negative %D) but much larger FHDs (positive %D). The same basic pattern obtains in almost all contrasts of human lower limb and lumbosacral joint size, but not all human upper-limb articular dimensions are predictably smaller than their non-human counterparts, especially in male–male comparisons.

Table 1	Narrow allometry of hominoid joint size in selected individuals						
Sex	Human	Chimpanzee	% difference in joint size	Human	Orang-utan	% difference in joint size
Males						
Mass	70·9 kg	70 kg		75·9 kg	75 kg	
AW	46·7 mm	51·8 mm	−11%	47·6 mm	53·0 mm	−11%
FHD	47·9 mm	40·2 mm	+16%	50·4 mm	39·8 mm	+21%
Females						
Mass	50·5 kg	50 kg		46·4 kg	45·5 kg	
AW	40·3 mm	46·0 mm	−14%	39·8 mm	41·9 mm	−5%
FHD	42·2 mm	32·5 mm	+23%	41·4 mm	31·3 mm	+24%

% difference = (human–ape)/human*100.
AW, articular width of distal humerus.
FHD, average diameter of femoral head.

The geometrically adjusted measures of joint size converted to Z-scores are provided in Table 2. Negative values indicate variables smaller than predicted by isometry, whereas positive values indicate joint dimensions larger than predicted by a model of geometric similarity. Only three of the sixteen adjusted variables exhibit a significant correlation with body mass (HHD, TRW and MCW), and these three dimensions scale allometrically with body mass in the same sample (Jungers, in press). In other words, size-related differences of an allometric nature tend to be preserved. The male values are presented in graphic form (Figure 1) in a simple attempt to look for trends. Values in Table 2 and the high frequency of cross-cutting lines in Figure 1 indicate that no single species can characterized as possessing uniformly large or small *upper* limb joints. For example, lesser apes have relatively small distal radii and humeral trochleae, but possess relatively large humeral heads and capitula and radial head diameters. In contrast, gorillas and orang-utans tend to have relatively large glenoid cavities and trochleae but small capitula. Orang-utans also possess the relatively largest ulnar head diameters. Both radial elements of the elbow (CPW and RHD) are relatively small in humans as is the humeral head and ulnar head. The pattern in the *hindlimb* and *lumbosacral region* is quite different. With the exception of the medial condyle dimension (MCW) of gorillas, the hindlimb joints and LSW of humans are relatively the largest of all hominoids and usually deviate markedly from the other species. In fact, the difference between humans and the other species is so great that most of the hindlimb joints (and LSW) of non-human hominoids exhibit negative departures from isometry; this is especially apparent in orang-utans (both sexes), male chimpanzees and female siamang.

The sex-specific regression-adjusted or "allometrically corrected" measures of relative joint size are summarized in Table 3. Numerous similarities to the patterns seen above for

JOINT SIZE RESIDUAL TO ISOMETRY (Z-SCORES)

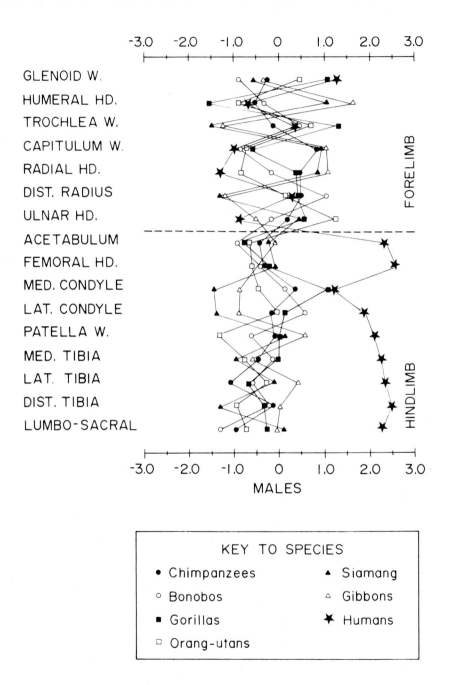

Figure 1. A plot for male hominoids of relative joint size computed as positive and negative departures from geometric similarity (and standardized by conversion to Z-scores). Note the hindlimb values for human males.

Table 2

Geometrically adjusted articular variables converted to Z-scores

Group	GLW	HHD	TRW	CPW	RHD	DRA	UHD	ACT	FHD	MCW	LCW	PAT	MTB	LTB	DTB	LSW
Chimp (male)	−0·26	−0·52	−0·15	0·97	0·46	0·49	0·18	−0·41	−0·36	0·37	−0·15	−0·10	−0·47	−1·07	−0·16	−0·97
Chimp (female)	−0·84	−0·19	0·35	1·17	0·72	0·64	0·31	−0·66	−0·52	0·62	0·00	−0·79	0·04	−0·58	−0·08	−0·08
Bonobo (male)	−0·82	−0·33	0·37	−0·83	−0·19	1·02	−0·21	−0·92	−0·53	0·16	0·57	−0·63	−0·14	−0·57	−0·20	−1·34
Bonobo (female)	0·46	−0·01	0·71	0·97	0·63	1·71	0·75	0·09	0·12	0·62	0·76	−0·41	0·28	−0·06	0·24	−0·39
Gorilla (male)	1·07	−1·56	1·32	−0·68	0·43	0·47	0·49	−0·76	−0·23	1·12	0·09	−0·03	−0·01	−0·60	−0·28	−0·25
Gorilla (female)	0·55	−0·81	1·03	−0·20	−0·26	0·44	0·27	−0·47	0·09	1·04	−0·01	−0·19	−0·03	−1·07	0·10	0·05
Orang (male)	0·48	−0·88	0·70	−0·72	−0·86	0·17	1·24	−0·69	−0·66	−0·49	−0·05	−1·60	−0·79	−0·25	−0·95	−0·75
Orang (female)	1·72	−0·12	1·08	−0·54	−0·55	0·59	1·72	−0·09	−0·25	−0·43	0·09	−1·01	−0·73	−0·47	−0·29	−0·56
Siamang (male)	−0·54	1·04	−1·49	0·94	0·85	−1·29	0·46	−0·25	−0·12	−1·44	1·41	0·13	−0·96	−0·11	−1·32	0·12
Siamang (female)	−2·08	1·05	−1·73	0·18	0·44	−1·70	−0·18	−0·43	−0·89	−1·94	−1·39	−0·18	−0·97	−0·45	−1·08	−0·45
Gibbon (male)	−0·28	1·68	−1·26	0·99	1·08	−1·23	−1·02	−0·17	−0·54	−0·83	−0·87	0·56	−0·60	0·45	0·01	−0·09
Gibbon (female)	−0·03	1·75	−0·65	0·80	0·97	−1·05	−1·03	0·35	−0·62	−0·82	−1·12	0·36	0·07	0·68	−0·11	0·76
Human (male)	1·24	−0·71	0·36	−0·99	−1·30	0·31	−0·93	2·35	2·56	1·12	1·84	2·05	2·25	2·32	2·48	2·28
Human (female)	−0·67	−0·40	−0·65	−2·06	−2·41	−0·57	−2·06	2·07	1·93	0·90	1·66	1·86	2·06	1·78	1·64	1·78

Table 3

Standardized residuals from sex-specific regressions of joint size on body mass

Group	GLW	HHD	TRW	CPW	RHD	DRA	UHD	ACT	FHD	MCW	LCW	PAT	MTB	LTB	DTB	LSW
Chimp (male)	−0·85	−0·97	−1·34	1·86	0·76	0·55	0·06	−0·26	−0·39	0·23	−0·42	−0·08	−0·47	−0·94	−0·14	−0·69
Chimp (female)	−0·63	−1·12	0·12	1·20	1·02	0·42	0·23	−0·93	−0·99	0·48	−0·43	−0·79	−0·17	−0·51	−0·40	−0·23
Bonobo (male)	−1·59	−1·69	0·84	−1·05	−0·27	1·89	−0·39	−0·71	−0·51	0·10	0·59	−0·59	−0·08	−0·51	−0·14	−0·99
Bonobo (female)	0·48	0·98	0·68	0·93	0·79	1·60	0·60	0·01	0·14	0·74	0·96	−0·37	0·19	0·06	0·23	−0·65
Gorilla (male)	0·61	0·36	0·58	0·38	1·65	−1·03	0·04	−0·68	−0·54	0·35	−1·01	0·18	−0·47	−0·49	−0·53	−0·07
Gorilla (female)	0·36	0·62	0·45	0·39	·0·65	−0·27	0·10	−0·82	−0·62	0·25	−1·39	−0·12	−0·59	−1·04	−0·58	−0·17
Orang (male)	0·08	0·63	0·42	−0·40	−0·87	−0·59	1·39	−0·52	−0·75	−1·65	−0·57	−1·43	−0·95	−0·17	−0·93	−0·51
Orang (female)	1·47	1·13	1·01	−0·29	−0·18	0·42	1·33	−0·22	−0·52	−1·03	−0·20	−1·03	−1·08	−0·40	−0·68	−0·86
Siamang (male)	−0·19	0·75	−1·47	0·44	0·36	−1·07	1·25	0·11	0·12	−1·15	−0·97	−0·15	−0·43	−0·28	−0·88	0·23
Siamang (female)	−1·39	−0·96	−1·38	−0·31	−0·13	−1·11	0·08	−0·64	−0·59	−1·83	−1·07	−0·16	−0·98	−0·77	−1·29	−0·86
Gibbon (male)	0·91	0·02	1·62	−0·24	0·28	0·00	−0·93	−0·06	−0·26	1·01	0·32	0·22	0·17	0·25	0·55	0·01
Gibbon (female)	0·99	0·65	0·90	−0·12	−0·19	0·28	−0·67	0·72	0·76	1·75	0·59	0·70	1·10	0·74	1·34	1·30
Human (male)	1·41	1·09	−0·24	−0·97	−1·59	−0·05	−1·54	2·21	2·20	1·31	1·91	1·92	2·16	2·15	2·06	2·07
Human (female)	−1·07	−1·15	−1·61	−1·92	−2·06	−1·44	−1·98	2·08	2·06	0·15	1·55	2·07	1·86	2·09	1·77	1·93

geometrically adjusted variables are evident, no doubt due in large part to the fact that many hominoid joints scale (empirically) rather close to isometry (Jungers, in press). Potentially significant differences do exist, however. For example, siamang and gibbons are now quite unlike each other with respect to most forelimb and hindlimb joints; in these cases the regression lines have cleaved these two groups of small hominoids. Cross-cutting trends still tend to characterize the forelimb variables in general. In sharp contrast to the lack of well-defined patterns in the forelimb, hindlimb and lumbosacral joint size residual to body mass exhibits a clear-cut trend (Figure 2) with respect to the expression of human relative joint size. Humans are again extreme outliers with very large hindlimb joints (and LSW) for their body size. This method of creating measures of relative joint size also consistently places the gibbon hindlimb dimensions closest to those of humans. Among males at least, orang-utans tend to lie at the opposite extreme with relatively small hindlimb joints.

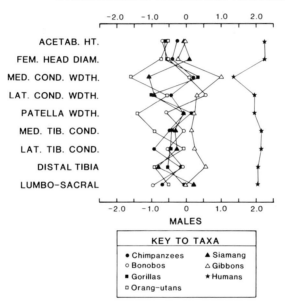

Figure 2. A plot for male hominoids of relative hindlimb joint size computed as the standardized residuals from the best-fitting least squares relationship of joint size as a function of body mass. Note the values for human males.

UPGMA clustering based on average taxonomic distances derived from the geometrically adjusted measures of relative joint size yields the phenogram in Figure 3. The cophenetic correlation between the original distance matrix and the distances implied by this clustering is 0·965. With the exception of bonobos, males and females of each sex cluster together before joining other clusters. A chimpanzee and bonobo cluster is joined by gorillas to produce an African ape grouping; this is joined in turn by the orang-utan to create a "great ape" cluster. Gibbons and siamang are more similar to each other than to any other single group; hence, the formation of a lesser ape cluster. The lesser ape cluster joins the great ape group to form a non-human hominoid assemblage. Only then are humans linked to this non-human cluster.

The three-dimensional principal coordinate ordination of the same distance matrix based on isometrically corrected variables is illustrated in Figure 4. These first three axes account for almost 90% of the total variance, with the first axis alone explaining over 50%. The correlation between the original distance matrix and the distances implied by this ordination is 0·992, suggesting that three dimensions adequately summarize the phenetic structure embodied in the original intergroup distances. Three obvious clusters are again apparent: great apes, lesser apes, and, far-removed from all others, modern humans. Correlations between the original geometrically adjusted variables (Table 2) and the three principal coordinate axes are presented in Table 4. Axis 1 not only accounts for most of the total variance but is also the one primarily responsible for isolating humans from the other species in ordination space. *All* of the hindlimb variables and LSW have highly significant negative correlations with Axis 1; CPW, RHD, and UHD show lower but significant positive correlations. This clearly corroborates the earlier findings (e.g., Figure 1) that *what makes humans unique is their relatively very large hindlimb joints (and LSW)*. Variable correlations with Axis 2 indicate that forms with relatively large humeral heads, small trochleae, and small joints of the distal forearm (lesser apes) are separated from those species with the opposite conditions (most of the great apes). Humans tend to occupy an intermediate position along this axis. The radial elements of the elbow (CPW and RHD) are significantly correlated with Axis 3, but the positions of taxa along this relatively minor axis are difficult to interpret.

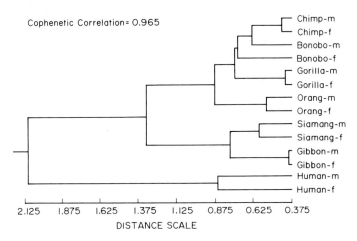

Figure 3. A phenogram of the UPGMA clustering of taxa based on average taxonomic distances computed from the geometrically adjusted joint variables.

The UPGMA clustering of the empirical, sex-specific regression-adjusted joint dimensions results in the phenogram depicted in Figure 5. The cophenetic correlation in this case is 0·901. As in the cluster analysis of geometrically adjusted variables, the sexes of each species are linked first with the exception of male and female bonobos. Subsequent clusters make less functional or systematic sense, with the noteworthy exception that humans are again the last group to join the other agglomerations. Chimpanzees are linked to gorillas but this aggregate is combined next to siamang. Only then do orang-utans join the cluster, followed by a combination of gibbons and female bonobos. Male bonobos are last to be added before the final distant linkage with humans. Use of the total sample rather than sex-specific standardized residuals complicates matters further by separating the males from females in siamang and orang-utans in addition to bonobos; the distinctive isolation of humans does remain intact, however.

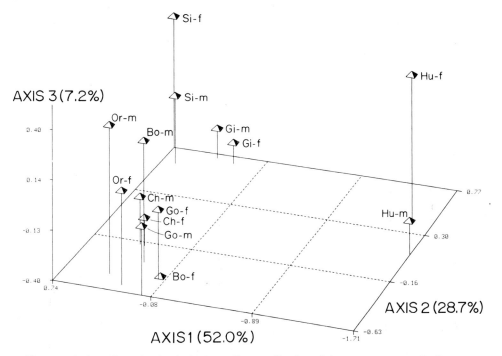

Figure 4. A three-dimensional principal coordinate ordination of the average taxonomic distances computed from geometrically adjusted joint variables. Abbreviations: m, male; f, female; ch, chimpanzee; bo, bonobo; go, gorilla; or, orang-utan; si, siamang; gi, gibbon; hu, human.

The three-dimensional, principal coordinate ordination of the sex-specific measures of relative joint size can be seen in Figure 6. These first three axes account for slightly over 85% of the total variance, and the matrix correlation between implied and actual distance matrices is a respectable 0·979. Humans are again sequestered from the non-human hominoids along Axis 1. Gibbons, especially females, approach humans along this axis more closely than any other species (recall Figure 2). Siamang are set off somewhat from the other groups (except female humans) and contrasted most noticeably with gibbons and female bonobos along Axis 2. Correlations between the allometrically adjusted variables (from Table 3) and the three ordination axes are presented in Table 5. Once again, the first and most important axis (56·3% of the variance explained by this axis alone) orders species/sexes primarily according to relative hindlimb joint (and LSW) size and somewhat on the basis (but in the opposite direction) of CPW, RHD and UHD. Relative size of the glenoid cavity and humeral trochlea are quite highly correlated with Axis 2 whereas humeral head diameter is the variable most highly correlated with Axis 3.

Table 4   **Correlations between geometrically adjusted articular variables and principal coordinate axes 1–3 (based on average taxonomic distances)**

Variable	Axis 1	Axis 2	Axis 3
GLW	−0·326	−0·599*	−0·295
HHD	0·373	0·824*	−0·219
TRW	−0·191	−0·946*	−0·116
CPW	0·657*	0·232	−0·669*
RHD	0·768*	0·184	−0·571*
DRA	−0·172	−0·884*	−0·248
UHD	0·564*	−0·671*	0·007
ACT	−0·901*	0·339	−0·057
FHD	−0·963*	0·057	−0·028
MCW	−0·643*	−0·612*	−0·247
LCW	−0·856*	−0·430	0·039
PAT	−0·783*	0·525	−0·149
MTB	−0·973*	0·035	−0·106
LTB	−0·813*	0·460	−0·050
DTB	−0·953*	0·005	−0·211
LSW	−0·829*	0·416	−0·148
Eigenvalue	6·748	3·726	0·940
% variance explained	52·0	28·7	7·2

* $P < 0.05$.

Table 5   **Correlations between sex-specific regression adjusted articular variables and principal coordinate axes 1–3 (based on average taxonomic distances)**

Variable	Axis 1	Axis 2	Axis 3
GLW	−0·216	−0·760*	−0·582*
HHD	−0·065	−0·571*	−0·781*
TRW	0·146	−0·835*	0·050
CPW	0·605*	−0·299	0·154
RHD	0·711*	−0·376	0·191
DRA	0·146	−0·581*	0·590*
UHD	0·833*	−0·088	−0·387
ACT	−0·963*	0·076	−0·167
FHD	−0·964*	0·094	−0·133
MCW	−0·568*	−0·620*	0·405
LCW	−0·855*	−0·185	0·226
PAT	−0·912*	0·447	0·017
MTB	−0·974*	−0·075	0·140
LTB	−0·965*	0·016	−0·144
DTB	−0·958*	−0·212	0·156
LSW	−0·918*	0·033	−0·201
Eigenvalue	7·556	2·335	1·547
% variance explained	56·3	17·4	11·5

* $P < 0.05$.

Discussion

One observation that emerges conspicuously from these results, regardless of statistical methods (narrow allometry, geometric adjustment and allometric adjustment), is that modern humans possess exceptionally large hindlimb and lumbo-sacral joints for their body size. Full-time terrestrial bipedality obviously precludes the sharing of weight support and propulsion with the forelimbs of humans, and this fundamental difference from the other hominoids is accurately reflected in the relative size of the hindlimb articulations (and LSW). In other words, because humans are obligate bipeds, they repeatedly subject these joints to much larger loads on average than do quadrupeds or suspensory species of comparable body size. The highly derived, relatively very large joints represent the biomechanical solution to this unique form of hominoid locomotion and posture. It follows that if it were possible to evaluate relative joint size in extinct early hominids, important new insights and inferences might emerge about the nature of their locomotor repertoire. Aspects of this problem are considered further on in the discussion.

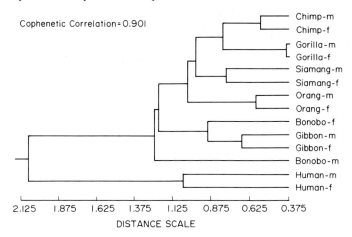

Figure 5. A phenogram of the UPGMA clustering of taxa based on average taxonomic distances computed from the sex-specific allometrically adjusted joint variables.

Examination of the geometrically adjusted measures of relative joint size revealed no non-human species that even remotely approached humans with respect to hindlimb joint enlargement. The allometrically adjusted ones, however, suggested that gibbons were consistently closer to humans than any other species in this respect. It is tempting to interpret this apparent trend as consistent with field reports that gibbons use hindlimb-dominated types of locomotion (bipedalism and leaping) more in travel than any other non-human hominoid (Fleagle, 1976; Gittins, 1983). However, and as was noted above, least squares regression tended to cleave gibbons from siamang because these are the two smallest taxa and relatively far-removed from the remainder of the hominoids. With respect of hindlimb joint size, this results in siamang appearing to have much smaller dimensions than gibbons despite a relatively high frequency of hindlimb-dominated positional behaviors in both species (Fleagle, 1976). Based on overall functional and systematic sorting of species/sexes, I regard the geometric correction to be easier to interpret and less prone to artifacts of the gibbon/siamang type (also see Creel, 1986). The overwhelming aspects of pure size differences among taxa are removed, and allometric information is retained (as was detailed above). This latter type of information about joint size is clearly important functionally in its own right (Jungers & Susman, 1984; Jungers, in press), and its inclusion in subsequent analysis and interpretation seems highly desirable. The remaining discussion will focus, therefore, on measures of joint size residual to isometry.

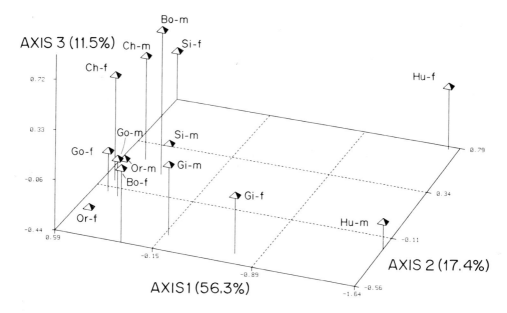

Figure 6. A three-dimensional principal coordinate ordination of the average taxonomic distances computed from sex-specific allometrically adjusted joint variables.

The results presented here indicate that lesser apes are characterized by relatively large humeral heads. Large, globular humeral heads are linked functionally to the requirements for great mobility and large degrees of excursion at the shoulder joint. A similar configuration can be found in New World monkey brachiators (Erikson, 1963). Similar requirements for enhanced mobility at the elbow may partially account for the relatively large radial elements (CPW and RHD) in lesser apes. If so, it does not appear to be reflected in any obvious fashion at the distal radioulnar joint (small UHDs in lesser apes) or the proximal surface of the radiocarpal joint (small DRAs in lesser apes); mobility-related aspects of hylobatid morphology seem more likely in the mid-carpal joint (Jenkins, 1981). It is also possible that the elbow joint is loaded fundamentally differently (i.e., more radially) in suspensory locomotion and postures than in pronograde positional behaviors. Relatively large humeral trochleae tend to sort great apes from lesser apes; it has been suggested that the load-bearing role of this element increases with increasing size in great apes (Jungers & Susman, 1984). This suggestion is consistent with the results of this study in that gorillas and orang-utans possess the largest values for TRW. It is difficult to account for the exceptionally large UHD of orang-utan in functional terms (greater load-bearing? greater mobility at the distal radio-ulnar joint?), but the tendency for African apes to possess larger distal radial articular breadths (DRAs) is consistent with conclusions reached by Jenkins & Fleagle (1975) from their cineradiographic analysis of wrist function. Medial femoral condyle size is absolutely and relatively larger in gorillas than in all other species; this is also reflected in the positively allometric scaling of this joint in African apes alone (Jungers & Susman, 1984) and in all hominoids (Jungers, in press). Based on biomechanical models of the ape knee (Preuschoft, 1970), these data suggest that the load-bearing role of this part of the knee joint increases with body size in species with a *genu varum* orientation (Jungers & Susman, 1984). The fact that bonobos exhibit significant sexual dimorphism in body mass but relatively little dimorphism in linear skeletal dimensions (Jungers & Susman, 1984) is reflected in their failure to cluster with each other and in the considerable distance between them in ordination space.

Returning to the issue of relative joint size and locomotion in early hominids, an attempt was made to place "Lucy" (A.L. 288-1, *Australopithecus afarensis*) into a size-related context similar to that discussed above for extant species. Twelve of the sixteen variables (six forelimb, five hindlimb and LSW) could be included in the analysis: GLW, TRW, CPW, RHD, DRA, UHD, ACT, FHD, MTB, LTB, DTB, plus LSW. All measurements were taken on the reference casts at the Cleveland Museum of Natural History. Geometric adjustments were then re-done for all taxa/sexes including Lucy, average taxonomic distances were computed, and the clustering and ordination procedures were repeated. To include Lucy in such an analysis obviously requires some estimate of body mass. McHenry (in press) has recently completed an extensive analysis of body mass estimates in Plio-Pleistocene hominids, and computes a value for A.L. 288-1 of almost 30 kg from measurements taken on the femoral diaphysis. A slightly more conservative value of 27·3 kg (60 lbs) is used here (Jungers, 1982). UPGMA clustering of the resulting 15 × 15 symmetric matrix is shown in Figure 7. The cophenetic correlation is quite high at 0·953. The overall picture is similar in many respects to that seen in Figure 3, but some alterations are evident. Males and females of the same species cluster first (with the exception of bonobos again). The common and pygmy chimpanzees form their own cluster, but gorillas link with orang-utans before joining the chimpanzee group. Hylobatids form their own well-defined cluster that is then joined to the great ape agglomeration. Humans are once again far removed from the other clusters. Note that Lucy affiliates with the non-human hominoids rather than with modern humans, although the linkage is not an especially close one. The three-dimensional principal coordinate ordination of the distance matrix is presented in Figure 8. These three axes account for almost 90% of the total variance, and the correlation between the distance implied by the ordination and the original distance matrix is 0·992. Along Axis 1, which alone explains over 52% of the variance, Lucy occupies an intermediate position between modern humans on the one hand and the rest of the non-human hominoids on the other. Lucy's location along Axis 2 places her in closest

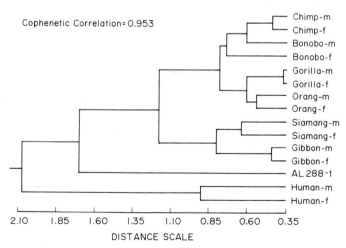

Figure 7. A phenogram of the UPGMA clustering of extant hominoid taxa and "Lucy" (A.L. 288-1) based on average taxonomic distances computed from geometrically adjusted joint variables.

proximity to the lesser apes, but she is unlike any extant species along Axis 3. Correlations between the geometrically adjusted joint variables and the three axes (Table 6) indicate that Axis 1 is once again primarily an axis of relative hindlimb joint (and LSW) size; RHD, CPW and UHD do contribute significantly in the opposite direction but less than any of the hindlimb variables. Lucy's position along Axis Three reflects relatively small radial elements of the elbow.

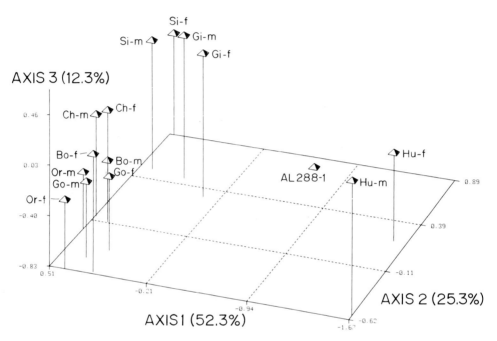

Figure 8. A three-dimensional principal coordinate coordination of the average taxonomic distances computed from geometrically adjusted joint variables in extant hominoids and A.L. 288-1.

A simple narrow allometric comparison has also been made between Lucy and two groups of modern humans (Table 7). Total pelvic height has been used as a body mass surrogate for reasons presented elsewhere (Jungers & Stern, 1983). Compared to either humans of European descent or to human pygmies, Lucy possessed a relative articular width of the humerus that is well within the range of either modern group. In contrast, relative femoral head size in Lucy is very small and far outside the range observed in either human group.

These results add to the growing list of features that distinguish Lucy and her relatives from modern humans (Stern & Susman, 1983; Schmid, 1983; Susman et al., 1984, 1985; McHenry, 1986). The intermediate position between humans and apes has been documented in other aspects of her postcranial skeleton, including interlimb and other body proportions (Jungers, 1982; Jungers & Stern, 1983). The data on relative size of the hindlimb joints in A.L. 288-1 indicate that some modest degree of enlargement had taken place if relative size in non-human hominoids can be taken as representative of the primitive condition. However, the highly derived, greatly enlarged hindlimb joints that characterize modern humans had not yet been achieved at this time in human evolution.

Table 6   **Correlations between geometrically adjusted articular variables in A.L. 288-1 and extant hominids and principal coordinate axes 1–3 (based on average taxonomic distances)**

Variable	Axis 1	Axis 2	Axis 3
GLW	−0·62	−0·861*	0·085
TRW	0·136	−0·930*	−0·269
CPW	0·572*	0·232	−0·669*
RHD	0·648*	−0·131	0·727*
DRA	0·165	−0·893*	−0·099
UHD	0·723*	−0·550*	−0·025
ACT	−0·904*	−0·303	0·094
FHD	−0·965*	−0·016	0·139
MTB	−0·936*	−0·231	−0·009
LTB	−0·856*	−0·023	0·394
DTB	−0·865*	−0·382	0·155
LSW	−0·926*	0·087	0·177
Eigenvalue	7·329	3·547	1·723
% variance explained	52·3	25·3	12·3

* $P < 0.05$.

Table 7   **Relative joint size in modern humans and "Lucy" (A.L. 288-1, *A. afarensis*)**

Group	Distal humerus articular width Total pelvic height		Average femoral head diameter Total pelvic height	
Humans of	x	0·206	x	0·214
European descent	s.d.	0·015	s.d.	0·011
(*n* = 18)				
Human pygmies	x	0·209	x	0·214
(*n* = 10)	s.d.	0·021	s.d.	0·010
"Lucy"				
(A.L. 288-1)		0·192*		0·180†

x, mean; s.d., standard deviation.
* Within observed range of both human groups.
† Completely outside the observed range of both human groups.

When on the ground, Lucy no doubt walked bipedally. That the frequency of terrestrial bipedality was almost certainly much greater than in any extant ape is probably linked directly to her somewhat greater relative joint size in the relevant anatomical regions. However, the decidedly non-human degree of relative joint size in A.L. 288-1 implies that this adaptation was far from complete and scarcely full-blown or functionally equivalent to that seen in modern humans (also see Stern & Susman, 1983; Susman *et al.*, 1984, 1985). A larger body size estimate of 30 kg would result in Lucy's hindlimb joints being relatively smaller still, and therefore even less human-like.

Relatively small hindlimb joints (e.g., femoral head diameter) in other early hominids have been noted before (Napier, 1964; McHenry & Corruccini, 1976; Corruccini & McHenry, 1978), and it seems highly likely that a Lucy-like locomotor adaptation was a stable one that persisted perhaps until the time of *Homo erectus* (McHenry, 1986; Johanson *et al.*, 1987). In other words, relatively long hindlimbs coupled with relatively large joints represent a major change in later hominid evolution, linked mechanically perhaps to the advent of longer distance travel and prolonged repetitive loading of the hindlimb joints. Energetic costs and joint loads would be increased further if weights of any kind were being carried (Miller & Stamford, 1987). Accordingly, the recent model proposed by Sinclair *et al.* (1986) that suggests the *origins* of bipedalism are to be found in an adaptation to long distance travel to scavenge from migratory ungulate herds seems most implausible in view of these results on the evolution of hindlimb joint size and on the basis of other independent lines of evidence (Leutenegger, 1987).

Acknowledgements

I wish to thank M. Dagosto and E. Strasser for the invitation to participate in their symposium at the American Association of Physical Anthropologists (New York City, 1987). I also wish to express my sincere appreciation to the many curators who made specimens available for study. Special thanks go to J. Kelly for typing the tables and bibliography, and to L. Jungers for preparing the figures. This research was supported by NSF grants BNS 8519747 and BNS 8606781.

References

Alexander, R. McN. (1980). Forces in animal joints. *Engin. Med.* **9**, 93–97.

Alexander, R. McN. (1981). Analysis of force platform data to obtain joint forces. In (D. Dowson & V. Wright, Eds) *An Introduction to the Biomechanics of Joints and Joint Replacements*, pp. 30–35. London: Mechanical Engineering Publishers, Ltd.

Anderson, M., Messner, M. B. & Green, W. T. (1964). Distribution of lengths of the normal femur and tibia in children from one to eighteen years of age. *J. Bone Joint Surg.* **46-A**, 1197–1202.

Corruccini, R. S. (1975). Multivariate analysis in biological anthropology: some considerations. *J. hum. Evol.* **4**, 1–19.

Corruccini, R. S. (1978). Morphometric analysis: uses and abuses. *Yrbk phys. Anthrop.* **21**, 134–150.

Corruccini, R. S. & McHenry, H. M. (1978). Relative femoral head size in early hominids. *Am. J. phys. Anthrop.* **49**, 145–148.

Creel, N. (1986). Size and phylogeny in hominoid primates. *Syst. Zool.* **35**, 81–99.

Currey, J. D. (1980). Skeletal factors in locomotion. In (H. Y. Elder & E. R. Trueman, Eds) *Aspects of Animal Movement*, pp. 27–48. Cambridge: Cambridge University Press.

Erikson, G. E. (1963). Brachiation in New World monkeys and in anthropoid apes. *Symp. zool. Soc. London.* **10**, 135–164.

Fleagle, J. G. (1976). Locomotion and posture of the Malayan siamang and implications for hominoid evolution. *Folia primat.* **26**, 245–269.

Fleagle, J. G. (1978). Size distributions of living and fossil primate faunas. *Paleobio.* **4**, 67–76.

Gittins, S. P. (1983). Use of the forest canopy by the agile gibbon. *Folia primatol.* **40**, 134–144.

Gower, J. C. (1966). Some distance properties of latent root and vector methods used in multivariate analysis. *Biometrika* **53**, 325–338.

Hodge, W. A., Fijan, R. S., Carlson, K. L., Burgess, R. G., Harris, W. H. & Mann, R. W. (1986). Contact pressures in the human hip joint measured *in vivo. Proc. Nat. Acad. Sci. USA* **83**, 2879–2883.

Jenkins, F. A., Jr (1981). Wrist rotation in primates: a critical adaptation for brachiators. In (M. H. Day, Ed.) *Vertebrate Locomotion*. Symposia of the Zoological Society of London, No. 48, 429–451.

Jenkins, F. A., Jr & Fleagle, J. G. (1975). Knuckle-walking and the functional anatomy of the wrists in living apes. In (R. H. Tuttle, Ed.) *Primate Functional Morphology and Evolution*, pp. 213–227. The Hague: Mouton.

Johanson, D. C., Masao, F. T., Eck, G. G., White, T. D., Walter, R. C., Kimbel, W. H., Asfaw, B., Manega, P., Nolessokia, P. & Suwa, G. (1987). New partial skeletal of *Homo habilis* from Olduvai Gorge, Tanzania. *Nature* **327**, 205–209.

Jungers, W. L. (1982). Lucy's limbs: skeletal allometry and locomotion in *Australopithecus afarensis. Nature* **97**, 676–678.

Jungers, W. L. (1984*a*). Scaling of the hominid locomotor skeleton with special reference to the lesser apes. In (H. Preuschoft, D. Chivers, W. Brockelman & N. Creel, Eds) *The Lesser apes: Evolutionary and Behavioral Biology*, pp. 146–169. Edinburgh: Edinburgh University Press.

Jungers, W. L. (1984*b*). Aspects of size and scaling in primate biology with special reference to the locomotor skeleton. *Yrbk phys. Anthrop.* **27**, 73–97.

Jungers, W. L. (1985). Body size and scaling of limb proportions in primates. In (W. L. Jungers, Ed.) *Size and Scaling in Primate Biology*, pp. 345–381. New York: Plenum Press.

Jungers, W. L. (in press). Scaling of postcranial joint size in hominoid primates. *Human Evol.*

Jungers, W. L. & Stern, J. T., Jr (1983). Body proportions, skeletal allometry and locomotion in the Hadar hominids: a reply to Wolpoff. *J. hum. Evol.* **12,** 673–684.

Jungers, W. L. & Susman, R. L. (1984). Body size and skeletal allometry in African apes. In (R. L. Susman, Ed.) *The Pygmy Chimpanzee: Evolutionary Morphology and Behavior*, pp. 131–177. New York: Plenum Press.

Kapandji, I. A. (1970). *The Physiology of the Joints*. New York: Churchill Livingstone.

Leutenegger, W. (1987). Origin of hominid bipedalism. *Nature* **325,** 305.

MacConaill, M. A. & Basmajian, J. V. (1969). *Muscles and Movements: A Basis for Human Kinesiology*. Baltimore: Williams & Wilkins.

McHenry, H. M. (1986). The first bipeds: a comparison of the *A. afarensis* and *A. africanus* postcranium and implications for the evolution of bipedalism. *J. hum. Evol.* **15,** 177–191.

McHenry, H. M. (in press). New estimates of body weight in early hominids and their significance to encephalization and megadontia in robust australopithecines. In (F. E. Grine, Ed.) *Evolutionary History of the Robust Australopithecines*. New York: Aldine.

McHenry, H. M. & Corruccini, R. S. (1976). Fossil hominid femora and the evolution of walking. *Nature* **259,** 657–658.

Miller, J. F. & Stamford, B. A. (1987). Intensity and energy cost of weighted walking vs. running for men and women. *J. appl. Physiol.* **62,** 1497–1501.

Napier, J. R. (1964). The evolution of bipedal walking in the hominids. *Arch. Biol. (Liege)* **75,** 673–708.

Neff, N. A. & Marcus, L. F. (1980). A Survey of Multivariate Methods for Systematics. New York (privately published).

Norkin, C. & Levangie, P. (1983). *Joint Structure and Function: A Comprehensive Analysis*. Philadelphia: F. A. Davis.

Preuschoft, H. (1970). Functional anatomy of the lower extremity. In (G. Bourne, Ed.) *The Chimpanzee*, Vol. 3, pp. 221–294. Basel: Karger.

Radin, E. L. (1982). Mechanical factors in the causation of osteoarthritis. In (M. Schattenkirchner, Ed.). *Rheumatology*, Vol. 7, pp. 46–52. Basel: S. Karger.

Radin, E. L., Orr, R. B., Kelman, J. L., Paul, I. L. & Rose, R. M. (1982). Effect of prolonged walking on concrete on the knees of sheep. *J. Biomech.* **15,** 487–492.

Reed, R. B. & Stuart, H. L. (1959). Patterns of growth in height and weight from birth to eighteen years of age. *Pediatrics* **24,** 904–921.

Reyment, R. A., Blackith, R. E. & Campbell, N. A. (1984). *Multivariate Morphometrics* (2nd edition). London: Academic Press.

Rohlf, F. J. (1972). An empirical comparison of three ordination techniques in numerical taxonomy. *Syst. Zool.* **21,** 271–280.

Rohlf, F. J., Kishpaugh, J. & Kirk, D. (1986). *NT-SYS: Numerical Taxonomic System of Multivariate Statistical Programs*. Stony Brook: State University of New York.

Rose, M. D. (1983). Miocene hominoid postcranial morphology. Monkey-like, ape-like, neither or both? In (R. L. Ciochon & R. S. Corruccini, Eds) *New Interpretations of Ape and Human Ancestry*, pp. 405–417. New York: Plenum Press.

Schmid, P. (1983). Eine Rekonstruktion des Skelettes von A.L. 288-1 (Hadar) und deren Konsequenzen. *Folia primatol.* **40,** 283–306.

Schmidt-Nielsen, K. (1984). *Scaling: Why Is Animal Size So Important?* Cambridge: Cambridge University Press.

Sinclair, A. R. E., Leakey, M. D. & Norton, M. (1986). Migration and hominid bipedalism. *Nature* **324,** 307–308.

Smith, R. J. (1980). Rethinking allometry. *J. theor. Biol.* **87,** 97–111.

Smith, R. J. (1984*a*). Determination of relative size: the "criterion of subtraction" problem in allometry. *J. theor. Biol.* **108,** 131–142.

Smith, R. J. (1984*b*). Allometric scaling in comparative biology: problems of concept and method. *Am. J. Physiol.* **15,** R152–R160.

Sneath, P. H. A. & Sokal, R. R. (1973). *Numerical Taxonomy*. San Francisco: W. H. Freeman & Co.

Sokal, R. R. & Rohlf, F. J. (1981). *Biometry* (2nd ed.). San Francisco: W. H. Freeman.

Stern, J. T., Jr & Susman, R. L. (1983). The locomotor anatomy of *Australopithecus afarensis*. *Am. J. phys. Anthrop.* **60,** 279–317.

Susman, R. L., Stern, J. T. Jr & Jungers, W. L. (1984). Arboreality and bipedality in the Hadar hominids. *Folia primatol.* **43,** 113–156.

Susman, R. L., Stern, J. T., Jr & Jungers, W. L. (1985). Locomotor adaptations in the Hadar hominids. In (E. Delson, Ed.) *Ancestors: the Hard Evidence*, pp. 184–192. New York: Alan R. Liss, Inc.

Thorpe, R. S. (1980). A comparative study of ordination techniques in numerical taxonomy in relation to racial variation in the ringed snake *Natrix natrix*. *Biol. J. Linn. Soc.* **13,** 7–40.